PREPARATIVE GAS CHROMATOGRAPHY

Preparative Gas Chromatography

Edited by

ALBERT ZLATKIS

Department of Chemistry
University of Houston, Houston, Texas

and

VICTOR PRETORIUS

Department of Physical and Theoretical Chemistry
University of Pretoria, Pretoria, South Africa

WILEY-INTERSCIENCE

a Division of John Wiley & Sons, Inc.
New York • London • Sydney • Toronto

Library of Congress Catalogue Card Number: 74   147238

ISBN 0   471   98384   5

Printed in the United States of America.

10 9 8 7 6 5 4 3 2 1

8/7/72 - 19.25

Preface

Gas chromatography is undoubtedly one of the outstanding scientific developments of the past decade. The success of the technique is certainly largely associated with chemical analysis. This is surprising when one realizes that the real strength of gas chromatography resides in a remarkable ability to separate chemical compounds rapidly and efficiently and only incidentally in identification and quantification. Even more surprising is the fact that this situation should have arisen in the face of a persistent demand—in both scientific and commercial circles—for increasingly refined techniques for isolating, recovering, or collecting pure chemical compounds. Nevertheless, it is a fact that an overwhelmingly large portion of the scientific effort that has been expended on gas chromatography has been directed toward exploiting and developing the technique for analysis and not for separation per se.

Possibly the main reason for this is the fact that at an early stage in the evolution of gas chromatography it became apparent that the efficiency with which compounds can be separated (resolution), the time needed for the separation (speed), and the quantity of material that can be handled (capacity) are linked, and that any one of these can only be enhanced at the expense of at least one of the others. Early optimism, which led to the coining of terms such as "production" and "plant-scale" gas chromatography, turned to disappointment as attempts to scale-up the capacity resulted in such losses in resolution that the separating potential of the technique became less attractive in comparison with other methods of purification. Talents were directed into more promising directions and preparative gas chromatography—as it had become known—faded and lay dormant.

As the processes underlying chromatographic separation became better understood, a reappraisal of preparative gas chromatography gradually took place. A variety of approaches, ranging from purely theoretical to purely empirical, have contributed to this progress. If nothing spectacular has emerged and if, as is clearly evident from the contributions to this volume, it is not yet certain whether preparative gas chromatography can be better exploited as a batch or as a continuous technique or whether long, narrow columns are preferable to shorter, wider ones, then at least it has been established that the technique can be competitively employed in judiciously selected areas. As things stand at the moment, we do know that preparative chromatography is most usefully applied where relatively difficult separations are involved, where grams rather than kilograms are handled per day, and where the emphasis is on the recovery of compounds that are valuable from a scientific or a commercial viewpoint. The technique is closer to the laboratory than to the plant, and the question as to whether or not it can be scaled-up to typically industrial dimensions can only be answered when much more is known about the scientific, technological, and economic factors involved.

The literature on preparative gas chromatography has become extensive and scattered. This volume is the outcome of a widely felt need to gather together the main threads. Since the technique is still in the formative stages, it was felt that this could best be reflected by a collection of relatively independent contributions covering the various aspects, rather than by a conventional monograph on the subject.

ALBERT ZLATKIS
VICTOR PRETORIUS

Houston, Texas
Pretoria, South Africa
March 1971

Contents

PREPARATIVE GAS CHROMATOGRAPHY

CHAPTER 1

Basic Theory

V. Pretorius and K. de Clerk, *Department of Physical and Theoretical Chemistry, University of Pretoria, Pretoria, South Africa*

I. Introduction

The purpose of a theory of preparative chromatography is to derive an expression relating a preparative efficiency to parameters that describe the processes involved and the way in which the latter may be affected by various conditions so that the optimum conditions for maximum preparative efficiency can be predicted. From a practical point of view, the theory becomes useful only if the parameters chosen are those that are experimentally convenient.

Theoretical treatments have an inborn tendency to become rapidly involved with a multitude of mathematical expressions which, if beautiful to the theoretician, are perhaps less so to the more practical-minded exponent of chromatography. It has therefore been deemed wise, before embarking on the final development, to attempt to sort out in a qualitative fashion the broad issues involved, so that the points at which the mathematical expertise are directed and the reasons for doing so can be appreciated.

A completely general theory of preparative chromatography would not be mathematically feasible at this time; in fact, it would almost certainly be so cumbersome as to defeat its own purpose. Consequently, this review primarily concerns itself with a relatively simple model from which the essential principles underlying the technique clearly emerge. In the first instance attention is confined to chromatography in columns (continuous forms are dealt with by Barker in Chapter 10). Second, a relatively simple efficiency parameter has been chosen, namely, the mass of a particular component of a stated purity that can be recovered per unit time. As a starting point, the theory is concerned with elution development used in the least complicated circumstances, i.e., two-component, equimolar solute; linear distribution isotherm; isothermal, isobaric operation, etc. Extensions to more complex situations are dealt with in a more-or-less qualitative fashion in Section V as perturbations of the simple model. Frontal analysis techniques are briefly covered in Section VI by using elution development as an analogy.

Because all theories of preparative chromatography are involved in one way or the other with the theoretical plate concept, a fair amount of attention is devoted to this aspect (Section II), and in particular to evolving a suitable plate height expression which correctly defines the functional dependence on practically important parameters such as column diameter.

Section III is concerned with inlet effects, and Section IV presents a detailed treatment of preparative elution chromatography for the simplified circumstances outlined.

A basic requirement of any chromatographic process, including preparative chromatography, is that the desired separation be obtained. In the case of an equimolar, two-component mixture, the purity can be expressed in terms of the resolution R (1–3) defined by

$$R = \frac{\Delta x}{4\sigma_{t1}} \tag{1.1}$$

where Δx is the distance between the peak maxima and σ_{t1} is the total peak variance of the least retarded component. Δx may be rewritten in terms of parameters that relate it to the thermodynamics of the separation process (4):

$$R = \frac{(\alpha - 1)}{4} \frac{k_1}{(1 + k_1)} \frac{l}{u\sigma_{t1}} \tag{1.2}$$

where σ_{t1} is measured in units of length and

where α = relative volatility = K_2/K_1

$K_j = C_{js}/C_{jm}$ = concentration distribution coefficient of the j-th component

C_{js} = concentration of j-th component in stationary phase

C_{jm} = concentration of j-th component in mobile phase

$k_j = m_{js}/m_{jm}$ = mass distribution coefficient where m_{js} and m_{jm} are the masses of the j-th component in the respective phases

l = length of column

u = carrier gas flow velocity (gas compressibility is neglected for the present purpose).

The total bandwidth σ_t measured at the outlet, but inside the column, may be regarded (5) as the sum of two contributions,

$$\sigma_t{}^2 = \sigma_c{}^2 + \sigma_{ii}^2 \tag{1.3}$$

namely, the variance at the column inlet as measured within the column σ_{ii}^2 and the variance $\sigma_c{}^2$ produced by the band-broadening processes within the column itself. The latter quantity is customarily expressed (6) in terms of the plate height H as

$$\sigma_c{}^2 = Hl \tag{1.4}$$

Pressure corrections attributable to gas compressibility are neglected for the moment and H is regarded as being adequately described by means of an empirical Van Deemter–type (4) expression, namely,

$$H = A + B/u + Cu \tag{1.5}$$

The coefficients A, B, and C are actually complex functions of, e.g., the geometry of the phases. (A more detailed examination of these matters is deferred to Section II.) The above form highlights the dependence of $\sigma_c{}^2$ on u and it is seen that there exists a certain velocity, namely $u = \sqrt{B/C}$, at which $\sigma_c{}^2$ and thus H is a minimum for a specific column (Fig. 1.1).

The inlet width of the sample is obviously determined by the volume of sample, the mode of introduction of the sample, and the column diameter. The mass contained within such a sample is simply related to the sample volume and the concentration. Consider, for argument's sake, that a sample containing two components A and B of masses m_A and m_B, respectively, is introduced in the form of a Gaussian concentration distribution of variance

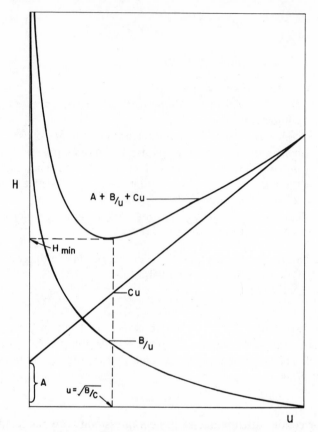

Fig. 1.1. Plate height vs. carrier flow velocity.

σ_{ii}^2 within the column at the inlet. This is illustrated in Fig. 1.2. It is important to note that the width of the sample is reduced to $1/(1 + k)$ of its value outside the column if equilibration between phases is considered as being instantaneous (see Section III). As the sample travels down the column, it broadens but the components gradually become separated since the peak separation processes are faster than those causing the peak broadening. The resolution as expressed by R therefore gradually increases until a point is reached where the R-value conforms to the required purity. For a given column the distance traveled to this point is least for a carrier velocity corresponding to H_{min}.

The mass rate at which a component, e.g., component 1, of total mass m_1, is produced at the column outlet per unit sample injected is given by (10)

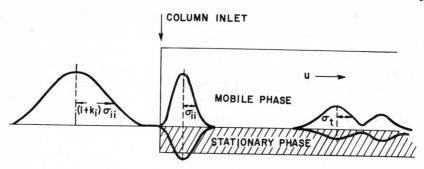

Fig. 1.2. Illustration of the sample contraction effect at the inlet.

$$E_p = \frac{(m_1 - \Delta m_1)\, u_o}{w_{to}(1 + k_1)} \tag{1.6}$$

where Δm_1 = mass discarded during fraction cutting

  u_o = carrier velocity at the outlet

  w_{to} = total chromatogram width per sample at the outlet inside the

column. This quantity is conveniently expressed in units of σ_t at the outlet as is explained in Section III, e.g.,

$$w_{to} = n\sigma_{to} \tag{1.7}$$

where n is a suitably chosen constant.

   The obvious way to increase E_p is to increase the numerator in Eq. 1.6, i.e., by increasing either $m_1 - \Delta m_1$ or u_o or both. m_1 can be increased by increasing the sample volume and/or the concentration, while u_o is increased by simply increasing the inlet pressure. Unfortunately, any such increases usually also result in an increase in the denominator because of the dependence of σ_{to} on these variables, namely,

$$\sigma_{to}{}^2 = Hl + \sigma_{ii}^2 \tag{1.8}$$

This statement requires some amplification. First, it is true only for u_o in excess of the value that corresponds to the minimum in H, which is invariably the case. Second, it is true that σ_{to} is not directly dependent on concentration; in any case not for the linear region. However, if the concentration is increased into the nonlinear region, peak distortion causes σ_{to} to become effectively dependent on the concentration.

   The price that must be paid for increasing σ_{to}, if the purity requirements are to be retained, is always in terms of column length, as can be seen from Eq. 1.2; if R must remain constant on increasing σ_{to}, l must be increased accordingly.

A more interesting way of framing the basic problem in preparative work now suggests itself. Accepting that an increase in preparative efficiency is paralleled by an increase in column length, the problem may be phrased as follows. Given a column length in excess of that required to merely effect separation, how should one avail oneself of the excess resolution in order to obtain the maximum gain in preparative efficiency? The situation is illustrated in Fig. 1.3. The problem is thus to determine how to fill most effectively the

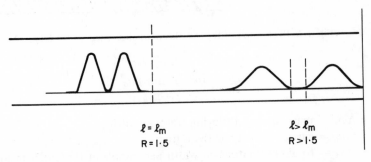

$\ell = \ell_m$ $\ell > \ell_m$
$R = 1.5$ $R > 1.5$

Fig. 1.3. Illustration of the superfluous resolution obtained with column lengths in excess of l_m.

open spaces created by the additional length. Should one increase the bandwidth by increasing the flow velocity or the inlet volume or both? The question is complicated by the fact that these are not the only variables involved. A most important one is the column diameter d_c. Increasing d_c affords a potentially attractive means for enlarging E_p since the inlet volume can thereby be increased without increasing the inlet bandwidth. However, the fact that H is also dependent on d_c interrelates these effects in such a complex manner that only a complete mathematical analysis could possibly unravel the finer details.

It is nevertheless instructive to assess the influence of flow velocity and inlet volume in a qualitative fashion. Consider the case in which a column operates with a high analytical efficiency, i.e., the inlet width σ_{ii} is negligible in comparison with σ_c, while the operating parameters, in particular u_o, are set at values designed to minimize σ_c. This implies that the column length is the shortest possible for the required separation. Let this length be designated by l_{am}. The preparative efficiency of this arrangement is obviously very low. In order to increase it one knows that column length must be increased. Let us assume that this has been done without altering the values of the optimized parameters, e.g., flow velocity is held constant by a suitable adjustment of the inlet pressure. This means that superfluous resolution becomes available which can be used to increase the preparative efficiency. If the velocity is held constant, this can be done by increasing the inlet volume until the peaks emerging at the outlet have just the required resolution. These new conditions yield a higher preparative efficiency but the question remains whether or not this is the optimum arrangement. Could an improved efficiency perhaps be obtained as a compromise between increased velocity and increased inlet volume? It is evident that the velocity increase is limited by the chosen length since a velocity is eventually reached where the superfluous resolution is consumed by the increases in $\sigma_c^2 = Hl$, necessitating an inlet volume that tends to zero. At this flow velocity the preparative efficiency therefore also tends to zero. Evidently, between the optimum velocity for the analytical efficiency and the velocity limited by the column length, there exists an optimum flow velocity for preparative chromatography as suggested in Fig.1.4.

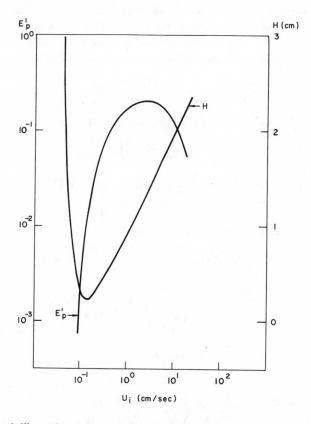

Fig. 1.4. Graph illustrating the extrema in the plate height and production rate vs. flow velocity curves.

The inlet volume corresponding to this velocity can then be computed. It is important to note that the optimum preparative flow velocity shifts to higher values as the column length increases, thereby increasing the preparative efficiency.

II. The Plate Height

The purpose here is to adopt from the confusing array of plate height expressions, one that clearly defines the role of parameters that could be expected to affect the preparative efficiency. Column radius is particularly important.

A useful starting point is the basic theory of the local plate height $H(x)$, which is defined (11) at constant local pressure as

$$H(x) = d\sigma^2(x)/d\langle x \rangle \tag{1.9}$$

i.e., the change of the variance of the concentration distribution with the position $\langle x \rangle$ of the distribution mean. This quantity is related to the effective diffusion coefficient D_{eff} contained in the basic differential equation (12) describing the passage of the distribution along the column axis.

$$\frac{\partial C(x)}{\partial t} = -\frac{u}{1+k}\frac{\partial C(x)}{\partial x} + D_{\text{eff}}\frac{\partial^2 C(x)}{\partial x^2} \tag{1.10}$$

On operating on both sides of this equation with the operator

$$\int_{-\infty}^{+\infty}(x - \langle x \rangle)^2\,dx \Big/ \int_{-\infty}^{+\infty} C(x)\,dx$$

and assuming suitable boundary conditions, one finds (12)

$$d\sigma^2/dt = 2D_{\text{eff}} \tag{1.11}$$

where σ^2 is the variance defined as usual by

$$\sigma^2 = \int_{-\infty}^{+\infty} C(x)(x - \langle x \rangle)^2\,dx \Big/ \int_{-\infty}^{+\infty} C(x)\,dx \tag{1.12}$$

Similarly, the operator

$$\int_{-\infty}^{+\infty} x\,dx \Big/ \int_{-\infty}^{+\infty} C(x)\,dx$$

yields

$$\frac{d\langle x \rangle}{dt} = \frac{u}{1+k} \tag{1.13}$$

so that $H(x)$ follows from Eqs. 1.9, 1.11, and 1.13 as

$$H(x) = \frac{d\sigma^2}{d\langle x \rangle} = \frac{d\sigma^2}{dt} \cdot \frac{dt}{d\langle x \rangle} = \frac{2(1+k)}{u} D_{\text{eff}} \tag{1.14}$$

The validity of replacing the exact differential equation which includes the transport in the radial directions by Eq. 1.10 has been demonstrated in various studies, e.g., Refs. 12, 13, 15, 16. It is well to note, however, that a certain initial time must elapse after sample introduction before Eq. 1.10 becomes fully operative. An estimate of this time, during which the long-time expression for $H(x)$ does not apply, has been made by Taylor (13,18), Giddings (19), and Levenspiel (20). Their results are in essential agreement; Levenspiel predicts that the lower limit to the transient duration t_{tr} is given approximately by

$$t_{tr} = 0.04(d_c^2/D_R) \tag{1.15}$$

where D_R is the radial dispersion coefficient, approximately equal to 0.6 D_m at low flow velocities. That this has a bearing on preparative work is apparent from the dependence on the column diameter; the actual functional dependence of H during this time is discussed again later.

The limiting expression for $H(x)$ can be found directly from Eq. 1.14 provided D_{eff} is known. This quantity is found to be the sum of three terms

$$D_{eff} = \gamma D_m + D_p + D'_{eff} \tag{1.16}$$

where γD_m is the usual molecular diffusion coefficient D_m corrected for packing effects by the labyrinth factor γ (21), D_p represents the dispersion attributable to the packing itself, and D'_{eff} corresponds to the effective diffusion that results from the random exchange of molecules between regions of different axial flow velocity in the column cross section. This transport takes place through both the mobile and stationary phases, and it has been shown (22) that D_{eff} can be written as the sum of contributions from the two phases

$$D_{eff} = D_{eff}(m) + D_{eff}(s) \tag{1.17}$$

An analogous separation for $H(x)$ is

$$H(x) = H_m(x) + H_s(x) \tag{1.18}$$

where $H_m(x)$ is now regarded as including plate heights attributable to the first two terms in Eq. 1.16. The other terms that contribute to $H_m(x)$ are designated by $H'_m(x)$ and correspond to $D'_{eff}(m)$. These are the most troublesome terms and have been calculated by various means, e.g., Giddings' nonequilibrium theory (23), random walk theories (24), and an integral method attributable to Aris (22). Of these, the Aris integrals probably provide the most direct insight into the factors that determine their values. Consider the mobile phase contribution in the open capillary case as illustration. D'_{eff} in this case is given (22) by

$$D'_{eff} = \frac{1}{1+k} \int_0^{r_c} \frac{r_c^2 dr}{2r D_R(r)} \left\{ \int_0^r \frac{r^2}{r_c^2} \frac{2r'dr'}{r^2} \left[u(r') - \frac{u}{1+k} \right] \right\}^2 \tag{1.19}$$

where $D_R(r)$ = the local value of the radial dispersion coefficient
    $u(r')$ = radial velocity profile

Equation 1.19 is actually a rearrangement of the Aris integral in order to facilitate the comparison with the expression for D'_{eff} as given by the simple random walk theory (26), namely,

$$D'_{eff} = \Lambda[(\Delta x')^2/\Delta t] = \Lambda v^2 \Delta t$$

where $\Lambda \equiv$ probability factor
    $\Delta x' \equiv$ effective axial length of the random step
    $\Delta t \equiv$ effective time for a random step
    $v \equiv$ effective axial velocity of random walk

In the present case the axial random motion is generated by the superposition of a radial diffusion (i.e., a radial random motion) on the flow profile. The random exchange of molecules between streamlines of different velocities give rise to the random steps relative to the mean flow. A complete interpretation of these matters is given elsewhere (27). For the present purpose a few general remarks suffice. It is obvious that Eq. 1.19 is a summation over all the radial elements of thickness dr in which resistance to radial mass transfer is experienced, the expression for a single cylindrical element being given by

$$dD'_{eff} = \frac{1}{1 + k} dt\, v^2$$

with

$$dt = \frac{r_c^2 dr}{2r D(r)} \tag{1.20}$$

$$v = \int_0^r \frac{2r'\, dr'}{r^2} \frac{r^2}{r_c^2} \left[u(r') - \frac{u}{1 + k} \right] \tag{1.21}$$

and $\Lambda = 1/(1 + k) \equiv$ probability to be in the mobile phase. The effective velocity is seen to be an average velocity relative to the mean velocity $u/(1 + k)$, weighted at each r by the probability r^2/r_c^2 of finding a representative molecule on that particular side of dr. v therefore gives a measure of the effective step length per unit time of a representative molecule as a result of the barrier presented by dr, assuming equal probability for this molecule to be anywhere in the mobile phase cross section. dt is a measure of the time required for a number of molecules, equal to that contained in the cross section, to have diffused through dr.

This "physical" interpretation of the Aris integrals is admittedly heuristic; nevertheless, it is very useful when one attempts to form a picture of the factors involved in the evaluation of plate heights. In particular, the roles of *dt* and *v* should be noted. Very roughly, the integration over *t* yields a time which will be larger as the dimensions of the column increase, corresponding to the larger distances over which the molecules must diffuse. The actual value of $H(x)$ is, however, still determined by the effective velocity factor, the value of which is in turn determined by the velocity profile. It is important to note that $H(x)$ can always be decreased by increasing the radial transport, i.e., $D_R(r)$.

These arguments are later applied to the radially dependent term of packed columns, i.e., a velocity profile extending over the whole column cross section. First, however, it is useful to summarize the present state of knowledge concerning plate heights by writing these expressions down explicitly for the two cases of interest, open tubular columns and particle-packed tubular columns.

A. Open Tubular Columns

For laminar flow in coated open tubular columns, the Golay equation (28) is applicable and is for convenience rewritten in terms of dimensionless variables

$$\frac{H(x)}{r_c} = h_m(x) + h_s(x) \tag{1.22}$$

$$h_m(x) = \frac{B}{Re} + C_m Re \tag{1.23}$$

$$h_s(x) = C_s \, Re \tag{1.24}$$

$$B = \frac{4(1 - \theta)}{Scm} \tag{1.25}$$

$$C_m = \left[\frac{(1 + 6k + 11k^2)}{24(1 + k)^2}\right] \frac{Scm}{2(1 - \theta)} \tag{1.26}$$

$$C_s = \frac{1}{12} \frac{k}{(1 + k)^2} \frac{\theta^2(2 - \theta)^2}{(1 - \theta)^5} \frac{v_m}{v_s} Scs \tag{1.27}$$

where $Re \equiv$ Reynolds number $= 2r_c u / v_m$
  $Scm \equiv$ Schmidt number for mobile phase $= v_m / Dm$
  $Scs \equiv$ Schmidt number for stationary phase $= v_s / D_s$
  $v_m \equiv$ kinematic viscosity for mobile phase
  $v_s \equiv$ kinematic viscosity for stationary phase
  $\theta = d_f / r_c$

The terms representing the longitudinal diffusion in the stationary phase and the resistance to mass transfer at the interface have been neglected.

As soon as turbulence sets in at $Re \sim 2100$ a number of changes with potential advantages for preparative work set in. These are a flattening of the velocity profile and enhanced radial dispersion attributable to the development of turbulent eddies. Both these effects tend to lower the reduced plate height to values below those predicted by Eq. 1.11 at the same flow velocity. These effects have been extensively investigated by Smuts (29), who numerically evaluated the Aris integrals for this specific case. By curve fitting the results he was able to obtain a useful analytical expression for the plate height valid for $2300 < Re < 20,000$. His results are summarized in Fig. 1.5.

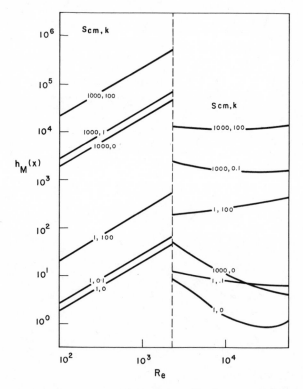

Fig. 1.5. Reduced plate height variations with Reynolds number in the transition from laminar to turbulent flow.

As can be seen from this figure, h_m decreases sharply when turbulence sets in. An unexpected feature is the strong dependence of h_m on k in the turbulent region.

B. Packed Columns

Zone dispersion in packed columns is not easy to describe because of the complexity of the flow dynamics in such systems. The stationary phase contribution is satisfactorily dealt with by assuming, as Giddings (30) did, that there is some typical or average configuration for the stationary phase, in which case its contribution may be represented by

$$H_s = \frac{1}{4} \frac{k}{(1 + k)^2} \frac{d_f^2 u}{D_s} \tag{1.28}$$

where d_f is the thickness of the stationary phase and the factor $1/4$ is regarded as a typical coefficient.

The mobile phase terms are less simply dealt with, however, and are even now a matter of controversy. Since they are so intimately related to the flow dynamics, there is merit in considering the latter in some detail. It is well-known (31) that packed-column flow dynamics can be described by the Ergun equation which is reproduced in Fig. 1.6. Inspection of this graph shows that the flow may be roughly divided into three different regions, namely, laminar, transient, and turbulent.

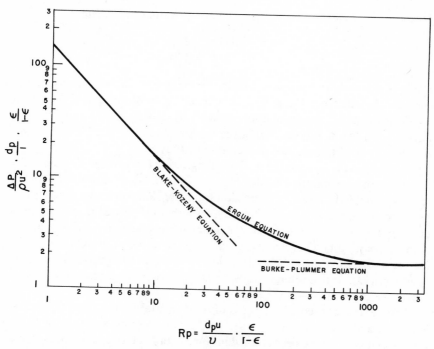

$$Rp = \frac{d_p u}{\upsilon} \cdot \frac{\epsilon}{1-\epsilon}$$

Fig. 1.6. A log–log plot of the Ergun equation.

Laminar flow is characterized by the existence of steady streamlines and persists approximately to a value of about 5 for the Reynolds number for packed columns defined by

$$R_p = \frac{d_p u}{\nu} \frac{\epsilon}{1 - \epsilon} \tag{1.29}$$

This picture is in accordance with the flow around a spherical body of diameter d_p immersed in a fluid flowing with a relative velocity u (32). Rough calculations show that for $R_p \lesssim 5$ the flow is mainly of the creeping type and that eddying behind the particles is still insignificantly developed.

In transient flow the chaotic flow pattern characteristic of turbulent flow develops progressively with increasing R_p until, at a value of $R_p \sim 2000$, turbulent eddies are present in practically all the interstices. This picture is substantiated by comparing this point with the onset point for turbulence in open tubular columns. If ϵ is taken as 0.4 and R_p as 2000, the velocity is given by

$$u = \frac{3000 \, \nu}{d_p}$$

This velocity would cause turbulence in an open tube of diameter

$$d_c = \frac{2100 \, \nu}{u} \sim \frac{2}{3} d_p$$

which is certainly of the correct order of magnitude of the interstices and suggests that the interstitial flow be likened to flow through a bundle of capillaries with a diameter of approximately d_p. Differences of opinion exist, however, on the flow types during the transient period (32a, 32b, 32c, 32d), but this is not of importance in the present study.

Mobile phase terms are, apart from longitudinal molecular diffusion, the result of two mechanisms, namely, the finite rate of lateral transport between regions of different flow and dispersion by the packing structure itself. These two mechanisms give rise, respectively, to terms proportional to, and independent of, the velocity. The best manner for classification of the various terms appears to be the scale on which random walks occur. On this basis, the terms may be grouped under the following headings as suggested by Giddings (33); transchannel, transparticle, short-range interchannel, long-range interchannel, and transcolumn.

Of these, the transcolumn term is of fundamental importance to preparative work and is considered separately. The contributions attributable to the other mechanisms have been studied in some detail in terms of a random walk model by Giddings (34) and de Clerk (35). Smuts (36) used material from the nonchromatographic literature in conjunction with the integrals of Aris, to obtain a plate height expression in terms of phenomenological variables. The details of these calculations are not essential to the present discussion. An interesting result is the fact that the k dependence of the terms diminishes as the scale of observation becomes larger until it eventually disappears. This is easily understood since D_{eff} in Eq. 1.14 is slowed down by the factor $1/(1 + k)$ when the flow variations extend over a number of particle diameters. The most remarkable feature of the results is, however, the fact that despite the complexity of the underlying mechanisms the behavior of the total plate height with R_p is surprisingly simple. The main reason for this fortunate result appears to reside in the properties of the radial dispersion coefficient D_R. This coefficient is made up of two contributions, the usual molecular radial diffusion and a contribution attributable to flow dispersion D_R'

$$D_R = (\gamma D_m + D_R')/(1 + k) \tag{1.30}$$

D_R' is a function of R_p and eventually becomes proportional to R_p as can be seen from Table 1.1, which contains the semiempirical results of Bischoff and Levenspiel (20) fitted to the form $A'R_p^{const}$. It is evident from Eqs. 1.14 and 1.19 that this tends to reduce complex plate height contributions to terms that are practically constant in the higher R_p regions which will be shown to be relevant for preparative work. This result has also been obtained by Giddings in a more restricted sense by means of his coupling theory (37).

TABLE 1.1

	$D_{R/\nu}' = \gamma D_{m/\nu} + D_{R/\nu}''$	
Gases $(Sc = 1.5)$	$0.1 < R_p < 0.94$	$0.48(R_p)^{0.11}$
	$0.94 < R_p < 15$	$0.52(R_p)^{0.46}$
	$R_p > 15$	$0.13R_p$
Liquids $(Sc = 1000)$	$0.001 < R_p < 4.8$	$0.013(R_p)^{0.83}$
	$4.8 < R_p < 37.5$	$0.001(R_p)^{2.4}$
	$R_p > 37.5$	$0.13R_p$

To summarize, the reduced plate height may be regarded as composed of four contributions

$$\frac{H}{d_p} = A + \left[\frac{1.2\,\epsilon}{Scm(1 - \epsilon)}\right]\frac{1}{R_p} + \left[\frac{1}{4}\frac{k}{(1 + k)^2}\theta_p^2\frac{\nu_m}{\nu_s}\frac{1 - \epsilon}{\epsilon}Scs\right]R_p + h_c \tag{1.31}$$

The fourth contribution, h_c, is the crucial term for preparative chromatography and therefore merits a detailed discussion.

The basic mechanism responsible for axial dispersion is once again the interplay between column-wide velocity variations and radial random walks. Quantitatively, this is conveniently interpreted in terms of an adaptation of the Aris integrals to this particular situation (38).

$$H_c = \frac{2}{\langle u \rangle} \int_0^{r_c} \frac{r_c^2[1 + k(r)]}{2rD(r)} \left\{ \int_0^r \frac{2r'dr'}{r^2} \frac{r^2}{r_c^2} \left[\frac{u(r')}{1 + k(r')} - \langle u \rangle \right] \right\}^2 \quad (1.32)$$

The adaptation, which assumes the correctness of the physical interpretation outlined above, consists of setting $\Lambda = 1$ since all molecules participate, slowing $D_R(r)$ down by a factor $1/[1 + k(r)]$, and using the actual local velocity $u(r')/[1 + k(r')]$ instead of the mobile phase velocity. $\langle u \rangle$ is the value of the local velocity averaged over the cross section. H_c is thus determined by the functional dependence of the velocity profile which is in turn a function of the mechanisms that cause these variations. The mechanisms that have been suggested include (1) packing techniques, (2) temperature profiles, and (3) stationary phase variations.

Of these, packing difficulties seem to be the main contributing factor. Giddings (41) has proposed a size-sorting hypothesis according to which larger particles pack preferentially toward the rim when a mountain (42) packing technique is employed. He has also evaluated the plate height contribution by means of the nonequilibrium theory as (43)

$$h_c = G_2 \frac{r_c^2 u}{96\gamma D_m} \quad (1.33)$$

with G_2 a constant.

In this analysis the column-wide flow profile was assumed to be quadratically dependent on the column radius. This implicitly assumes that all mechanisms are scaled up in proportion to the radius, which is certainly not generally true. The result therefore must be treated with reservation.

A packing defect that appears to be inherent in particle-packed tubular columns with circular cross sections has become known in the literature as the wall effect (44,45). This effect actually has two distinct facets. The first results from the fact that any surface parallel to the flow direction exerts a frictional drag proportional to the surface area. The flow in the space bounded by the wall and the outermost particles is therefore retarded relative to the flow in the packing proper because of the larger surface area of the wall. In the remainder of the cross section, flow is governed by the particles and the interstitial velocity should be constant provided that the packing is perfect. A second defect which is unavoidable with tubular columns of circular cross section, is the misfit in the vicinity of the wall. This results from the attempt to pack spheres, which tend to pack in square (cdp) or hexagon (hdp) units, into a cylindrical container. It is evident from Fig. 1.7 that the additional area

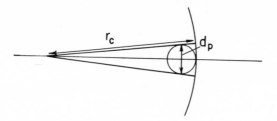

Fig. 1.7. Dependence of wall effect on particle–column diameter ratio.

available for flow in the vicinity of the wall depends on the angle subtended by the outermost particle at the center of the column. However, this angle is simply equal to the ratio of the particle diameter to the tube radius, i.e., d_p/r_c. It is therefore expected that the maximum flow velocity deviation should be a function of $\rho = d_p/d_c$. The expected flow profile is illustrated in Fig. 1.8. The above model was used as basis for an analysis (46) in terms of the Aris integrals. The following semiempirical expression for h_c was obtained

$$h_c = (m/2\rho^2 D_R)\, d_p u$$

$$m = (1/100) \exp\left(-\frac{1}{10\rho}\right) \tag{1.34}$$

According to this study, the plate height first increases with d_c (at constant ρ), reaches a maximum at $\rho \sim 0.05$ and then decreases with increasing radius.

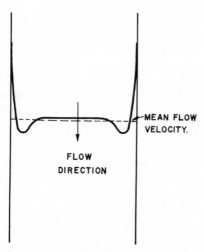

Fig. 1.8. Flow profile in packed column.

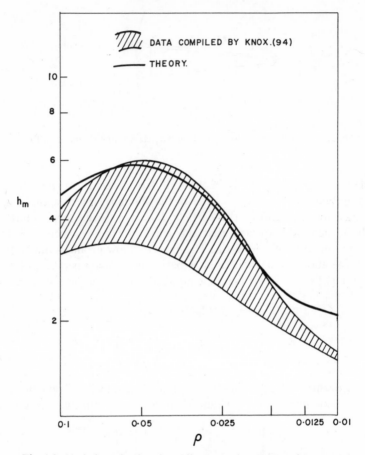

Fig. 1.9. Variation of reduced mobile phase plate height with $\rho = d_p/d_c$.

[See Fig. 1.9 in which data from the nonchromatographic literature ($k = 0$) is also presented.] It should be emphasized that these results are for $k = 0$ and do not necessarily apply to retarded velocity fronts which need not be a replica of the carrier gas profile. A thorough experimental investigation is obviously needed. It is interesting to note, however, that the results of Spencer and Kucharski (47) give experimental support to the above hypothesis. However, variations in stationary phase loading over the cross section can have a profound effect on the velocity profile and consequently on the plate height. In this connection the unexpected experimental results of Hupe (48) could be significant. He found that densification of coated packing material could lead to an accelerated front, completely contrary to what would be expected from the functional dependence of k on V_m. [$k = K(V_s/V_m)$]. A satisfactory explanation of this phenomenon is still lacking.

Other theoretical treatments of transcolumn effects include those of Higgins and Smith (49) and Huyten et al. (50). This work has been reviewed by Rijnders (51), who also begins with the Aris integrals to obtain

$$h_c = 2\Re(r_c{}^2 u / D_R) \tag{1.35}$$

from which it is inferred that h_c should increase with the square of the column radius. This is not necessarily true, however, since $\Re$ is in general a function of ρ (and therefore r_c) because of the probable dependence of the velocity profile on ρ. This dependence may even be the overriding factor and can result in a decrease of h_c with increasing r_c as noted above.

Higgins and Smith also derived an expression for h_c ,but they were more concerned with the dependence of the effective radial diffusion on the packing structure than with the dependence of h_c on the form of the flow profile. The latter was inferred semiempirically by means of a measured parameter x. Further details may be found in the cited references.

In summary, flow variations over the column cross section occur but their exact functional dependence, especially on d_c, and therefore the transcolumn contribution to the plate height, is a debatable matter. In particular, the problem is obscured by the difficulty of separating the fortuitous contributions resulting from avoidable packing defects from the unavoidable contributions resulting from inherent transcolumn effects. It is obvious that, irrespective of the cause of these variations, the plate height contribution can always be reduced by enhancing the radial mixing. To this end a variety of baffle systems have been proposed (53–55), and there is no doubt that these radial mixers have contributed greatly to the improvement of the efficiency of preparative chromatographic systems.

Another effect that requires discussion is the contribution to the plate height during the initial time when Eq. 1.10 has not yet become operative. Giddings (19) contends that this contribution is given by

$$H = G^2 l / 16 \tag{1.36}$$

where

$$G = \Delta u / u$$

with $\Delta u =$ difference between the extremes of velocity variation. He later (57) incorporated this into a general plate height expression by means of his coupling theory. The plate height vs. flow velocity curve then assumes the form given in Fig. 1.10. As yet, experimental evidence (57) is insufficient to conclusively substantiate these predictions, and more theoretical and experimental work is required.

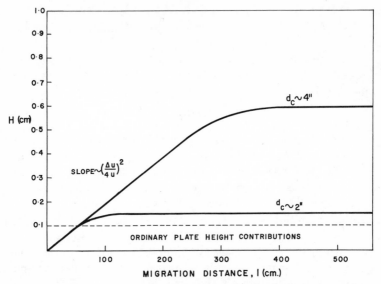

Fig. 1.10. Examples of the transient phenomena in wide-bore columns according to the theory of Giddings.

The effect of temperature fluctuations and coil bending was also studied by Giddings (19). The migration velocity depends on the local temperature, so that thermal fluctuations across the column cross section give rise to an additional plate height contribution. Giddings found this to be given approximately by

$$H_t(x) = \frac{\alpha_t(\Delta T)^2}{900} \frac{r_c^2 u}{D_m} \qquad (1.37)$$

where α_t is a constant with a value of about 0.004 and ΔT is the temperature difference between the center and the wall. (D_m should more correctly be replaced by D_R.)

For bent columns, Giddings found

$$H_b(x) = \frac{4\alpha_b}{R_0^2} \frac{r_c^4 u}{D_m} \qquad (1.38)$$

which results from the fact that molecules travel faster or slower depending on whether they are on the inside or the outside of the bend of radius R_0. The value of α_b is of the order of 0.1.

It should be emphasized that the above discussion pertains to the local plate height $H(x)$. The effect of the pressure gradient over the column, however, should be taken into account in the actual calculations of Section IV.

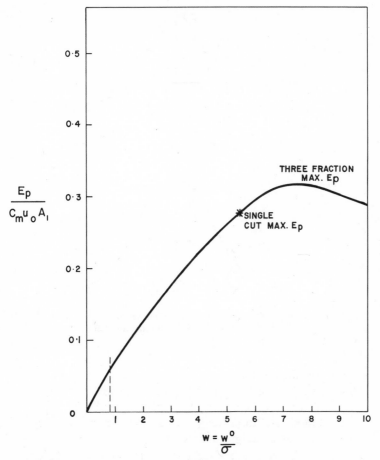

Fig. 1.11. The amount of the solute 1 which may be recovered per sample $(M_r)_1$, calculated as a function of $w = w^0/p$ for equimolar peaks of equal bandwidths, $(\eta_1^w)^0 = 0.0001$, $(\eta_1^c)^0 = 0.01$, and $R_i = 2$ (76a). The asterisk and dotted line, respectively, indicate the corresponding values for a single cut and a sample volume of $w = 1/\sqrt{2}$. The latter criterion was suggested by Glueckauf (61) and is analogous to that of Sawyer and Purnell.

The last aspect to be considered here is the concentration of the inlet sample. It is well known (68) that peak distortion results from operating in the nonlinear part of the distribution isotherm because of the fact that different parts of the zone assume different values of k. This effect has been investigated by Sawyer and Purnell (69), who found that the solution-limited volume is given by

$$V_{sl} = \left(\frac{X}{1-X}\right)\left(\frac{l_f}{l}\right)\left(\frac{\rho_p w_l}{M_l}\right)\left(\frac{\pi r_c^2 l M_s}{\rho_s}\right) \tag{1.43}$$

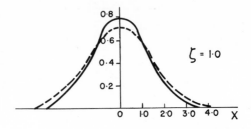

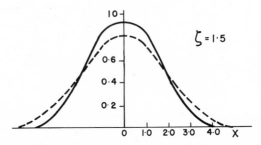

Fig. 1.12. Comparison of actual elution curves for plug inlet (solid line) with the elution curves for an equivalent Gaussian inlet.

where X = the mole fraction of solute in the solution
  l_f/l = the fractional length of column into which injection is made
  ρ_p = density of the composite packing
  w_l = fraction of total weight of packing present as solvent
  M_l = molecular weight of solvent

Nonlinearity can occur at values of X corresponding to less than 1 mole %, and in general linearity does not extend beyond about 10 mole %. Since l_f/l is determined by the feed-volume restrictions, and X by nonlinearity, the liquid sample volume can be uniquely determined from Eq. 1.43.

IV. Elution Development

The discussion in this section is based on a simplified model developed by de Clerk (70). In this model it is assumed that:

1. The mixture contains only two equimolar components, one of which has to be recovered in purified form.
2. The equivalent Gaussian inlet is applicable.
3. Only double fractions are recovered.

4. Flow is restricted to the laminar region.

5. No temperature or flow programming is employed.

6. Concentrations are restricted to the linear part of the distribution isotherm.

The first task, then, is to formulate a simplified analytical expression for preparative efficiency in terms of experimentally meaningful parameters. The parameters that are usually practical constraints in the laboratory are column length and pressure gradient, and it is therefore useful to retain these as running variables and eliminate the rest as far as is possible by means of optimization. For this purpose the definition of preparative efficiency given in Section I is retained, but in the actual presentation of the results the aim is to conform to the slightly modified definition:

The efficiency of a preparative column is measured by its production rate at a given purity, column length, and pressure gradient.

A. The Simplified Basic Equations

1. Derivation

The mass rate at which a component 1 of total mass m_1 is produced at the column outlet per unit sample injected is given by Eq. 1.6, namely,

$$E_p = \frac{(m_1 - \Delta m_1)\, u_o}{w_{to}(1 + k_1)} \tag{1.44}$$

This equation also applies to the continuous production rate if, as is assumed throughout, samples are introduced repetitively (Fig. 1.13). The task is now to transcribe this equation into variables that have practical significance. The troublesome parameters are $m_1 - \Delta m_1$ and w_{to}.

It is evident that Δm_1 can arise from two sources, the overlap with other components within a specific sample and the overlap between successive samples. The actual values of Δm_1 and w_{to} depend of course on the position of the cuts themselves and on the separating ability of the column. It is thus difficult to generalize about m_1 and w_{to} since they can differ widely according to the number of components contained in the sample. However, for the purpose of studying the influence of various variables on the inherent efficiency of a preparative column, it is sufficient to consider the case of an equimolar, two-component mixture. Estimates of the preparative efficiencies in other cases may then be made from the results for this special one.

The question of the measuring units for w_{to} has already been dealt with in Section III. The applicability of the equivalent Gaussian inlet is first tentatively assumed in the present development, and the shape of the peaks at the outlet is assumed to be sufficiently Gaussian so that there is a well-defined

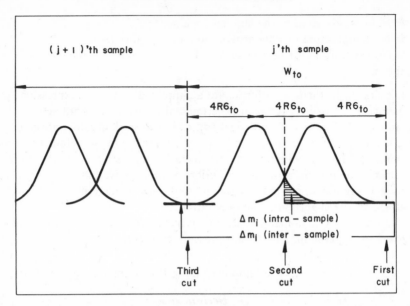

Fig. 1.13. Figure illustrating the resultant elution curves for a two-component sample when a repetitive inlet is employed.

relationship between the number of σ_{to}-values between the peak maxima within the sample and the purity. (σ_{to}^2 is here taken to be the total variance of the first peak at the outlet.) In the same way, cross contamination between samples can be specified in terms of the distance between them as measured in units of σ_{to}. For example, in the special equimolar, two-component case being considered here, a distance of $4\sigma_{to}$ between peaks leads to a contamination of about 2%, while a $6\sigma_{to}$ interval between samples gives an impurity of about 0.15%. It is always assumed in the following discussion that the frequency of sample injection has been regulated to make the cross contamination negligible in comparison with the overlap within the sample. This is ensured by taking the distance between samples as equal to twice the distance between the two peaks within the sample, i.e.,

$$w_{to} = (2R + R)\,4\sigma_{to}$$

$$= 12R\sigma_{to} \tag{1.45}$$

An expression for $m_1 - \Delta m_1$ is now easily found as

$$m_1 - \Delta m_1 = (m_1/2)(1 + erf\sqrt{2}R) \tag{1.46}$$

The mass m_1 itself can be related to the equivalent Gaussian variance. Consider a plug input located in an identical extension of the column but with impenetrable stationary phase.

$$m_1 = C_i V_i \qquad (1.47)$$

where C_i is the concentration, at the inlet pressure, of m_1 in the volume V_i. From the definition of the EG inlet it follows that

$$w_{ii} = \sqrt{2\pi} \, \sigma_{ii} \qquad (1.48)$$

where w_{ii} is the width of the plug within the column at the inlet. It is assumed that the time of introduction is short enough for peak form deviations attributable to plate height effects to be negligible. If the porosity is defined as the void fraction in the column

$$\epsilon = \frac{\text{volume of mobile phase}}{\text{total volume available for packing}}$$

the expression for m_1 can be written as

$$m_1 = (1 + k_1) \, C_i \pi r_c^2 \epsilon \sqrt{2\pi} \, \sigma_{ii} \qquad (1.49)$$

where r_c is the internal radius of the column. E_p thus becomes

$$E_p = \frac{\sqrt{2}\pi^{3/2} C_i r_c^2 \epsilon \sigma_{ii} u_o (1 + erf\sqrt{2}R)}{24R\sigma_{to}} \qquad (1.50)$$

It remains to relate σ_{to} and σ_{ii} to useful practical parameters. This can be effected by means of the resolution function R, which for equimolar peaks, is given by

$$R = \frac{\text{distance between peak maxima}}{4\sigma_{to}} \qquad (1.51)$$

A general expression for R, which takes variations along the column axis into account, can be derived (71) as

$$R = \frac{(\alpha - 1) \, k_1 u_o \int_0^l [dx/u(x)]}{4(1 + k_1)\left[\int_0^l [P^2(x)/P_o^2] \, H(x) \, dx + p^2\sigma_{ii}^2 \right]^{\frac{1}{2}}} \qquad (1.52)$$

where $P(x) =$ pressure at x
$P_i =$ inlet pressure
$P_o =$ outlet pressure
$p = P_i/P_o$
$\alpha =$ relative volatility.

When the indicated integrations are carried out and the resulting equation is solved for σ_{ii}, one finds

$$\sigma_{ii} = \beta \left\{ \left[\frac{(\alpha - 1)\, lk_1}{4R(1 + k_1)} \right]^2 - Hl \right\}^{\frac{1}{2}} \tag{1.53}$$

where

$$\beta = 2(p^3 - 1)/3p(p^2 - 1)$$

is a pressure correction associated with σ_{ii} and H is the HETP including pressure corrections. Specific expressions for these quantities are used in the actual calculations.

The bracket factor in Eq. 1.52 is just σ_{to}, so that this quantity follows directly as

$$\sigma_{to} = \frac{\delta(\alpha - 1)\, lk_1}{4R(1 + k_1)} \tag{1.54}$$

with the pressure correction δ given by

$$\delta = \frac{u_0}{l} \int_0^l \frac{dx}{u(x)} = p\beta = \frac{2(p^3 - 1)}{3(p^2 - 1)} \tag{1.55}$$

Substitution of Eqs. 1.53 and 1.54 in Eq. 1.50 yields the required expression for E_p as

$$E_p = \left[\frac{\sqrt{2\pi}\, C_i \epsilon \pi r_c^2 u_i}{24R} \right] \left[1 - \frac{16(1 + k_1)^2 R^2 H}{k_1^2(\alpha - 1)^2 l} \right]^{\frac{1}{2}} \left[1 + erf\sqrt{2}R \right] \tag{1.56}$$

$$= [f_A][f_B][f_C]$$

2. Discussion

Several interesting deductions may be made from the general properties of Eq. 1.56. In particular, the roles played by the variables in the present situation may be compared to those in the analytical case.

a. The Plate Height (H). It is evident from Eq. 1.56 that H plays only a secondary role in preparative work in contrast to its dominant place in the analytical efficiency function. Theories of preparative chromatography based on the plate height as efficiency function should therefore be considered inadequate.

b. The Concentration (C_i). Since E_p is directly proportional to C_i, the maximum concentration should be used. This maximum is determined by the linearity of the distribution isotherm if operation is restricted to linear chromatography. This does not imply that it would be necessarily deleterious to operate the column in the nonlinear region, since a fair amount of skewing may be tolerated in view of the gain in mass. In fact, preliminary theoretical investigations indicate that such a procedure may considerably enhance the efficiency of a preparative column. A complete discussion of these effects is, however, not possible at the moment.

c. The Concentration Distribution Coefficient (K). The remarks made in connection with the plate height also apply to K. It might at first be thought that an increase in K would lead to a higher efficiency because of the shorter initial plug length. This effect is, however, exactly off-set by the reduction in the velocity at the column outlet so that f_A is independent of K.

d. The Column Length (l). Inspection of Eq. 1.56 shows that a critical length l_a must be reached before production can start. This should correspond to zero inlet volume, i.e., l_a should be equal to the analytical length. That this is indeed the case, is seen by equating the bracketed term to zero and solving for l_a. The result is

$$l_a = \frac{16R^2H(1 + k_1)^2}{(\alpha - 1)^2 k_1^2} \tag{1.57}$$

which is simply the expression for the analytical length. When l is increased beyond l_a, larger inlet volumes may be used and E_p increases.

e. The Velocity (u_i). The velocity appears both explicitly and implicitly (through H) and these dependences have opposite effects on the efficiency. This leads to an optimum flow velocity for preparative chromatography, the value of which exceeds that for the corresponding analytical case. This may be seen by noting that the shortest critical length l_{am} is obtained by making H a minimum. The flow velocity at which this occurs is the usual optimum flow velocity for analytical work. For every $l > l_{am}$ there is therefore an optimum flow velocity for preparative work which exceeds that of the analytical value by a factor which increases with increasing length.

f. The Inlet Volume (V_i). The existence of an optimum velocity makes it possible to test the assumption of an equivalent Gaussian inlet. Since this velocity is in excess of that which makes H a minimum, H may for sufficiently long columns be approximated by

$$H = Cu_i \tag{1.58}$$

where the complete expression has been taken to be of the Van Deemter form. E_p may then be written as

$$E_p = E_p' u_i \left[1 - \frac{16(1 + k)^2 R^2 C u_i}{k^2(\alpha - 1)^2 l} \right]^{\frac{1}{2}} \tag{1.59}$$

where E_p' is independent of the velocity, and pressure effects have been neglected. The latter is a good approximation since H has been written in terms of the inlet velocity (72). The optimum inlet velocity is then found by differentiation as

$$u_i(\text{opt}) = \frac{k_1^2(\alpha - 1)^2 l}{24(1 + k_1)^2 R^2 C}$$ (1.60)

from which the ratio ξ follows as

$$\xi^2 = (\sigma_{ii}^2/Hl) = 0.5$$ (1.61)

so that $\xi \sim 0.7$. The inlet width is thus even slightly less than that produced by the column so that the equivalent Gaussian inlet appears to be a valid assumption provided that the optimum flow velocity is used.

The actual functional dependence of the inlet volume is found by substituting Eqs. 1.49 and 1.53 into Eq. 1.49.

$$V_i = (1 + k_1)\,\pi r_c^2 \epsilon \sqrt{2\pi\beta} \left\{ \left[\frac{(\alpha - 1)\,lk_1}{4R(1 + k_1)} \right]^2 - Hl \right\}^{\frac{1}{2}}$$ (1.62)

It is evident that the determination of V_i requires, apart from l and u_i which are, respectively, given and optimized, the specification of the column radius which may or may not be fixed by an optimum.

g. The Column Radius (r_c). The effect of the column radius is more complicated to assess because of the uncertainties in the dependence of H upon the radius. If H is explicitly radius dependent, the role of r_c appears to be analogous to that of u_i. Actually, this is more subtle since u_i and r_c are connected via the pressure gradient equation and the details of the optimization with respect to the radius is deferred to the specific treatment of open and packed columns.

h. The resolution (R). R appears in all three factors of Eq. 1.56. In f_C it expresses the fact that an increase in R leads to an increase in the amount of substance recovered with a corresponding increase in the efficiency. Its presence in f_B indicates that a price must be paid for this increase in terms of an increase in length (the other variables are considered as remaining constant). The R in the numerator of f_A merely takes into account the increased width of the sample (and consequently reduced E_p), with increasing R.

i. The Effect of the Stationary Phase Loading. The effect of the stationary phase loading is reflected in the preparative efficiency via ϵ, k, and H. In the plate height it effects both the mobile and stationary phase contributions, which complicates its interpretation to such an extent that it has been found expedient to assess its influence numerically. The results are given in the specific treatments of the open and packed columns, the interesting general result being that there is an optimum in the stationary layer thickness which depends slightly on the value of K.

j. The Temperature (T). Temperature does not appear explicitly in Eq. 1.56 but influences E_p in a complex fashion through its effect on a number of the variables. An extensive theoretical analysis of its influence is still lacking at the present time. It is interesting, however, to take note of the experimental results of Rose et al. (73). These investigators found a definite optimum temperature for the preparative efficiency, which appears to be independent of the carrier velocity. They attribute this to a balance between sample separation (resolution is usually better at lower temperatures, thus allowing larger samples to be used) and elution rate (higher temperatures reduce the retention time).

The general theory outlined above is now applied to the two main types of columns used in chromatography, namely, coated open tubular columns and particle-packed columns. The reasons for treating open tubular columns in any detail despite their inherent low capacities require some explanation. First, capacity demands are usually less stringent for difficult preparations for which the excellent resolution of such columns could be used to advantage. Pressure problems are also more easily avoided. In fact, it has been shown (74) that the ideal preparative column lies somewhere in between particle-packed and open columns, i.e., where the pressure advantages are combined with the capacity of packed columns. Since the flow dynamics in open columns are better understood and the mathematical treatment is simpler, open columns present a natural starting point for such an analysis.

3. Preparative Efficiencies in Open Tubular Columns

The plate height equation for open columns is given by Eq. 1.22, while the preparative efficiency expression can be rewritten in terms of Re as

$$E'_p = \frac{E_p}{C_i} = FRer_c\left(1 - \frac{G_1 H}{l}\right)^{\frac{1}{2}}$$

where

$$F = \frac{\pi\sqrt{2\pi}\, \nu_m(1 - \theta)(1 + erf\sqrt{2R})}{48R} \tag{1.64}$$

and

$$G_1 = \frac{16(1 + k_1)^2 R^2}{k_1^2(\alpha - 1)^2} \tag{1.65}$$

Once the purity has been specified by fixing R, the significant variables that remain are Re, r_c, l, p, and θ. Since these are coupled via the Poisseuille equation

$$p^2 - 1 = \frac{8\mu\nu_m plRe}{(1 - \theta)^3 r_c^3 P_o} \tag{1.66}$$

only four of them are independent. (P_o is considered to be fixed.) Of these, Re, r_t, and θ apparently exhibit optima with relation to the preparative efficiency, so that either l or p can be chosen as the independent variable. However, the optimization with respect to Re and r_t leads to identical equations so that optimization fixes only two variables, e.g., Re and θ, leaving two variables to be specified from the group r_t, p, and l. Since p and l are the most convenient pair in this respect, they are used as running variables while r_t is determined at optimum Re and θ from the Poisseuille equation.

The optimization with relation to Re is easily carried out if it is assumed that the optimum Re will so exceed that of the analytical case that H may be taken as $H = Cr_cRe$

$$Re(\text{opt}) = \left\{ \frac{(\alpha - 1)^2 l k_1{}^2}{24 R^2 C (1 + k_1)^2} \left[\frac{(1 - \theta)^3 (p^2 - 1) P_o}{8 \mu \nu_m p l} \right]^{\frac{1}{3}} \right\}^{\frac{2}{4}} \tag{1.67}$$

When this is substituted into E'_p, one finds

$$E'_p = \frac{2Fl(\alpha - 1)^2}{3\sqrt{3} G_3 C} \tag{1.68}$$

where

$$G_3 = G_1(\alpha - 1)^2$$

Optimization with relation to θ is complicated and has been performed numerically. Its value is fortunately independent of α, p, and l but depends on K, becoming smaller and less critical as K increases. This is shown in Fig. 1.14.

The radius corresponding to the optimum E_p is given by

$$r_c(\text{opt}) = \left\{ \frac{l^2 \mu_m \nu_m p k^2 (\alpha - 1)^2}{3(1 - \theta)^3 (p^2 - 1) P_o (1 + k)^2 R^2 C} \right\}^{\frac{1}{4}} \tag{1.69}$$

Several interesting deductions may be made from the above equations.

1. The preparative efficiency is directly proportional to the length of the column. Columns that are too long for operation in series may therefore be run in parallel without loss of efficiency.

2. E'_p is also proportional to $(\alpha - 1)^2$ so that considerable simplification in the presentation of the results is possible by introducing the variable

$$E''_p = \frac{E'_p}{(\alpha - 1)^2}$$

3. The efficiency itself is practically independent of the pressure gradient. This does not mean that the pressure is altogether unimportant. In fact, it is a very useful parameter to control r_c and Re within realistic limits as is evident from Eqs. 1.67 and 1.69.

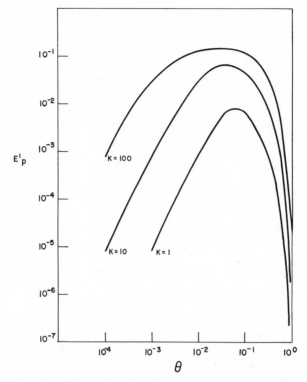

Fig. 1.14. Illustration of the optimum in $\theta = d_f/r_c$ for preparative GLC in open tubular columns (laminar flow).

E_p'' has been computed for various values of K for the following values of the constants. $\nu_m = 0.2$; $Scm = 1$; $R = 1.5$; $\nu_s = 1.0$; $Scs = 1000$; $\mu_m = 0.0002$.

For the calculations θ_{opt} was chosen as 0.01. The inlet volume (at 1 atm pressure) was calculated by means of Eq. 1.62.

The results are given in Figs. 1.14 to 1.17.

Figure 1.14: E_p' vs. θ for different values of K.

Figure 1.15: $E_p'' = \dfrac{E_p}{C_i(\alpha - 1)^2}$ vs. l for different values of p

Figure 1.16: $r_c'(opt) = \dfrac{r_c(opt)}{(\alpha - 1)^{\frac{1}{2}}}$ vs. l for different values of p

Figure 1.17: $V_i' = \dfrac{V_i}{(\alpha - 1)^2}$ vs. l for different values of p

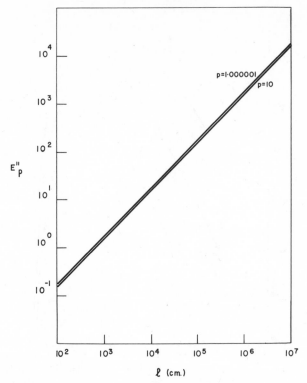

Fig. 1.15. Graph of $E_p = E_p/[C_i(-1)^2]$ against column length l for $K = 100$ and various values of p in open tubular columns.

For design purposes a suitable stationary phase is first chosen. This fixes the K-value, which in turn allows a θ-value to be chosen by means of the first graph. The next graph gives the length required for the desired production rate, while the radius and inlet volumes are found, respectively, from the final two graphs. The pressure gradient is selected to yield convenient values for p and V_i.

An indication of the yield to be expected is found by assuming $C_m = 0.001$ g/ml. For $\alpha = 1.1$ and a total column length in the range 10^3–10^5 cm, the preparative efficiency lies in the range 1.6–160 g/day. Preparative open tubular column work is thus seen to be restricted to the small-scale preparative field, as was anticipated.

4. Preparative Efficiencies in Packed Columns

The treatment of packed columns is complicated by the uncertainties in the plate height expression, and the numerical results presented here should be seen in this light.

The plate height is taken as given by

$$H = A + (B/R_p) + CR_p \qquad (1.70)$$

where $A = 5d_p$

$$B = \frac{2\gamma}{Scm}\left(\frac{\epsilon}{1-\epsilon}\right)d_p$$

$$\gamma = 0.6$$

$$C = C_s + C_c$$

$$C_s = 0.25 \frac{k}{(1+k)^2} \theta_p{}^2 d_p \frac{\nu_m}{\nu_s} Scs \frac{(1-\epsilon)}{\epsilon}$$

$$C_c = C_c' r_c{}^y$$

$$R_p = \left(\frac{d_p u_i}{\nu_m}\right)\frac{\epsilon}{1-\epsilon}$$

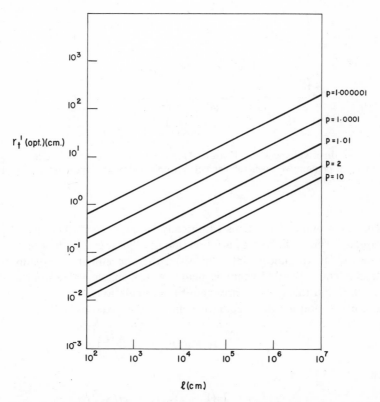

Fig. 1.16. Graph of the optimum in $r_c' = r_c/(\alpha - 1)^{1/2}$ vs. column length for various values of the pressure ratio p.

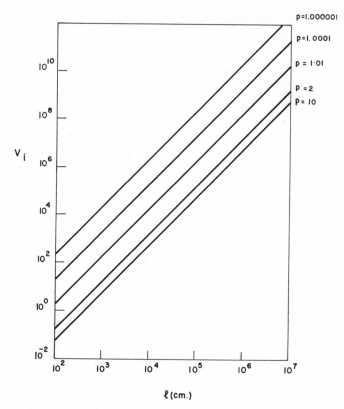

Fig. 1.17. Inlet volume $V' = V_i/(\alpha - 1)^2$, at inlet pressure, as a function of column length for a preparative open tubular column operated at the optimum values of θ, r_c, and u ($K = 100$).

The transcolumn term C_c is empirically taken as $C'_c r_c^y$ where y is a fitting parameter. The coefficient C'_c is a function of at least d_p and R_p (via D_R), but because of all the uncertainties involved the convenient approximation of setting it equal to a constant is made. A reasonable value appears to be $C_c = 0.01$ and this is used throughout the calculations.

The expression for the preparative efficiency is given by

$$E'_p = \frac{E_p}{C_i} = F_1 r_c^2 R_p \left(1 - \frac{G_1 H}{l}\right)^{\frac{1}{2}} \tag{1.71}$$

with

$$F_1 = \frac{\pi \sqrt{2\pi}(1 - \epsilon)\, v_m (1 + erf\sqrt{2R})}{24 R d_p}$$

and

$$G_1 = \frac{16(1 + k)^2 R^2}{k^2(\alpha - 1)^2}$$

An approximate expression for ϵ in terms of θ_p has been derived by assuming the liquid coating to be excluded from contact points between spherical particles with an average of three contact points per particle. If the particles are considered to be impenetrable, one finds

$$\epsilon = \epsilon_0 - (1 - \epsilon_0)\{(1 + 2\theta_p)^3 - 1\} + 6\theta_p^2(1 - \epsilon_0)(3 + 4\theta_p)$$

where $\epsilon_0 = 0.4$ is taken as the void fraction for uncoated particles. This expression predicts clogging at about $\theta_p = 0.2$, which seems reasonable. In the case of completely porous particles, ϵ is initially independent of θ_p; the dependence of ϵ on the stationary phase loading in actual packings is very difficult to describe theoretically. For the present purpose the nonporous model is considered adequate.

The flow must satisfy the pressure gradient equation. Since flow is considered to be restricted to the laminar region, a good approximation is to use the Blake-Kozeny equation (75), suitably corrected for gas compressibility. i.e.,

$$l = D/R_p \tag{1.72}$$

where

$$D = \frac{(p^2 - 1)\, P_o d_p^3 \epsilon^3}{300 p \mu \nu_m (1 - \epsilon)^3}$$

If the pressure corrections to the plate height are neglected, E_p' is independent of the pressure when written in the form of Eq. 1.56. The optima in the flow velocity and in the column radius can then easily be computed by setting

$$\left(\frac{\partial E_p'}{\partial R_p}\right)_{l, r_c, d_p, \theta_p, y} = 0$$

and

$$\left(\frac{\partial E_p'}{\partial r_c}\right)_{l, R_p, d_p, \theta_p, y} = 0$$

This yields

$$R_p(\text{opt}) = \frac{2(l - G_1 A)}{3 G_1 C} \tag{1.73}$$

and

$$r_c(\text{opt}) = \left(\frac{2C_s}{C_c(y - 2)}\right)^{1/y} \tag{1.74}$$

This illustrates an important point, namely, that simultaneous optima in R_p and r_c need not exist, e.g., for $y \leq 2$ in the above case. For the actual calculations it was found expedient to substitute l in Eq. 1.72 from Eq. 1.71. Subsequent optimization with respect to R_p at constant pressure leads to

$$R_p(\text{opt}) = -\frac{3A}{8C} + \left(\frac{9A^2}{64C^2} + \frac{D}{2G_1C}\right)^{\frac{1}{2}} \qquad (1.75)$$

from which p may be solved as

$$p = \frac{Z + (Z^2 + 4)^{\frac{1}{2}}}{2} \qquad (1.76)$$

with

$$Z = \frac{2G_1C}{C_3}\left(R_p{}^2 + \frac{3AR_p}{4C}\right)$$

and

$$C_3 = \frac{P_o d_p{}^3 \epsilon^4}{300\mu_m \nu_m (1 - \epsilon)^3}$$

The behavior of E_p with respect to θ_p was first investigated numerically. This is illustrated in Fig. 1.18. It is seen that E_p first increases with θ_p, then levels off, and should eventually decrease and go through zero when the column becomes clogged. For the present calculations $\theta_p = 0.01$ was used. p was next calculated from Eq. 1.76 for a given R_p value and used to calculate l from Eq. 1.72. These values then correspond to those at which the velocity is optimized. The results are presented in Fig. 1.19. As an example of the actual yield that can be expected provided that the various assumptions are satisfied, consider a component to be regained from a binary mixture with $\alpha = 1.1$ in which its concentration is, e.g., 0.001 g/ml. The above equations then predict a production rate of about 250 g per day when a pressure of 32 atm is applied to a 420-cm column packed with 0.01-cm-diameter spherical particles and 108-ml samples ($\sim$3 ml liquid) are injected repetitively. This fairly low rate is to be expected in view of the narrow-bore column used, namely, $r_c = 1.0$ cm.

The efficiency decreases with increasing d_p, but since the required pressure gradient also diminishes, particle diameter becomes an important variable for determining the balance point between efficiency and pressure gradient. Pressure gradient appears to be the limiting practical constraint in packed-column preparative work and the results clearly indicate the desirability of investigations into the possibility of alternative packing materials which retain capacity but with less flow resistance than particles, e.g., foams and perforated plates.

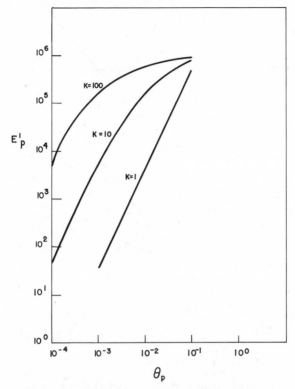

Fig. 1.18. E_p (inarbitrary units) plotted as a function of $\theta_p = d_p/r_c$ where a nonporous support has been assumed.

V. Modifications of the Simplified Theory

The simplifying assumptions that underly the above treatment were stated explicitly. The present section attempts to assess the effect of removing these constraints.

A. Multicomponent Mixtures

The presence of more than two components always results in a lower preparative efficiency because of the increased total bandwidth. This should not affect the optima in the case in which a single component must be recovered since this is determined by the nearest neighbor component. The only major change in such cases would be in the frequency of injection. However, the range of k-values may become so large that programming becomes necessary.

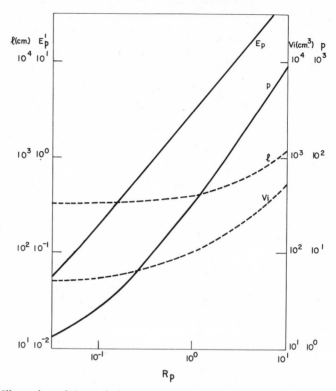

Fig. 1.19. Illustration of the variation of E_p, p, l, and V_i with $R_p = (d_p u / v_m)[\epsilon/(1 - \epsilon)]$ for an optimized packed preparative column.

The consequences for the preparative efficiency in such cases have not been explored theoretically.

B. Nonequimolar Peaks

The effect of nonequimolar peaks may be conveniently studied by regarding its influence as reflected in a higher R-value required to effect the separation to the same purity. Such an analysis has not yet been carried out in the idiom of Section IV, but Said (76) has developed graphs that can profitably be used for this purpose.

C. The Three-Fraction Technique

The three-fraction technique is a natural extension of the two-fraction method. Consider a two-fraction column operating at optimum. It is apparent that the efficiency can be improved by increasing the inlet volume while simultaneously ensuring that the purity specifications are maintained by making two cuts instead of one. Eventually, however, the inlet volume passes through an optimum. This problem has been extensively investigated by Gordon (77–81) from whose data Fig. 1.20 has been compiled. This figure shows the efficiency increase that could be expected by employing the three-fraction technique. It is seen that substantial gains are possible at high purification but that the increase at lower purities is small. Whether or not this increase would be financially justified, in view of the additional cost entailed by the more complex fraction cutter and the recirculation or discarding of the middle portion, is a matter that would require careful consideration for each specific case.

Gordon's work is important in its own right, however, since it represents the first systematic analysis of the efficiency of a preparative process. It is also in a

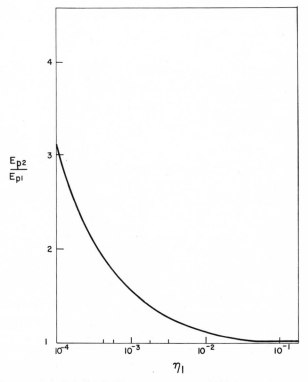

Fig. 1.20. An illustration of the ratio of the efficiencies which can be expected by employing a double or a single cut (equimolar peaks, equal bandwidth. η_1 gives the impurity of the first component as a result of overlap within the band.

form that lends itself directly to a comparison with an extensive analysis of preparative chromatography by frontal analysis (82) so that it is worthwhile to state some of the main results here.

The first task in this study was to find an expression for the preparative efficiency at the optimum inlet volume.

$$E_p(W_{opt}^0) = \frac{C_i u_i A_1}{2}\left[1 - \frac{f(\eta_1^E)^0}{R_i}\right] \tag{1.77}$$

where

$$f(\eta_1^E)^0 = \frac{[36 + 22 \log_{10}(1/\eta^E_1)^0]}{125}$$

and $\eta_1^E \equiv$ the impurity fraction of the specific component.

The resemblance of this equation to that of the simplified model is striking. It should be pointed out, however, that the R_i used in this equation is the resolution for an infinitesimally small sample and thus differs from the R defined in Eq. 1.1. The influence of the following variables were then successively assessed:

1. Flow velocity (78).
2. Column diameter (79).
3. Column length and particle size (80).
4. Stationary liquid loading and column packing (81).

The plate height expression used in these studies include the initial effects according to Giddings, i.e.,

$$H = A + A_d l + B/u_i P_i + (C_g P_i + C_l)\, u_i \tag{1.78}$$

when

$$u_i \geq (u_i)_{tr}$$

and

$$H = A + B/u_i P_i + (C_g P_i + C_l + C_d d^2 P_i)\, u_i$$

when

$$u_i \leq (u_i)_{tr}$$

where $(u_i)_{tr}$ is defined as that velocity at which the two transcolumn terms become equal, namely, $A_d l = C_d d^2 u_i P_i$.

The use of these expressions introduces certain subtleties into the optimization, the significance of which is difficult to evaluate at this stage because of the uncertainties in the transcolumn contribution. In the present review, these are largely disregarded and attention is restricted to the general role played by the various variables.

1. Flow Velocity

The general conclusion is that preparative efficiency increases with flow velocity and ultimately reaches a maximum velocity which is the limiting velocity at which the required resolution can still be met. There is also a theoretical optimum in the flow velocity but this lies at a higher value than the above maximum. This differs from the result for the simplified single-cut model and probably results from the fact that the three-fraction technique effectively adds to the actual column length, thereby shifting the value of the optimum flow velocity to a higher value.

2. Column Diameter

The role of the column diameter is intimately related to the plate height expression used and this is reflected in the results obtained in this study. It is shown that at the limiting value of the flow velocity, for fairly large α-values, the preparative efficiency increases linearly with the square of the column diameter, whereas for smaller values a limiting value of the preparative efficiency with respect to the diameter is reached. The latter corresponds to the plate height with $h_c = Cr_c^2 u$, in which case the column diameter plays a role mathematically analogous to the flow velocity. In the former case H is independent of r_c so that there is no restraining influence and the result is intuitively evident.

3. Column Length

It is found that when operating the column at the maximum flow velocity, the preparative efficiency increases almost linearly with column length. This is in accord with the results of the single-cut model. In the process of increasing the column length, the pressure required to maintain the desired flow velocity is of course also increased and the preparative efficiency therefore eventually becomes limited by the available pressure. This may be overcome by increasing d_p, but this again occurs at the expense of efficiency.

4. Stationary Liquid Loading and Column Packing

This effect was extensively investigated by Gordon, to whose original paper (81) the interested reader is referred for details. In general, it was found that the percentage of liquid should be the maximum that the packing can hold, again in essential accord with the predictions of the simplified model.

D. Turbulent Flow

This subject has not been extensively investigated, but a preliminary survey (83) indicates that there are probably no significant advantages despite the fact that the mobile phase contribution to the plate height is less than it would have been for laminar flow at the same velocity. In addition, many practical difficulties are foreseen, e.g., excessively long columns, high inlet pressures, and stripping of the stationary phase at the high flow velocities.

E. Nonlinearity

Consider a preparative column that is optimized with respect to a single cut. The mass production can now be still further increased by increasing the inlet concentration. However, when this concentration reaches the nonlinear part of the distribution isotherm, the peaks become skewed and it becomes necessary to increase the column length to conform to the purity requirements. The question is whether or not this additional length could be used more profitably by increasing flow velocity and inlet volume. This question has not been answered in general terms.

F. Thermal Effects

In the simplified model isothermal operating conditions were assumed. This is, however, not true in practice because of the combined effects of the heat of solution of the larger samples and the finite rate of heat transport across the larger-diameter columns. Scott (84) has studied the temperature effects resulting from the passage of a solute through a theoretical plate and concludes that the excess heat generation increases with (1) increasing flow rate (2) increasing sample size and (3) decreasing k-values. The first two factors are of obvious significance for preparative work, and the experimental results of Rose et al. (73) indicate that the deleterious effects are attendant on these temperature variations and heat transfer properties could be of primary importance in determining the design of preparative columns. The effects on the peak shape are similar to those produced by nonlinear effects since the larger heat production at higher concentrations leads to a front-sharpening resulting from an acceleration of the zones of higher local temperature.

The temperature lag between the center and the wall of the column attributable to the resistance against heat transfer is also of importance when temperature programming is used. Giddings (86) investigated this phenomenon and concludes that its effect can be incorporated in terms of an additional contribution to the plate height, namely,

$$H_{tp}(x) = \frac{\alpha \beta^2 u r_c{}^6}{14,400 k^2 D_m} \tag{1.79}$$

where β = heating rate
$\alpha \sim 0.004$

VI. Frontal Analysis

Preparative chromatography by frontal analysis differs from its elution counterpart in that the column is so overloaded that separate peaks are not present at the column outlet; recovery of the least retarded component can consequently only be effected at the peak front and not also from the tail of the peak as is the case in elution chromatography.

In order to avoid confusion, however, it should be pointed out that the characteristic flat-topped peaks of frontal analysis may be produced in two distinct ways, namely, by means of either a single-step input function or a double step. This is illustrated schematically in Fig. 1.21.

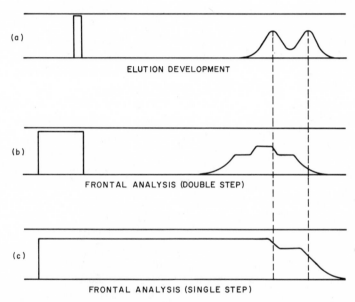

Fig. 1.21. Schematic comparison of some techniques in preparative chromatographic work.

According to the strict definition of frontal analysis, sample introduction is continued until the cut point is reached (Fig. 1.21c), after which the column is completely flushed out before introduction of the next sample. The application of this technique to preparative work has been extensively investigated by Krige (87), who has compared his results to those of Gordon on elution chromatography in a series of papers. The preparative efficiency in this case is given by

$$E = \frac{(M_r)_1}{t_i + t_F{}^1} \tag{1.80}$$

where t_F^l denotes the time required to flush the sorbed solutes from the column, as measured from the moment that flushing is commenced. The time t_i is simply the total time during which the sample mixture is continuously introduced at the column inlet. In the case of a dilute solution, Eq. 1.80 may to a good approximation be related to Gordon's corresponding expression for elution development at the optimum sample inlet volume (Eq. 1.77) by

$$E \simeq \left[\frac{\overline{V}_b - \overline{V}_a}{\overline{V}_b} \right] E(W_{\text{opt}}^0) \qquad (1.81)$$

The functional similarity of these equations leads one to expect practically identical optimum operating conditions for the two modes of operation and this is indeed the case. As regards the actual production rate, it is evident from this equation that a higher preparative efficiency can be obtained from repetitive injection of plug samples than from frontal analysis, provided that the injected samples are highly diluted. The superiority of elution development becomes progressively more pronounced as the difference between the retention volumes of the components increases. However, the possibility is not excluded that the situation may be reversed in some cases in which high inlet concentrations are utilized, since frontal analysis is in general less sensitive to nonlinear effects. The treatment in such cases becomes involved, however, and comparisons between the two methods are not readily made. For a treatment of the phenomena that accompany high-concentration preparative work by frontal analysis, the interested reader is referred to Krige's original publications (88).

An obvious limitation in frontal analysis by the single-step technique is the fact that only the least retarded solute can be directly recovered in a pure form. In order to circumvent this difficulty, a natural extension would be the repeated processing of a double-step input (Fig. 1.21b) in which the least and most retarded remaining components after a specific run are, respectively, purified by front and tail cutting in the consecutive run. Schemes for effecting this have been elaborated by Reilly and co-workers, but the relation to elution development in terms of a suitably defined preparative efficiency has not been clarified. In view of Krige's results on the single-step procedure, it also seems unlikely that such procedures will be competitive with elution development, at least not for dilute samples.

VII. Conclusions

The problem of preparative chromatography is essentially that of the scale-up of the analytical procedures to a desired production level, and the theoretical studies outlined above can be considered to have demonstrated—albeit in a semiquantitative fashion—the role of the key parameters involved in this process. It has been shown that an increase in length is always required when an increased production is effected by means of an increase in one or more of the parameters: column diameter, flow velocity, and sample inlet volume. In particular, the existence of an optimum represented by a simultaneous increase of flow velocity and sample volume has been demonstrated. Increase in column length and flow velocity is accompanied by an increase in pressure requirements and in practice a convenient compromise between column length and pressure will probably be found by adjusting the particle diameter (for packed columns). The preparative application of the conventionally coated open tubular columns, because of their inherent low capacity, will probably be restricted to difficult separations of costly substances. It is to be expected, however, that new open-type column packings will be evolved which will combine the permeability of open tubular columns with the capacity of particle-packed columns. In this respect prefabricated, rigid foam packings appear especially promising because of their lower cost and the precision with which the porosity can be controlled. This could to a considerable extent help to eliminate the difficulties involved in diameter scale-up. Even at present, these difficulties can apparently be effectively minimized by incorporating radial mixers. The optimum column diameters will therefore probably be determined as a balance between capital and running costs.

In conclusion, it should once again be stressed that the results reviewed above were obtained on the basis of a simplified criterion of merit, namely, production rate. In actual practice this is but one of the parameters—albeit an important one—that determine the design of a commercial installation. It is to be expected that the aim in practice will be to maximize profits for a production rate as defined by the market. In addition, it is important to realize that a complete optimization must include auxiliary equipment, in particular a carrier gas recovery plant. One can therefore expect future applied preparative chromatographic theory to acquire a distinct operations research character.

However, much remains to be done in the fundamental theory, especially on the thermodynamic aspects that govern the choice of phases, nonlinearity, thermal effects, and multicomponent separations, and these must be thoroughly understood if the full potential of preparative chromatography is to be realized.

Symbols

A Parameter defined by Eq. 1.5
A_1 Cross-sectional area of column
A_d Parameter defined by Eq. 1.78

α Relative volatility
α_b Parameter defined by Eq. 1.38
α_t Parameter defined by Eq. 1.37
B Parameter defined by Eq. 1.5
β Heating rate Eq. 1.79
C Parameter defined by Eq. 1.5
C_m Mobile phase contribution to C
C_s Stationary phase contribution to C
$C(x)$ Concentration at x
C_i Concentration at the inlet
C_{jm} Concentration of j-th component in mobile phase
C_{js} Concentration of j-th component in stationary phase
C_g Parameter defined by Eq. 1.78
C_d Parameter defined by Eq. 1.78
D_m Molecular diffusion coefficient
D_R Radial dispersion coefficient
D_R' D_R, excluding D_m
D_s Molecular diffusion coefficient in stationary phase
D_{eff} Effective longitudinal diffusion coefficient (axial dispersion coefficient)
D_{eff}' Axial dispersion coefficient resulting from random exchange of molecules between regions of different axial flow velocity (resistance to mass transfer term)
d_c Column inside diameter
d_f Thickness of stationary phase liquid layer
E Production rate for solute i by frontal analysis
$E(W_{\text{opt}}^0)$ Production rate for solute i by elution development at the optimum equivalent sample inlet volume for the three-fraction technique
E_p Production rate of i-th component by elution development for two-fraction technique

$$E_p' = \frac{E_p}{c_i}$$

$$E_p'' = \frac{E_p}{C_i(\alpha - 1)^2}$$

ϵ Void fraction
ϵ_0 Void fraction for uncoated particles
η_i^E Impurity fraction of i-th component in elution development
$(\eta_i^w)^0$ Equivalent impurity ratio for overlap between samples
$(\eta_i^c)^0$ Equivalent impurity ratio for overlap within a sample
F Parameter defined by Eq. 1.64
G Parameter defined by Eq. 1.36
G_1 Parameter defined by Eq. 1.65
G_2 Parameter defined by Eq. 1.33
γ Labyrinth factor
H Height equivalent to a theoretical plate (HETP)
$H(x)$ Local HETP
$H_m(x)$ Mobile phase contribution to local HETP
$H_s(x)$ Stationary phase contribution to local HETP
$H_t(x)$ Plate height contribution attributable to thermal fluctuations
$H_b(x)$ Plate height contribution attributable to column bending

$H_{tp}(x)$ Plate height contribution attributable to temperature programming

h Reduced plate height $= H/r_c$ for open columns $= H/d_p$ for packed columns

h_m Reduced plate height for mobile phase

h_s Reduced plate height for stationary phase

$K = C_s/C_m$ Concentration distribution coefficient

$K_j = \dfrac{C_{js}}{C_{jm}}$ Concentration distribution coefficient of the j-th component

$k = m_s/m_m$ Mass distribution coefficient

$k(r)$ Local value of k at radial distance r

α Parameter defined by Eq. 1.35

l Column length

l_a Column length required for the separation of an infinitesimally small sample

l_{am} Minimum of l_a

$l_{f/l}$ Fractional length into which injection is made

λ Probability factor

$(M_r)_1$ Amount of solute recovered per run in frontal analysis

M_l Molecular weight of solvent

M_s Average molecular mass of sample

m Parameter defined by Eq. 1.34

m_j Total mass of j-th component in sample

m_{js} Mass of j-th component in stationary phase

m_{jm} Mass of j-th component in mobile phase

Δm_j Mass discarded of component j during fraction cutting

μ Viscosity

μ_m Mobile phase viscosity

ν_m Kinematic viscosity for mobile phase

ν_s Kinematic viscosity for stationary phase

P_i Inlet pressure

P_o Outlet pressure

$p = P_i/P_o$

R Resolution $= \Delta x/4\sigma$

R_i Resolution for infinitesimal sample

r Radial coordinate

r_c Inside radius of column

$Re = \dfrac{2r_c u}{\nu_m}$ Reynolds number for open column

$R_p = \dfrac{dpu}{\nu} \dfrac{\epsilon}{1 - \epsilon} =$ Reynolds number for packed column

R_0 Radius of bend in column

$p = d_p/d_c$

ρ_s Average density of sample

ρ_p Density of composite packing

$Sc = \nu/D =$ Schmidt number

$Scm = \nu_m/D_m$

$Scs = \nu/Ds$

σ^2 Second moment (variance) of concentration distribution

σ_{ii}^2	Variance at inlet within column
σ_t^2	Total $\sigma^2 = \sigma_c{}^2 + \sigma_{ii}^2$
$\sigma_c{}^2$	Variance produced within column
σ_{to}^2	Total variance at column outlet within the column
t	Time variable
t_{tr}	Transient time required for effective axial dispersion to become a valid mode of description
Δt	Effective time for a random walk step
θ	d_f/r_c
θ_p	d_f/d_p
ΔT	Temperature difference between column wall and center (Eq. 1.37)
$\overline{t}$	Retention time of solute front in frontal analysis
t_i	Time for which a solute mixture is continuously introduced into a column
t_F'	Time required to flush the sorbed solutes from a column with an inert carrier, as measured from the moment flushing is commenced in frontal analysis
u_o	Linear carrier velocity at outlet
u_i	Linear carrier velocity at inlet
Δu	Difference between extremes of velocity variation
$(u_i)_{tr}$	Transition velocity defined by Eq. 1.78
$\langle u \rangle$	Sample velocity averaged over column cross section
v	Effective axial velocity of random walk
V_s	Volume of stationary phase
V_m	Volume of mobile phase
V_{fl}	Sample volume as a liquid
V_i	Inlet volume as a gas at the pressure existing at the inlet but without the column
$\overline{V}$	Retention volume of solute front in frontal analysis, i.e., the total volume of fluid that has passed through the column outlet from $t = 0$ to $t = \overline{t}$
W^0	Equivalent inlet volume of sample in volume of eluate as measured at the column outlet during the interval in which solute is introduced into the column
$W = W^0/p$	
W_{opt}^0	Optimum equivalent inlet volume of a sample
w_{to}	Total width of fraction cut out at outlet within column
w_e	Fraction of total weight of packing present as solvent
x	Axial coordinate
Δx	Distance between peak means
$\Delta x'$	Effective axial step length during random walk
X	Mole fraction of solute
y	Parameter defined by Eq. 1.70
Z	Parameter defined by Eq. 1.76
$\xi = \sigma_{ii}/\sigma_c$	

References

1. T. Ellerington, in *Gas Chromatography 1958*, D. H. Desty, Ed., Butterworths, London, 1958, p. 199.
2. W. L. Jones and R. Kieselbach, *Anal. Chem.*, **30**, *1590* (1958).

3. In *Gas Chromatography 1960*, R. P. W. Scott, Ed., Butterworths, London, 1906, p. 423.
4. J. C. Giddings, in *Dynamics of Chromatography*, Part I, Edward Arnold, London, 1965, p. 269.
5. J. C. Sternberg, in *Advances in Chromatography*, *Vol. 2*, J. C. Giddings and R. A. Keller, Eds., Marcel Dekker, New York, 1966, Chapter 6.
6. See Ref. 4, p. 24.
7. J. J. van Deemter, F. J. Zuiderweg, and A. Klinkenberg, *Chem. Eng. Sci.*, **5**, 271 (1956).
8. M. J. E. Golay, in *Gas Chromatography 1960*, R. P. W. Scott, Ed., Butterworths, London, 1960, p. 143.
9. P. C. Haarhoff, *Contributions to the Theory of Chromatography*, D.Sc. Thesis, University of Pretoria, Pretoria, South Africa, 1962, p. 182.
10. K. de Clerk, *Contributions to the Theory of Chromatography*, D.Sc. Thesis, University of Pretoria, Pretoria, South Africa, 1966, p. 111.
11. A. B. Littlewood, *Gas Chromatography*, Academic Press, New York, 1962, p. 192.
12. R. Aris, *Proc. Roy. Soc. (London)*, **A235**, 67 (1956).
13. G. I. Taylor, *Proc. Roy. Soc. (London)*, **A219**, 186 (A53).
15. See Ref. 4, Chapter 3.
16. See Ref. 9, Chapter 3.
18. G. I. Taylor, *Proc. Roy. Soc. (London)*, **A225**, 473 (1954).
19. J. C. Giddings, *J. Gas Chromatog.*, **1**, 12 (1963).
20. K. B. Bischoff and O. Levenspiel, *Chem. Eng. Sci.*, **17**, 257 (1962).
21. J. H. Knox and L. McLaren, *Anal. Chem.*, **36**, 1477 (1964).
22. R. Aris, *Proc. Roy. Soc.*, **A252**, 538 (1959).
23. See Ref. 4, Chapter 4.
24. See Ref. 4, Chapter 2.
26. K. de Clerk, T. W. Smuts, and V. Pretorius, *Separation Sci.*, **1**, 443 (1966).
27. K. de Clerk, to be submitted for publication.
28. L. S. Ettre, *Open Tubular Columns in Gas Chromatography*, Plenum Press, New York, p. 13.
29. V. Pretorius and T. W. Smuts, *Anal. Chem.*, **38**, 274 (1966).
30. See Ref. 4, p. 180.
31. R. B. Bird, W. E. Stewart, and E. N. Lightfoot, *Transport Phenomena*, John Wiley, New York, 1960, p. 198.
32. See Ref. 31, p. 193.
32a. K. R. Jolles and T. J. Hanratty, *Chem. Eng. Sci.*, **21**, 1185 (1966).
32b. V. G. Levich, *Physicochemical Hydrodynamics*, Prentice-Hall, Englewood Cliffs, New Jersey, 1962.
32c. A. E. Scheidegger, *The Physics of Flow through Porous Media*, University of Toronto Press, Toronto, 1960.
32d. H. S. Mickley, K. A. Smith, and E. I. Korchak, *Chem. Eng. Sci.*, **20**, 237 (1965).
33. See Ref. 4, p. 42.
34. See Ref. 4, Chapter 2.
35. See Ref. 10, Chapter 3.
36. T. W. Smuts, *A Fundamental Study of High Speed Chromatography*, D.Sc. Thesis, University of Pretoria, Pretoria, South Africa, 1967, p. 89.
37. See Ref. 4, p. 52.
38. K. de Clerk, to be submitted for publication.
41. J. C. Giddings and E. N. J. Fuller, *J. Chromatog.*, **7**, 255 (1962).

42. G. M. C. Higgins and J. F. Smith, in *Gas Chromatography 1964*, A. Goldup, Ed., Institute of Petroleum, London, 1965, p. 94.

43. J. C. Giddings, *J. Gas Chromatog.*, **1**, 38 (1963).

44. M. Golay, 2nd International Symposium ISA, June 1959, Preprints p. 5.

45. See Ref. 50.

46. See Ref. 36, p. 77.

47. S. F. Spencer and P. Kucharski, presented at 1966 Pittsburgh Conference F and M, Technical Paper No. 37.

48. K. P. Hupe, U. Busch, and K. Winde, in *Advances in Chromatography*, A. Zlatkis, Ed., Preston Technical Abstracts Co., Evanston, Illinois, 1969, p. 107.

49. See Ref. 42.

50. F. H. Huyten, W. Beersum, and G. W. A. Rijnders, in *Gas Chromatography 1960*, R. P. W. Scott, Ed., Butterworths, London, 1960, p. 224.

51. G. W. A. Rijnders, in *Advances in Chromatography*, **3**, Marcel Dekker, New York, 1966, p. 215.

53. E. Bayer, K. Hupe, and H. Mack, *Anal. Chem.*, **35**, 492 (1963).

54. R. F. Baddour, U. S. Patent 3,250,058 (1966).

55. R. W. Reiser, *J. Gas Chromatog.*, **4**, 390 (1966).

57. J. C. Giddings, *J. Gas Chromatog.*, **2**, 290 (1964).

58. See Ref. 11, p. 19.

59. C. N. Reilly, G. P. Hildebrand, and J. W. Ashley, Jr., *Anal. Chem.*, **34**, 1198 (1962).

60. W. J. de Wet and V. Pretorius, *Anal. Chem.*, **32**, 169 (1960).

61. E. Glueckauf, *Trans. Faraday Soc.*, **51**, 34 (1955).

62. W. J. de Wet and V. Pretorius, *Anal. Chem.*, **32**, 1396 (1960).

63. T. M. Reed, J. F. Walter, R. R. Cecil, and R. D. Dresdner, *Ind. Eng. Chem.*, **51**, 271 (1959).

64. P. T. Sawyer and H. Purnell, *Anal. Chem.*, **36**, 457 (1964).

65. P. C. Haarhoff, P. C. van Berge, and V. Pretorius, *Trans. Faraday Soc.*, **57**, 1838 (1966).

66. See Ref. 10, p. 18.

67. See Ref. 10, p. 18.

68. E. Glueckauf, *Trans. Faraday Soc.*, **51**, 1540 (1955).

69. D. T. Sawyer and H. Purnell, *Anal. Chem.*, **36**, 457 (1964).

70. See Ref. 10, p. 111.

71. See Ref. 10, p. 107.

72. See Ref. 9, p. 89.

73. A. Rose, D. J. Royer, and R. S. Henly, *Separation Sci.*, **2**, 257 (1967).

74. See Ref. 10.

75. See Ref. 31, p. 200.

76a. A. S. Said, *J. Gas Chromatog.*, **2**, 60 (1964).

76b. S. M. Gordon, *A Fundamental Study of Preparative Gas-Liquid Chromatography*, D.Sc. Thesis, University of Pretoria, Pretoria, South Africa, 1964.

77. S. M. Gordon and Victor Pretorius, *J. Gas Chromatog.*, **2**, 196 (1964).

78. S. M. Gordon, G. J. Krige, and V. Pretorius, *J. Gas Chromatog.* **2**, 241 (1964).

79. S. M. Gordon, G. J. Krige, and V. Pretorius, *J. Gas Chromatog.*, **2**, 285 (1964).

80. S. M. Gordon, G. J. Krige, and V. Pretorius, *J. Gas Chromatog.*, **2**, 246 (1964).

81. S. M. Gordon, S. J. Krige, and V. Pretorius, *J. Gas Chromatog.*, **3**, 87 (1965).

82. G. J. Krige, *Large-Scale Chromatography by Frontal Analysis*, D.Sc. Thesis, University of Pretoria, Pretoria, South Africa, 1964.

83. See Ref. 10.

84. R. P. W. Scott, *Anal. Chem.*, **35**, 481 (1963).
85. A. Rose, D. J. Royer, and R. S. Henly, *Separation Sci.*, **2**, 257 (1967).
86. J. C. Giddings, *J. Gas Chromatog.*, **1**, 12 (1963).
87. G. J. Krige and V. Pretorius, *J. Gas Chromatog.*, **2**, 115 (1964).
88. G. J. Krige and V. Pretorius, *Anal. Chem.*, **37**, 1186, (1965).
89. J. H. Knox, in *Advances in Gas Chromatography*, A. Zlatkis and L. Ettre, Eds., Preston Technical Abstracts Co., Evanston, Illinois, 1966.

CHAPTER 2

Inlet System

K. -P. Hupe, *Hupe + Busch, Karlsruhe, Germany*

I. Introduction

The use of gas–liquid chromatography (GLC) for the separation of larger quantities of material is noted for the fact that gas chromatographic columns cannot be loaded with as much sample as one likes because the resolving power diminishes with increased loading. Increase in column dimensions can compensate for this only within certain limits. When larger quantities must be separated, it therefore becomes necessary to divide the total quantity into smaller portions with which the column can deal and to subject each of these portions to separation. Preparative GLC as it is presently used in the laboratory is thus a discontinuous process. The separate sequential phases of this process, such as injection, evaporation, separation, and trapping of the materials, known collectively as "the separating cycle," must therefore be repeated until the required amount of purified material is obtained.

II. Injection Systems

A new separating cycle starts with sample injection. The distance between injections is either determined by time or by a recurring point in the separating cycle such as the occurrence of a certain peak in the chromatogram. The sample injection can naturally be made manually by syringe as in analytical work. This method is recommended for small quantities of sample requiring only a few cycles, and every preparative instrument has an injection port for manual injection. Similar to all other parts of the separating cycle, this process is made automatic for longer runs. The injection system has thus the task of transferring at a certain distance, on a given signal, a definite quantity of starting material from the storage vessel to the vaporizer. The following requirements must be complied with; they stem partly from the chromatographic process and partly from the expediency of the practical operation.

1. The quantity of material to be injected must be easily adjustable. It is generally sufficient for the accuracy of adjustment to be $\pm 10\%$, as the change in resolving power caused by differences in loading within these limits is hardly noticeable. An instrument for general laboratory use should be capable of injecting quantities in the range of 50 μl to 10 ml. The lower limit is set by correspondingly small quantities of starting material or by the use of narrow columns to obtain higher column efficiency. Quantities of this order are generally sufficient for most spectrometric methods. The purification of a solvent, however, generally requires large quantities of material at low column efficiency.

2. The reproducibility of repeatedly injected quantities of material requires great accuracy. This value should be below 1% if possible, in any case lower than 2%, as the change in quantity injected also alters, e.g., the retention time and the peak height in the chromatogram. As one of these quantities is usually necessary for the control of the run, their values must be kept as constant as possible.

3. The injection system should be so designed that the sample liquid is always kept under carrier gas atmosphere and is not affected by air or moisture from the surroundings. It should be possible to protect the system from light when separating light-sensitive materials. Facilities for cooling low-boiling liquids and warming viscous ones should be provided.

4. The geometry of the injection system should be uncomplicated and the distance to the vaporizer should be short so that no dead space exists in which residual sample material is prevented from reaching the vaporizer, or reaching it between injections. This also assures that the whole system can be cleaned easily.

5. The injection must be sharply defined at the start and especially at the end so that no dribbling occurs, and the material introduced in as concentrated a form as is possible into the vaporizer and from there into the column (see Section III). Between injections complete disconnection should be assured between injection system and vaporizer so that none of the sample reaches the vaporizer even during involuntary fluctuations in pressure, carrier gas flow, and temperature. A leakage of this sort usually ruins a run and contaminates materials already collected.

6. Finally, it is essential that the injection system guarantees trouble-free and reliable operation—a requirement that naturally must be fulfilled by all other parts of the instrument. A preparative gas chromatograph must work continuously for days, and sometimes weeks, to separate adequate quantities of material. As a fault in a run usually leads to contamination of material already separated and collected, all results obtained until then will be destroyed. Reliability is therefore a very important requirement.

The criteria put forward here that must be considered when choosing a suitable injection system are not always easily realized and may entail considerable expense. In the early stages of preparative gas chromatography (GC), this aspect posed many problems. They have been gradually solved, and several excellent inlets are described in the literature or are commercially available. The most important ones are considered in the following discussions.

In a method described by various authors (1–4), the injection is carried out using excess pressure. A typical example (2) is shown in Fig. 2.1: The sample

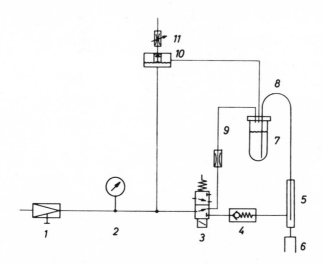

Fig. 2.1. Injection system. *1*, Fine-pressure regulator; *2*, pressure gauge; *3*, three-port solenoid valve; *4*, check valve; *5*, heated injection block; *6*, column; *7*, storage vessel; *8*, dosing capillary; *9*, pressure compensation line with flow restrictor; *10*, cutoff relay; *11*, stream contoller.

liquid is stored in a vessel *7*, which is connected with the vaporizer *5* and the column *6* via a capillary tube *8*. The sample vessel is vented to the atmosphere via the pressure controller *10* and needle valve *11*, which always causes a small stream of carrier gas to flow through the sample vessel from the column and thus prevents diffusion in the reverse direction. The carrier gas normally enters the vaporizer via regulating valve *1*, the three-way valve *3*, and the check valve *4*; a manometer shows the column inlet pressure.

When the three-way valve is turned, the carrier gas flows into the sample vessel instead of into the column. While the pressure rises in the sample vessel, it falls at the column inlet. A definite quantity of sample flows from the storage vessel to the vaporizer depending on the adjusted pressure difference, the time, as well as the length and diameter of capillary *8*. The injection is terminated by turning back the three-way valve. The sample still in capillary *8* is forced back to the sample vessel. When the pressure drops at the entrance to the pressure controller, the sample vessel is automatically degassed which prevents a pressure difference from being built up between sample vessel and vaporizer. Check valve *4* prevents the sample from entering the cold carrier gas inlet during injection. The storage vessel must be thermostated since the temperature of the liquid in the sample vessel changes with the temperature of its surroundings. The ensuing change in viscosity causes a change in quantity injected according to the Hagen-Poiseuille law.

This type of inlet complies with the essential requirements mentioned above. The construction is uncomplicated and therefore almost trouble free. The geometry is simple and the dead space small. As shown by Jentzsch, the sample inlet profiles obtained are almost rectangular and compare favorably with those obtained using manual injection (Fig. 2.2). The quantity of sample

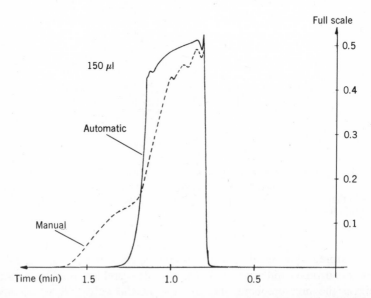

Fig. 2.2. Dosing profile. 150 μl *n*-pentane with automatic (solid line) and manual (broken line) dosing. Dosing capillary, 0.5 mm diameter, 70-cm length; injection block temperature, 150°C; $\frac{3}{8}$-in. column, 2-m length with uncoated Chromosorb W; carrier gas flow, 95 ml/min.

to be injected can be varied over a wide range between 2 μl and 5 ml either by using capillaries of different diameter and length, or by using injection times of different duration (1–31 sec). The reproducibility is always better than 2%. A disadvantage of the method is the fact that the quantity injected can be found only by comparing peak heights in a chromatogram for manually and automatically injected quantities of sample.

The open line between the liquid in the sample vessel and the vaporizer is not without problems; in spite of the constant gas stream between these two, erroneous injections are possible when large pressure changes occur, such as at the end of a temperature or flow-programmed cycle. In laboratory instruments it is unwise to close the dosage capillary with a magnetic valve as this involves relatively large dead spaces.

As in many other technical laboratory applications, injection can also be carried out by means of a piston pump. The cylinder head of this pump is connected on one side with the sample vessel (inlet) and on the other side with the vaporizer or column (outlet) (Fig. 2.3). The inlet and the outlet are

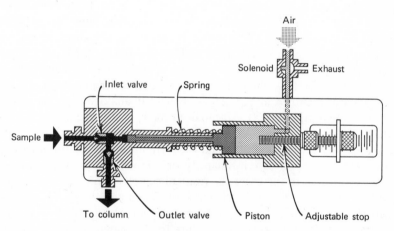

Fig. 2.3. Pneumatic piston pump. Discharging (see text).

closed with check valves. On one side of the piston is a spring (suction stroke), on the other side, air pressure (pressure stroke). The sample is drawn into the cylinder during the suction stroke and is pushed into the connecting tube to the vaporizer or column during the pressure stroke. The piston stroke and thus the volume of the injection can be varied by an adjustable stop. The three-way magnetic valve is usually in the position shown in Figure 2.3 so that air pressure is brought to bear on the piston which is pushed into the cylinder. The magnetic valve is activated by an electric pulse which decompresses the compression chamber, whereby the piston is moved to the right by the spring and the sample is sucked in. Return of the magnetic valve to its original position moves the piston to the left and injects the sample.

This method has been described here because it is so widely used. Unfortunately, it has several disadvantages. The presence of moving parts e.g., the piston and check valves, enhances the danger of leaking (6); susceptibility to dirt; relatively large dead space; complicated geometry, and thus difficulty in cleaning and changing the sample. When using viscous samples, there is a danger that the cylinder is filled only partly and not reproducibly. This is also true for low-boiling materials which partly evaporate during the suction stroke but the problem can be overcome by pressurizing the sample vessel. When the connection between the pump and the column is not purged with carrier gas after the injection stroke, dribbling of sample material may occur, which will certainly cause tailing of the leading edge. Only a few advantages can be set against these disadvantages; e.g., the ease of adjusting the sample volume over a wide range and the good reproducibility for materials of lower viscosity and low vapor pressure. Some of the disadvantages of this method can be offset by replacing the check valves with three-way valves, which usually have a closer fit, and by driving the sample into the cylinder with carrier gas instead of drawing it into the cylinder. Viscous and low-boiling materials can be injected better in this way (7).

The apparatus shown in Fig. 2.4 is somewhat analogous to the syringe used for manual injection (8). A geared motor _1_, pushes the piston _4_ into a cylinder _5_ which is connected to the vaporizer via a small tube _6_ and a Teflon capillary _7_. The spindle makes a certain number of turns for each dosage, which brings a certain quantity of sample into the vaporizer. The tube is then heated to force out the residual sample it still contains in order to obviate dribbling. The piston is made of Teflon and the cyclinder, which is also the storage vessel, is made from precision glass tubing to ensure a good seal between both parts. According to the manufacturer, the dosage lies between 200 μl and 2 ml. This apparatus basically fulfills all requirements for a reliable working system. It is still unsatisfactory, however, in that the vaporizer cannot be closed off from the sample reservoir and provision is not made for purging the connecting capillary with carrier gas between dosing.

Figure 2.5 (9) shows an apparatus that pays particular attention to the importance of a clear separation between sample liquid and vaporizer. The system consists essentially of a glass vessel of which one part _1_ serves as a storage vessel and the other part _2_ as a dosage vessel. A height-adjustable

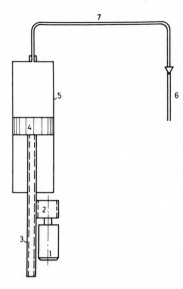

Fig. 2.4. Automatic syringe pump. *1*, Motor; *2*, worm; *3*, spindle; *4*, piston; *5*, cylinder; *6*, injection needle; *7*, connecting capillary.

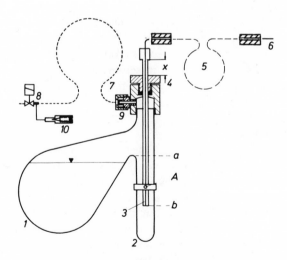

Fig. 2.5. Automatic dosing apparatus (see text).

tube *3* projects into the dosage vessel, which is held and sealed at *4*. This tube is connected to a metal capillary *6*, which leads to the vaporizer via a flexible Teflon capillary *5*. Flexible high-pressure piping *7*, which is normally closed by magnetic valve *8*, opens out into the upper part of the dosage vessel. The inlet of the magnetic valve is connected with the carrier gas inlet. The pressure at this point can be adjusted to 0.1–0.5 atm above the pressure in the vaporizer.

The whole system swivels at point *A* on a motor shaft. When the vessel swings ca. 45° to the right, the liquid runs from the storage vessel to the dosage vessel. After moving back to the initial position, the storage vessel is filled to the upper edge *a* with liquid. When the magnetic valve is opened, the incoming carrier gas flow presses the liquid between *a* and *b* through the tube into the vaporizer. Adjustment of tube *3*, e.g., alteration of distance *a* to *b*, continuously adjusts the quantity of liquid dosed. The dosage is standardized by filling the dosage vessel first to level *a* with liquid and then sucking up the desired quantity with a syringe. Level *b* is then found and the end of the tube is positioned there. Quantities between 50 μl and 10 ml can be dosed by using vessels of different sizes. To prevent evaporated starting material from diffusing in the vaporizer, a needle valve *10* is installed, through which a small stream of carrier gas from the vaporizer continuously escapes to the outside, thus counteracting possible diffusion. The dosage and injection apparatus are separated in this method as they are when using the piston pump.

A certain quantity of liquid is first measured volumetrically and then pushed pneumatically into the vaporizer; this ensures a reproducibility of better than 1%. As a negative meniscus forms at level *b* and the tube no longer enters the liquid, a complete separation between storage liquid and vaporizer is ensured so that no false injections are possible. Although the system must be moved to dose the sample, the liquid itself is not in contact with moving parts so that the advantage of the piston pump (real volumetric dosing) is combined with the advantages of other systems (simple geometry without moving parts, etc.).

The arrangement for dosing gases depends on whether or not the gas is under pressure. In the event that the gas is under pressure and the pressure is fairly constant, an arrangement as shown in Fig. 2.6 has been found to give very satisfactory results. Carrier gas stream *A* normally enters the column *1* via magnetic valve *2*, whereby valve *3* is closed. The gas mixture to be separated is stored in cylinder *4*. By means of pressure regulator *5* the gas pressure is set to a suitable value. Upon opening valve *3* and closing valve *2*, the sample gas *B* streams into the column. The quantity dosed depends on the set gas pressure and on the time during which valves *2* and *3* are actuated. This is carried out by a timer that can be set between 0.1 and 99 sec. Figure 2.7

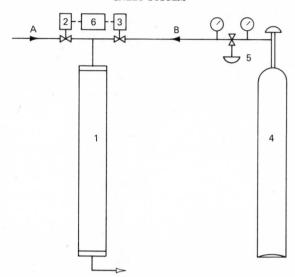

Fig. 2.6. Dosing arrangement for gases under pressure. *A*, Carrier gas; *B*, sample gas; *1*, column; *2* and *3*, solenoid valve; *4*, gas cylinder; *5*, pressure regulator; *6*, timer.

shows the chromatogram of the purification of propane which was carried out with the aforementioned arrangement.

In case the sample gas is under atmospheric pressure, an arrangement as shown in Fig. 2.8 is used. Upon opening of valve *3*, pump *5* is simultaneously actuated which sucks the gas out of vessel *4*. Preferably the pump is of the membrane type.

III. Evaporation

The injection systems described, apart from the dosing of materials that are gaseous in normal conditions, deliver the starting material in liquid form. Whereas the liquid in analytical GLC is often injected directly by syringe onto the column without previous evaporation, the relatively large quantities used in preparative GLC must first be evaporated. This is mainly for three reasons. First, the first part of the column that must hold the injected quantity does not have a heat capacity sufficiently large to heat the liquid from room temperature to column temperature and to evaporate a quantity corresponding to the partition coefficient. Second, it is very difficult to distribute liquid material homogeneously over a cross-sectional area of the column; this is nevertheless absolutely necessary on the basis of an even distribution of the concentration. This problem is much more easily solved when the starting material is already in the gas phase. Third, the direct

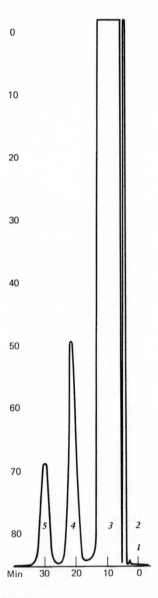

Fig. 2.7. Purification of propane. Column: $l = 4$ min, $d = 40$ mm, 20% SE 30 on Chromosorb P (60–80 mesh), 1.2 min^{-1} N$_2$, 25°C. Sample: commercially available propane (*1*, methane; *2*, ethane; *3*, propane; *4*, *n*-butane; *5*, isobutane); dosed quantity per cycle, 1.4 *l* (injection time, 4 sec; sample pressure, 1.5 atm; reproducibility, better than 1%).

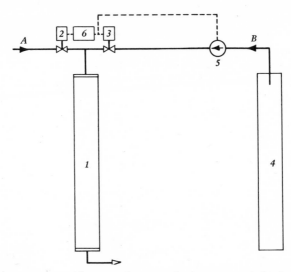

Fig. 2.8. Dosing arrangement for gas under atmospheric pressure. *A*, Carrier gas; *B*, sample gas; *1*, column; *2* and *3*, solenoid valve; *4*, vessel; *5*, pump; *6*, timer.

injection of liquid causes the liquid phase at the beginning of the column to be washed off the support and to be distributed elsewhere. The result is that after some time part of the column is not coated, while another part contains too much coating on the support, both of which unfavorably influence the separation power.

Satisfactory results have however, been reported (11) when the injected material is merely heated to the column temperature and then injected onto the column as a liquid. The carrier gas supply must be interrupted for a relatively long time when using this method in order to give the injected material sufficient opportunity to reach equilibrium (12,13). It is therefore doubtful that this method could be used universally and over long periods. On-column injection should therefore be limited to small amounts ($< 10 \ \mu l$) and only when it is feared that the material could change its chemical structure during evaporation. This may be the case with especially high-molecular-weight biochemical materials or with samples that tend to polymerize. In all other cases the substance is first evaporated and put onto the column in the gaseous phase.

Certain requirements for the evaporation process are determined by theoretical (14–17) and various practical considerations. It is known that the total peak width at the end of the column is compounded from the width it has at the beginning immediately after injection and from a widening effect which takes place during the chromatographic process and is caused

by various kinds of diffusion and the final velocity of mass transfer (18).
The initial width should therefore be as small as possible. When the height of
the initial peak is defined to be the maximum possible concentration, the
ideal peak shape at the beginning of the column should be a rectangle with a
small width and as great a height as possible. The area of the rectangle
represents the quantity of sample dosed.

Van Deemter has shown that the quantity introduced has no noticeable
influence on the width of the eluted peak as long as the following relation is
valid:

$$V < 0.5 \, V_R'/\sqrt{n}$$

where V = gas volume used to transport the substance in the column

V_R' = retention volume

n = column performance in theoretical plates

When V exceeds this value, column overload occurs with a subsequent
decline in column efficiency. Related to average retention time, this means
for a column of 20-mm diameter and 2-m length a dosage in the range of
100–200 mg. This value is generally exceeded in practical applications. This
equation can also be expressed in units of time:

$$t < 0.5 \, t_R'/\sqrt{n}$$

where t = time required to introduce the gas volume containing the sample
onto the column

t_R' = retention time for very small values of t.

In relation to the example given above, t should be *ca.* 5–10 sec. This, to
first approximation, is the time during which the evaporation process must
take place.

The maximum possible quantity of material to be dosed depends essentially
on the system, substance–solvent, which is characterized by the partition
coefficient. With increased solubility (increasing partition coefficient), more
material can be injected. According to Henry's law, the solubility of a substance is proportional to its partial pressure over the solvent. As the partial
pressure drops in the presence of the carrier gas, the maximum solubility
(and thus the maximum quantity of substance to be injected) is obtained
when the substance is not diluted with carrier gas. This process has been
investigated theoretically as well as experimentally (19), showing that the
column efficiency quickly drops with increased dilution of the injected
material. This is accentuated by increases in the injected quantity of the
substance.

In Fig. 2.9, n, the theoretical plate number, is plotted against K, the ratio
of the carrier gas volume (with which the sample is mixed upon entering the

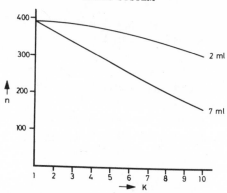

Fig. 2.9. Relation of theoretical plate number n to dilution factor K. Column: $d = 38.2$ mm; $L = 230$ cm; $T = 90°C$; stationary phase, 30% dibutylphthalate on Celite; substance, Toluene.

column) to the gas phase volume of the sample. It is noteworthy that smaller quantities of material can evidently stand quite considerable dilution ($K = 5$).

In spite of the simplifying assumptions that have been made, the following conclusions can be drawn.

1. There is a limiting value with regard to the quantity with which each column can be dosed; exceeding this limit causes overloading.

2. The quantity can be related to a time limit within which the evaporation and injection must take place.

3. The most favorable shape for the concentration profile of the initial plug is a rectangle.

4. The sample should not be diluted with carrier gas when introduced into the column.

5. The smaller the solubility of the substance in the stationary phase, the greater the significance of points *(1)–(4)*. The injection procedure is thus more important for quickly eluting substances than for those with long retention times.

When these results are taken as a guide for the construction of an evaporation chamber, the chamber must be able to conduct the necessary evaporation heat to the substance very quickly (short evaporation time), and the sample must be diluted with as little carrier gas as possible during this process (small dead space corresponding to the quantity to be evaporated). The same is valid for the transport of the substance from the vaporizer to the column, where it should enter with as small a rectangular profile as possible. The evaporation must obviously be complete, as a wet vapor injection has at least partly the same disadvantages as an injection in liquid form.

Apart from the fact that these requirements can be approximated only in practice, fast evaporation is limited by the fact that a large number of substances of interest to preparative GLC are sensitive to heat, especially when the heating takes place very quickly. A compromise must therefore be found between the quickest and the most careful evaporation.

Of all the constructions published in the literature and known from commercial instruments, a design has been approved in which the sample is introduced into a heated chamber filled with packing of a suitable material (20–24). The packing is heated to ca. 50°C above the boiling point of the highest boiling material and gives a part of its heat capacity to the injected material. Other ways of heating tried for this purpose, such as IR or induction heating, were eliminated as they could not supply adequate temperature distribution. Local overheating or undercooling should be avoided.

The size of the vaporizer is therefore established by the availability of a certain heat capacity (20). It is not necessary that the total vapor volume be located in the vaporizer. The ratio of vapor volume to vaporizer volume should be ca. 5% and this ratio is not very critical (25,26). The packing material particles should be small (0.5–1 mm ϕ) to ensure a large surface and a short heat transfer path. They should, moreover, have good thermal conductivity. Spheres or other shapes of copper, silver, or also steel particles are most suitable from this point of view, but as metals may have catalytic effects at high temperatures, glass beads are usually employed. These should also be deactivated as much as possible, and glass beads of this quality can be obtained commercially (27). Very sensitive substances, such as steroids, often require vaporizers constructed entirely of glass to exclude any contact with metal surfaces.

Finally, it is necessary to pay attention to the geometry of the evaporation chamber. Figure 2.10 shows an arrangement frequently used. A and B are the inlets for the sample and the carrier gas—the latter is spread as homogeneously as possible over the cross section by using a perforated plate in the lid. C is the outlet connected with the column. Thermocouples are placed at positions through 1–9. They show the changes in temperature during evaporation. Figure 2.11 shows the temperature pattern at various positions. The first thing noted is that the temperature at 2 is considerable lower than at the other positions. This is caused by the carrier gas flowing into the vaporizer without preheating. The temperature at positions 2 and 5 falls quickly, immediately after injection, and returns to normal only after ca. 15 min. The temperature of the other positions, however, changes only slightly. The often expressed idea that the vaporizing liquid spreads in clouds over the total cross section of the vaporizer immediately after touching the hot glass beads is therefore incorrect. The liquid obviously runs down along the shortest route so that the heat capacity of the glass beads is used only to a small

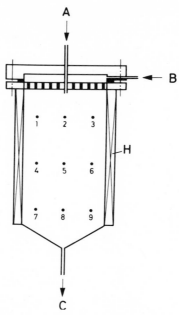

Fig. 2.10. Simple evaporation chamber. A, Sample inlet; B, carrier gas inlet; C, outlet; H, heating; 1–9, temperature measurements positions; d = 10 cm; L = 20 cm; 6 ml hexane.

extent. The radial temperature equilibrium is accomplished very slowly because of the poor thermal conductivity of the glass bead layer. This causes the sample to reach the column as a wide, very diluted band with a long tail, and the evaporation is furthermore incomplete. The chromatographic results are correspondingly bad. The cause of this is basically very trivial and has been noted by other authors (28–30); it results from not paying attention to essential thermodynamic laws. As the configuration given in Fig. 2.10 appears in many publications, it merits further discussion, particularly since it appears that the bad separations often obtained with preparative instruments are not caused by the column but by the evaporation chamber.

Figure 2.12 shows a vaporizer in which an attempt was made to eliminate some of the problems mentioned above. In chamber 1, surrounded by heating jacket 2, is placed a metal body 3 (silvered copper), which can be taken out when cold but which at higher temperatures fits closely against the inner wall of tube 1 because of thermal expansion and thus makes good heat contact (large heat capacity, good thermal conductivity). Holes are drilled in the metal body which vary in size for different quantities of sample. The drill holes are filled with packing material (short heat transport path, large

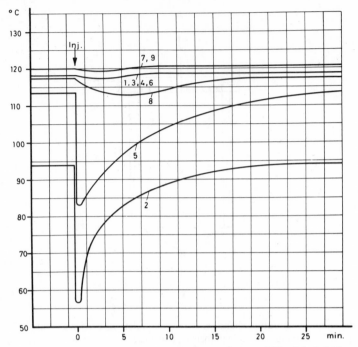

Fig. 2.11. Temperature distribution in evaporation chamber of Fig. 2.8. *1–9*, Temperature measurement positions.

surface). A cone is attached to the underside of the metal body, which is somewhat flatter than the cone of the bottom part. This leaves a gap through which the material and the carrier gas flow but through which the packing material cannot pass. This saves a sieve or sinter at this point which often lead to blocking. The already preheated carrier gas flows through a layer of packing material and establishes a complete temperature equilibrium. The carrier gas inlet contains a check valve *4*, which prevents substances from entering the inlet during evaporation with its related higher pressures. The vaporizer has a removable lid which is necessary for cleaning purposes and for removal of the metal body.

The evaporation of substances in the main carrier gas stream is called "dynamic evaporation." Although easily attainable instrumentally, it has the disadvantage that the vaporizing substance always reaches the column after considerable dilution with carrier gas. This effect can be reduced by fitting a solenoid valve *5* (Fig. 2.12), which is closed during the evaporation process. Check valve *4* (Fig. 2.12) only safeguards the solenoid valve against direct entry of substance.

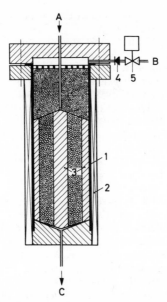

Fig. 2.12. Evaporation chamber. *1*, Tube; *2*, heating element; *3*, metal body; *4*, check valve; *5*, solenoid valve; *A*, substance inlet, *B*, carrier gas inlet; *C*, outlet.

The other possibility is the so-called "static evaporation" (31,32). Here the vaporizer is situated in a lateral branch of the carrier gas stream, immediately in front of the column. Its inlet and outlet are closed by valves. The sample is introduced through a further opening and has ample time to evaporate. The valves are turned for injection and the carrier gas stream purges the sample onto the column. As already noted, this method gives a homogeneous, saturated vapor which can be introduced into the column as an only slightly diluted plug if the carrier gas stream is directed correctly. Static evaporation has found little application in the field of preparative laboratory GLC because of the greater instrumental expense and especially because an automatic valve is necessary between the vaporizer and the column, e.g., at the hottest part of the apparatus.

References

1. E. Heilbronner, E. Kovats, and W. Simon, *Helv. Chem. Acta*, **40**, 2, 2410 (1957).
2. D. Jentzsch, *J. Gas Chromatog.*, **5**, 226 (1967).
3. H. Boer, *J. Sci. Instr.*, **41**, 365 (1964).
4. J. W. Frazer and C. J. Morris, University of California, Lawrence Radiation Laboratory, TID-4500, 46th ed. (1965).
5. Anonymous, Varian Aerograph, *Previews and Reviews*, No. 8, 10 (1968).

6. E. P. Atkinson and G. A. P. Tuey, in *Gas Chromatography 1958*, D. H. Desty, Ed., Butterworths, London, 1959, p. 194.
7. J. M. Kauss, J. Peters, and C. B. Euston, *Develop. Appl. Spectry.*, **2**, 383 (1962).
8. Anonymous, Nester/Faust, Sales Bulletin, No. 867 (1967).
9. K. -P. Hupe, *Chromatographia*, **1**, 462 (1968).
10. J. J Kirkland, in *Gas Chromatography*, V. J. Coates, H. J. Noebels, and I. S. Fugerson, Eds., Academic Press, New York, 1958, p. 203.
11. E. M. Taft and J. E. Booker, U. S. Pat. No. 3,366,149 (1968).
12. R. P. W. Scott, *Nature*, 198, 782 (1963).
13. S. J. Hawkes (discussion remarks), in *Gas Chromatography 1962*, M. V. Swaay, Ed., Butterworths, London, 1962, p. 65.
14. G. W. A. Rijnders, in *Advances in Chromatography*, J. C. Giddings and R. A. Keller, Eds., Vol. 3, Marcel Dekker, New York, 1966.
15. R. S. Henley, *J. Am. Die Chem. Soc.*, **42**, 673 (1965).
16. R. S. Henley, A. Rose, and R. F. Sweeny, *Anal. Chem.*, **36**, 744 (1964).
17. D. T. Sawyer and H. Purnell, *Anal. Chem.*, **36**, 457 (1964).
18. J. J. van Deemter, F. J. Zuiderweg ,and A. Klinkenberg, *Chem. Eng. Sci.*, **5**, 271 (1956).
19. W. J. de Wet and V. Pretorius, *Anal. Chem.*, **32**, 169 (1960).
20. A. B. Carel and G. Perkins, *Anal. Chim. Acta*, **34**, 83 (1966).
21. F. H. Huyten, W. v. Beersum, and G. W. A. Rijnders, in *Gas Chromatography* 1960, R. P. W. Scott, Ed., Butterworths, London, 1960, p. 224.
22. E. Bayer (discussion remarks), in *Gas Chromatography* 1960, R. P. W. Scott, Ed., Butterworths, London, 1960, p. 236.
23. P. A. Bushong, in *Lectures on Gas Chromatography* 1962, H. A. Szymanski, Ed., Plenum Press, New York, 1963, p. 247.
24. A. B. Carel, R. E. Clement, and G. Perkins, in *Advances in Chromatography 1969*, A. Zlatkis, Ed., Preston Technical Abstracts Co., Evanston, Illinois, 1969.
25. M. -B. Dixmier, B. Roz, and G. Guiochon, *Anal. Chim. Acta*, **38**, 73 (1967).
26. H. Eggert, Diplomarbeit, Lehrstuhl fur Thermodynamik, University of Karlsruhe, 1964.
27. Corning Glass Works, Corning, New York, Chromatography Products Group.
28. A. Rose, D. J. Royer, and R. S. Henley, *Separation Sci.*, **2**(2), 211 (1967).
29. R. B. Sumantri, Chem. Eng. Department, Notebook 0101, Pennsylvania State University, University Park, Pennsylvania, p. 8.
30. J. W. Amy and W. E. Baitinger, in *Lectures on Gas Chromatography 1962*, H. A. Szymanski, Ed., Plenum Press, New York, 1963, p. 19.
31. E. Bayer, K. -P. Hupe, and H. G. Witsch, *Angew. Chem.*, **73**, 525 (1961).
32. M. Modell and J. M. Ryan, U. S. Pat. 3,352,089 (1967).

CHAPTER 3

Preparative Column Technology

Raymond E. Pecsar, *Varian Aerograph, Walnut Creek, California*

I. Introduction

Chromatography naturally divides itself into two distinct classes: (*1*) analytical—in which the purpose is to obtain qualitative or quantitative information and (*2*) preparative—in which the goal is the collection and recovery of purified fractions for further identification, chemical synthesis, reference materials, or production of chemicals. The latter purpose is dealt with in much detail in later chapters of this volume. Our concern here is the attainment of sufficient pure material for the former uses. Generally, this requires milligrams to grams of the substance at most and is accomplished via laboratory-scale operations, i.e., with apparatus of analytical proportions. The needed amounts of material may usually be collected in short times even when high resolution is required.

At the heart of this technique is the column. Generally, the inlet and detector sections of the chromatograph are quite similar to the analytical device. However, the column must normally be scaled up from the conventional

$\frac{1}{8}$- to $\frac{1}{4}$-in. packed columns used in analytical studies so as to accommodate the characteristically larger preparative sample. It is in this area that difficulties are encountered and in which many of the developments in preparative column technology are centered. If the intrinsic efficiency of the column is reduced markedly, then much of the advantage of the chromatographic technique is sacrificed and the user will return to alternate separation methods such as distillation. Likewise, if the analysis time is extensively increased because of the larger sample size employed to obtain sufficient pure material, then the second major advantage of chromatography is lost.

From the above discussion it is obvious that in preparative chromatography even the most optimized column represents a compromise between speed of analysis, component resolution, and column sample capacity (1). This compromise is represented in Fig. 3.1. Any two of the desired goals may be had at the expense of the third. That is, for preparative purposes, a large capacity and a rapid analysis can be obtained only on an easily separable mixture, or a difficultly resolved mixture using large samples necessitates a lengthy analysis. The trade off between these parameters is always required no matter what the situation.

In the present discussion of preparative column technology, considerations are restricted to laboratory-scale columns, i.e., columns with diameters of 4 in. or less. The choice is defined by the commercially available apparatus operating in a laboratory environment presently. The separations likewise will be on a batch scale as no continuous laboratory apparatus is existent. Continuous and circular preparative chromatographic separations are discussed elsewhere in this book and are usually conducted on a larger scale. All of the subject matter is based on elution analysis, although many of the conclusions are equally applicable to frontal analysis. In fact, frontal analysis is one technique by which the sample throughput of the column may be increased. In addition, GLC is considered exclusively. Gas–solid chromatography can certainly be conducted on a preparative scale but extremely few examples have been recorded.

The following treatment assumes that the reader is conversant with the basic elements of GC especially those dealing with column technology. Many areas are discussed on a rudimentary level, but this approach is not universal. For those not familiar with the salient features of the method, reference is made to the excellent text of Ettre and Zlatkis (2) for a practicing viewpoint without the extreme rigors of mathematics. Likewise, for brief introductions to the preparative technique in general, worthwhile reading is provided by the reviews of Rijnders (3), Verzele (4), and Hupe (5). Frequent reference in the main text is made to these works.

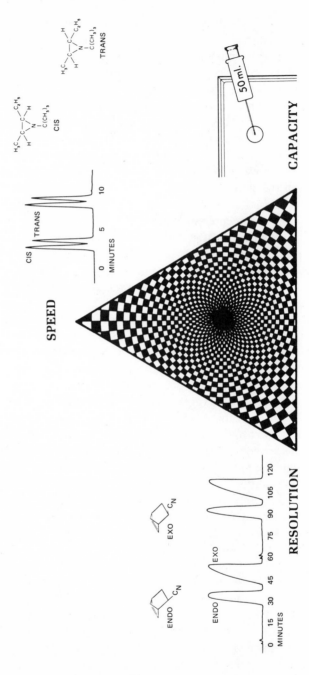

Fig. 3.1. Compromise between resolution, speed, and capacity.

II. Theoretical Implications

The theoretical development of preparative GC has been ably presented by Pretorius in Chapter 1. The intent here is to provide an interpretation of the theory and to point up the ramifications toward improved preparative columns.

A. Plate Equation

In order to express the ability of a column to separate a given mixture, we invoke a measure of efficiency called the number of theoretical plates. This efficiency parameter is a carryover from laboratory distillation, and in a chromatographic column the analogy is somewhat strained. When the length of the column and the number of plates are known, the ratio can be computed, and this is called the Height Equivalent to a Theoretical Plate, commonly written as HETP or h. Basically, the lower the plate height is the more efficient the column.

1. Standard Form

Much theoretical work has been done in order to express the behavior of a chromatographic column in terms of the HETP. The most well-accepted relation, in one form or the other, is the classic van Deemter equation.

$$h = A + B/u + C_g u + C_l u \qquad (3.1)$$

As h is proportional to the linear gas velocity as well as the inverse of the linear gas velocity, it may be intuitively concluded that an optimum velocity exists at which h is a minimum. Figure 3.2 illustrates the relation between h and u with a qualitative indication of the effect of the correlation constants A, B, C_g, and C_l. These terms are far from constant but in fact depend on a multitude of chromatographic variables. The expanded form of the plate height equation in terms of these variables is:

$$h = 2\lambda\, d_p + \frac{2\gamma D_g}{u} + \frac{8}{\pi^2}\left(\frac{k}{1+k}\right)^2 \frac{d_p{}^2}{D_g} u + \frac{2k}{3(1+k)^2}\frac{d_f{}^2}{D_l} u \qquad (3.2)$$

The term $2\lambda d_p$ is related to the component band broadening caused by gas flow irregularities as the carrier gas is forced around the support particles. The second term accounts for molecular diffusion in the axial direction in the gas phase. This term predominates at lower gas velocities, and in this region high-molecular-weight carrier gases are desirable to minimize h. At most practical velocities the rate of gas transport makes axial diffusion of minor significance. The last two terms are the primary contributions to plate height and are attributable to the finite rates of mass transfer in the

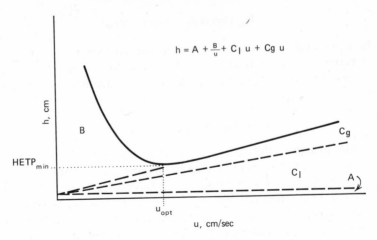

Fig. 3.2. Plate height dependence on gas velocity.

mobile and stationary phases. From Eq. 3.2 it can be deduced that more efficient columns can be obtained by using small support particles with light liquid loadings and high diffusivity (low viscosity) in the stationary phase. While these observations are correct in a theoretical way, other considerations concerning practical operating conditions dictate compromises which are delineated in subsequent sections.

2. Preparative Expanded Form

The van Deemter equation has provided an adequate representation of the behavior of a standard chromatographic column. However, for preparative columns which are characteristically of larger diameter the agreement is less desirable. This is in essence attributable to the lack of consideration of all the factors that influence band broadening, many of them being negligible in normal analytical situations. Giddings (6) has given a detailed discussion on 10 factors contributing to increased HETP in large columns. The most significant of these is gas flow nonuniformity at a given cross section. The source of this nonuniformity is support particle size segregation, i.e., larger particles seem to congregate near the wall with a decreasing size gradient as the center is approached.

This gas flow nonuniformity coupled with a lateral diffusion of the molecules in the gas phase is the main source of additional band broadening in preparative columns over that accounted for by the standard plate equation. Many approaches have been made for theoretically formulating the added contribution for this type of peak spreading. Giddings (7) has extended his generalized nonequilibrium theory to large-diameter columns. Huyten et al.

(8,3), in considering peak dispersion, used the concept of additivity of variances and derived an additional term for radial diffusion. Finally, Higgins and Smith (9) represented the random process of diffusion by a sequence of microscopic vectors without phase coherence. While the complexity of the above treatments varied considerably, all three methods generated an added term of the form

$$h_{\text{prep}} = \frac{Cr^2u}{D_g} \tag{3.3}$$

Combining Eqs. 3.2 and 3.3 yields the preparative extended form of the plate equation

$$h = 2\lambda d_p + \frac{2\gamma D_g}{u} + \frac{8}{\pi^2}\left(\frac{k}{1+k}\right)^2 \frac{d_p^2}{D_g} u + \frac{2k}{3(1+k)^2} \frac{d_f^2}{D_l} u + \frac{Cr^2}{D_g} u \tag{3.4}$$

Two investigations have been performed to test the validity of Eq. 3.4. Giddings (10) measured the plate height as a function of the linear carrier velocity for both analytical (¼-in. diameter) and preparative (2-in. diameter) columns, extrapolating all data to zero sample size. By preparing plots of h as a function of u/D_g and making cross correlations, he determined the value of C in Eq. 3.3 to be 0.8, which concurred well with the predicted value of 0.75. Both helium and nitrogen carrier gases produced the same value. Hargrove and Sawyer (11) compared ¼-in. analytical columns with 1-in. preparative columns. The gas phase mass transfer terms were measured and computed. The theoretically based values predicted an increase by a factor of 6 for the preparative over the analytical column while the actual experimental results showed an increase by a factor of 12. When considering the complexity of the theoretical formulation, the experimental agreement is certainly acceptable.

One further study in this area is worthy of mention. Hupe (12) has generated an added preparative term to the van Deemter equation using a statistical treatment. The resultant expanded form of the equation is

$$h = A + B/u + C_g u + C_l u + 2.83 \frac{r^{0.58}}{u^{1.886}} \tag{3.5}$$

While the cross-sectional velocity profile corresponding to this relationship has a somewhat unusual shape, the fit to the experimental results on a variety of preparative columns ranging from ½ to 4-in. in diameter was very good. However, because of the somewhat empirical appearance of Eq. 3.5, extension to other preparative systems is discouraged with preference being given to Eq. 3.4.

B. Operating Variables

As mentioned in the discussion of the van Deemter equation, a large number of chromatographic variables can influence the efficiency of the column. Readily apparent were the effects of the particle size, the loading level of the stationary phase, and the carrier gas linear velocity. Also, for preparative columns the influence of column diameter was clear. Less obvious but of definite importance are the influence of the column length, the sample size and injection technique, the nature of the solid support, the choice of carrier gas, and the thermal environment of the column. In analytical GC these variables are optimized to produce a minimum HETP. While this is often true in preparative work also, the separation may be alternately optimized for sample throughput, i.e., the maximum amount of material passing through the system per unit time. The effect of these important variables on each of the possible cases is now examined.

1. Optimization of HETP

The smaller the particle size, the lower the eddy diffusion term and the resistance to mass transfer in the gas phase. These considerations dictate the use of small-mesh particles. However, this process is eventually limited by the permeability of the column. Decreased permeability means increased pressure needed to generate flow and this becomes the practical limit. Often with long columns coarser supports are used to reduce pressure drop to a workable value. As the efficiency usually increases in direct proportion to the length, the HETP is independent of length (13). However, according to the theory of Giddings (6,10), a dependence of HETP on length for larger-diameter columns is proposed for short columns, after which HETP is again length independent. The transition length is dependent on column diameter.

The increased preparative sample size definitely degrades the plate height. This is attributable to a variety of reasons which are treated in later sections of this chapter. In order to compensate for these larger loads, the amount of liquid phase is often increased as a percentage of the packing material. This is in direct contradiction to the plate equation, as a thin liquid film is desired to reduce the liquid phase mass transfer resistance and so a compromise is required. If the gas phase mass transfer is limiting, then moderate increases in amount of liquid phase have little effect on efficiency and greatly increase sample capacity. Likewise, the method of sample introduction is critical and best efficiencies are obtained with plug injection (14). A poor injection system heavily penalizes column performance.

The choice of optimum carrier gas is very much dependent on the linear gas velocity. At low velocities where longitudinal diffusion is prevalent, a low diffusivity is desired and nitrogen is preferred. For higher velocities and more rapid analyses where gas phase mass transfer is predominant, the opposite is true and helium is preferred. Generally, at the optimum velocity a lower value of $HETP_{min}$ is obtained when using nitrogen in preference to helium.

Finally, the influence of temperature on preparative plate height can be expressed as complex and problematic. In order to handle large samples of wide-boiling mixtures, often temperature programming should be used. However, with large amounts of insulative packing material significant gradients exist. Also, the solution thermal effects associated with these large samples aggravate and amplify the temperature gradients. In practice the optimum column temperature is determined by experimentation and often small variations can have profound effects on plate height.

2. Optimization of Throughput

As the purpose of preparative gas chromatography is the isolation of pure materials, efficiency may be established based on throughput, i.e., the ratio of the sample size to the analysis time. The main evaluation of parametric effects on preparative throughput has been conducted by Gordon and Pretorius (15). They define the preparative efficiency E as the number of moles of a particular solute of a given purity that can be recovered per unit time. This efficiency definition is preferable to the plate height as it is much less sensitive to sample size variations. The effect of the operating variables on E depends on the difficulty in accomplishing a given separation as expressed by α, the relative volatility or relation retention value. For $\alpha > 1.2$, i.e., reasonably simple separations, the dependence of E is very similar to that of HETP on the variables being considered, while for $\alpha < 1.2$ a different behavior occurs.

A porous support, such as the diatomaceous earths, always gives a higher efficiency E (16). For simple separations, as the support particle size decreases the efficiency increases, but again pressure drop in the column is the limiting practical factor. For more difficult separations the efficiency is independent of particle diameter (17). In all cases, as the column length increases so does E with a concomitant increase in pressure drop.

The sample size is still an important variable relative to E, as an optimum sample inlet volume exists for each component in a mixture. However, the efficiency decreases quite slowly as the optimum value is exceeded. Here too, in order to handle large samples the stationary phase loading is increased. The preparative efficiency increases as the percentage liquid phase increases, but beyond a reasonably heavy load only slight additional advantage is gained (16).

An interesting behavior relative to variations in column diameter is observed. For easily resolved mixtures the preparative efficiency continues to increase approximately proportional to the square of the column diameter. However, as the resolution complexity increases, this diameter dependence holds to about 4-in. columns. Above this value no increase in E is gained by enlarging the column (18).

The carrier gas linear velocity shows an optimum value with respect to E similar to the effect on h, but $u^E_{opt} > u^h_{opt}$ although in practice this u^E_{opt} is difficult to achieve. As the column diameter increases, the optimum velocity decreases, but for a constant diameter the value of u^E_{opt} increases as the column length increases (19). In general, when optimizing on throughput more flexibility in carrier gas linear flow rate is possible.

C. Interrelation between HETP, Column Diameter, and Carrier Gas

The dependence of HETP on carrier gas linear velocity and on column diameter has been discussed in a qualitative manner in the previous section. At this point some quantitative data is presented to provide an indication of the magnitude of typical values. In Table 3.1 carrier gas flow rates for both helium and nitrogen are provided by way of comparison. These values are not necessarily the lowest HETP achievable but are representative practical results. All the columns were packed using conventional techniques, but with care, and no ancillary techniques or devices were used to decrease the above-mentioned velocity profile effects. In the larger-diameter columns the methods to be subsequently delineated can very markedly reduce the plate height by as much as 30%.

TABLE 3.1

Dependence of HETP on Carrier Gas and Column Diameter

Column diameter in.	Helium carrier flow, ml/min	Nitrogen carrier flow, ml/min	HETP min, mm
1/8	30	15	0.4
1/4	65	25	0.6
3/8	220	90	1.0
3/4	—	150	1.5
1–1/2[a]	2,100	—	1.7
4[b]	—	10,000	3.0
20[b]	—	—	4.2

[a]Ref. 20.
[b]Ref. 12.

An inspection of Table 3.1 shows the expected efficiency loss as the column diameter increases. However, it can be seen that for extremely large columns the increase in $HETP_{min}$ does not occur proportionately. Values of 3–4 mm for plate height are still respectable, and fairly difficult separations can be achieved with such a column. Likewise, as the column cross section increases, the carrier gas flow increases but again nonlinearly. A linear relation between column area and flow does exist up to ⅜-in. columns, but above this size the incremental flow increase is not proportional.

The variation of plate height with carrier gas velocity differs for each carrier gas because of the diffusion coefficient variations. Thus for a given column both $HETP_{min}$ and u_{opt} vary depending on the choice of carrier gas. In general, the higher the molecular weight of the carrier, the lower both u_{opt} and $HETP_{min}$. This behavior is depicted in Fig. 3.3. Here, for a ⅛-in.

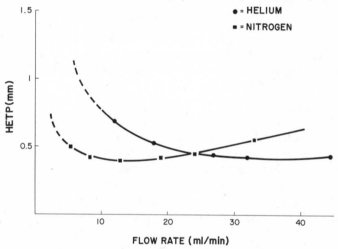

Fig. 3.3. Efficiency comparison with nitrogen and helium carrier gases.

column it can be seen that a slight improvement in $HETP_{min}$ can be obtained with nitrogen. However, a significant decrease in flow rate is required to accomplish this efficiency gain. By employing helium as the carrier gas, the flow rate can be greatly increased with only a slight penalty in efficiency above the optimum value. This permits a much more rapid analysis to be conducted without much compromise in resolution. Unfortunately, the economics of the situation often dictate employing the less costly nitrogen especially when considering the high flow rates associated with the larger-diameter columns listed in Table 3.1. Practically speaking, for the sample sizes usually accommodated with these larger columns, very little difference in resolution is gained by varying the carrier gas, and the type most readily available is the one used.

III. Inlet and Sample Size Effects

The column efficiency is very markedly decreased as the sample size is increased. As larger samples are employed, the bandwidth broadens as a result of selective thermal changes occurring in portions of the band arising from solution and dissolution in the stationary phase. Likewise, increased resistance to mass transfer occurs in the column. Attempts to increase the sample capacity by additional amounts of liquid phase only magnify the mass transfer resistance. For each column diameter there is a definite maximum sample size for practical chromatographic separations, but these limits are normally exceeded and the column operated in an overloaded condition. In these cases one is more often concerned with throughput maximization while maintaining a given product purity.

A. Sample Capacity

Sample size limitations in preparative columns are caused by two factors, the sample feed volume and the solution concentration level (21). In order to prevent an increase in h, the maximum feed volume allowable is

$$V_f \leq \frac{V_R}{2\sqrt{n}} \tag{3.6}$$

If this criterion is met, then the amount of solute in the sample must still be kept below 10 mole % (preferably 1 mole %) in order that a linear partition isotherm results. If larger samples are introduced, increased solution thermal effects will occur, nonlinear partition isotherms will result, and an increase in h will be manifested. The most common means of overcoming these limitations is by repeated injection of samples that conform to the above conditions.

1. Maximum Allowable Sample Size

Much work on the effect of sample size in relation to column performance has been done by Verzele (22). The decrease in efficiency for larger samples on a $\frac{3}{8}$-in. column is presented in Fig. 3.4. For finite sample sizes up to 500 μl, the decrease is extreme but then asymptotically levels off for larger samples. It is also clearly demonstrated that for preparative conditions of operation the carrier gas has an extremely small effect when comparing hydrogen, helium, and nitrogen. Similar reductions in efficiency result irrespective of the column diameter as shown in Fig. 3.5. Here the behavior of the 1-in.-diameter column is similar to that of the $\frac{3}{8}$-in. column in Fig. 3.4, while the 3-in. column exhibits a uniform but higher value of h (23). The results of this behavior is that for sample sizes below 7 ml the 1-in. column yields better separations, while above this sample size the 3-in. column is to be preferred.

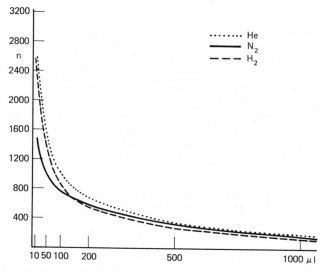

Fig. 3.4. Sample size effect on efficiency as a function of carrier gas (22).

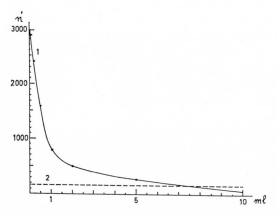

Fig. 3.5. Sample size effect on efficiency as a function of column diameter (23). *1*, 1-in. diameter; *2*, 3-in. diameter.

In order to quantify to some extent this sample size relationship with column performance, the required efficiency to affect a given separation must be considered. This can be expressed as (24)

$$n = 16R^2 \left(\frac{\alpha}{\alpha - 1}\right)^2 \left(\frac{k + 1}{k}\right)^2 \tag{3.7}$$

From this equation and an experimental plot such as that given in Fig. 3.4, the maximum allowable sample size can be formulated (25)

$$\text{MASS} = \frac{100S(n)}{x} \tag{3.8}$$

The utilization of Eqs. 3.7 and 3.8 is as follows. From the chromatogram the relative retention value for the least resolved pair may be computed as a ratio always greater than unity. Then the relative retention ratio between a peak of interest and an unretained peak produces a value for k. Finally, as a resolution R of unity is adequate for practical purposes, the required number of plates can be computed. By using this value the data in a plot such as Fig. 3.4 can be interpolated to yield a sample size S. As this sample size is for but a single component of the least resolved pair, the mole fraction of that member of the pair present in largest amount is used as x to determine the MASS.

Perusal of the preceding relations shows that the maximum allowable sample size is dependent on three factors, the separating power of the column, the difficulty of the separation, and the time required for the separation. The difficulty of the separation is expressed by α for the least resolved pair. As the term involving α grows rapidly as α approaches unity, for practical conditions values in excess of 1.1 should be used. If the value obtained from the chromatogram is below 1.1, then alternate stationary phases should be sought which provide greater selectivity on the component pair of interest and thus a larger α. The separation power of the column, expressed as n, is for the actual sample sizes employed and as such is the preparative plate number. Utilization of the number of plates for very small samples as expressed in analytical applications has no practical meaning. The last factor, which is the time required for separation, is couched in the term k. If the peaks are retained for only a short period, the value of k will be low and the sample size reduced. A working range of 3 to 10 is recommended (24).

A further analytical extension of the above approach has been suggested by Beeson (26). As a plot of h as a function of sample size yields a linear relation through the useful range of the column, the dependence can be expressed as

$$h = mS + b \tag{3.9}$$

where m and b are, respectively, the slope and intercept of the line. Utilizing a value of $R = 1$ and neglecting the dependence on k for relatively large values of k, Eqs 3.7 and 3.9 can be combined to yield

$$S = \frac{L(\alpha - 1)^2}{16m\alpha^2} - \frac{b}{m} \tag{3.10}$$

Graphs or tabulations for a given column can then be prepared expressing the dependence of S on α.

2. Overloading

Overloading is a condition occurring whenever the two criteria advanced for maximum sample capacity are exceeded and a decrease in the intrinsic column efficiency results. Practically speaking, overloading is the rule rather than the exception in preparative GC. Nonlinear partition isotherms are in force, and the peak shape undergoes transition from symmetrical to tailing with slight overloading and finally to leading with increased overloading. However, as the desire to obtain pure components is paramount, the peculiar peak shapes have little impact.

Five distinct contributions to overloading have been defined by Higgins and Smith (9). Condensation overloading occurs because of flash vaporization in the injection port with subsequent condensation in the head of the column. This increases the effective liquid film thickness and obstructs gas flow. This type of overloading can be minimized by reducing the vapor concentration below the partial pressure at the column temperature. A second contribution is feed volume overloading, previously discussed. For relatively large samples with well-designed injectors, this is the predominant overloading effect. Nonlinear isotherm loading is equivalent to the above-described solution concentration effect. For systems that exhibit low solubility and high activity coefficients, this type of overloading is enhanced and more dilute solutions are required for avoidance. This overloading generally produces the leading peaks attributable to the high amounts of sample. Counteracting this concentration effect is enthalpic overloading. This is caused by the heat of solution, which is generated as the solute dissolves in the liquid phase, resulting in a temperature rise. This temperature rise is dissipated via the heat capacity of the column packing. Even though this effect is partly compensatory with the preceding mechanism, a distorted peak still results. Finally, if the heat is dissipated through the column walls, a severe radial thermal gradient will exist causing radial profile overloading. This may be particularly severe in large-bore columns and can be minimized by insulating the walls carefully from the surroundings. This phenomenon further reduces the efficiency of large columns. It should be kept in mind that the above effects are interactive and minimization of one will most probably increase another.

B. Throughput Rate

The concept of throughput is very integrally tied to the overloading just discussed. For mixtures with very low α-values, high resolution is needed and overloading is to be avoided. Conversely, for most separations it is desired to separate the maximum amount of material in a given period of time and column overloading is common. In these instances the two peaks most difficult to resolve often are overlapping to a degree, and a technique referred to as heart cutting is employed (27,28). In this process when collecting the fractions three cuts are obtained, the center being the region of band overlap. This portion may then be recycled through the column. By this means a higher throughput rate is achieved.

Another method that can frequently increase the production rate is that of overlapping injections. If the sample is relatively clean and free of materials that elute after the desired components, and if a significant time interval exists between injection and elution of the peaks of interest, then multiple overlapped injections may be made. First, the time band in which all components to be collected elute is determined. A first injection is made, and following elapse of this time band another is made. In fact, a number of overlapped injections can often be accommodated. If, for example, the time band interval is 10 min and the retention time of the first desired peak is 32 min, then three injections can be interspersed. In this way the chromatographic throughput is maximized.

Sawyer and Purnell (21), in their evaluation of the effect of overloading on sample throughput, proposed the following five dependencies on column operating parameters:

1. The throughput should increase with the cross-sectional area.
2. Easily separable mixtures (high α) produce more material faster.
3. Operation at velocities higher than u_{opt} is desirable.
4. The more dense the sample, the lower the production rate.
5. No advantage is gained by increasing k above 10.

In addition to these general conclusions, if the inlet pressure is greater than 1 atm the system is solution concentration limited and a high support loading as well as high column temperatures enhance throughput. However, for low inlet pressures where feed volume is limiting, a high support loading is still favored but a reduced column temperature is beneficial.

Various investigations have confirmed these hypotheses. Verzele (24), as mentioned above, found no improvement by increasing k over 10. Carel et al. (29), in separating normal hydrocarbons, obtained a direct scale-up in throughput in going from a $\frac{3}{8}$-in. column up to a 12-in. column. By increasing the flow rate in a 1-in. column up to 4 liters/min, Rose et al. (30) observed a continually increasing production rate. The column temperature, however, exhibited a distinct maximum relative to throughput. This may be indicative of a transition from solution concentration- to feed volume limited-overloading. In actuality, although the controlling mode of overloading is not easily identified, by utilizing the suggested guidelines the throughput rate for any system can be considerably increased.

C. Injection Technique

In some preparative gas chromatographs, the sample is automatically delivered from a reservoir to the injector. These devices commonly employ a piston-driven pump or controlled-time application of a pressure on the storage container for sample transfer to the chromatograph. However, for most laboratory-scale separations, sufficient material is obtained rapidly enough so that manual syringe injection suffices. Thus two basic alternates are available, namely, flash vaporization and on-column injection. Theoretical considerations suggest that plug injection primarily via flash vaporization is the preferred method. Here again the practical limitations of the system often dictate the utilization of on-column sample introduction. As sample size increases, the normal injector is unable to "instantaneously" transmit sufficient heat to the sample to vaporize the entire amount in a discrete plug. In essence, the sample is volatilized out of the injector with a nearly exponential decay. Likewise, the buildup of back pressure may often generate a cessation of flow. This sequence of events actually spreads the band over a larger segment of the column than on-column injection directly into the initial coated section of the column.

The comparison between the two methods for a ⅜-in. column in a very overloaded condition was demonstrated by Dimick (31). Figure 3.6 shows

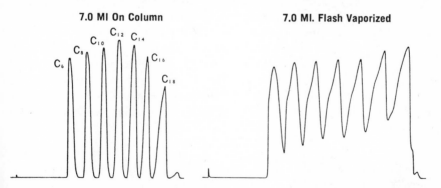

Fig. 3.6. Effect of syringe injection technique on column performance (31).

the resulting separation attainable in each case. Here again the effect on column performance of improper injection is vividly pointed up. Much loss in resolution occurs, as well as marked skewing of the peaks in the flash vaporization case. Verzele (4) also has pointed out this column decay with careful injections of decalin on SE-30. He further cautions that a stripping of the liquid phase may occur with on-column injection so that the technique should be used with awareness. For extremely large samples another technique consists of stopping the gas flow deliberately, followed by injection of the material and recommencement of flow. In this manner considerably less difficulty is encountered in introducing the sample, and reasonable peak shapes are obtained.

In order to transfer more rapidly the heat needed for flash vaporization and obtain plug injections, many workers have attempted to pack the injection volume with a high-heat-capacity material such as steel shot. The added thermal mass is then reheated via normal means while the separation is occurring in the column. Rose and co-workers (32), in their extensive study of preparative GC, investigated this method as well as a number of other variations in injection means relative to the maximum throughput rate attainable. In addition to studying flash vaporization injections into filled and empty cavities, they also evaluated a special rotary spray technique. With the standard method a slight improvement was gained by filling the injection cavity with steel shot. By using the rotary spray mode with an unfilled injector, the highest throughput was obtained, i.e., 19.5 ml/hr on a 2-in. column. The rotary spray is accomplished by bending the syringe needle and crimping the end. Following insertion through the septa, the syringe is rotated 360°, spraying the sample in a jet onto the hot walls. With the rotary spray as well as with the normal means, an optimum injection period existed. Overly rapid or slow injections gave decreased resolution and lower throughput rates. Injection times of 1.5–3 min produced acceptable results.

From the above discussion the primary point to be gleaned is that the injection method, while often overlooked as a minor factor, can very much influence column resolution and throughput. A column with a low apparent efficiency is frequently the victim of an improper injection technique.

D. Gradient Loading

With on-column injection the concentration of sample components in the introductory stages of the column is extremely high. This usually leads to solution concentration overloading. One means of overcoming this problem and accommodating increased sample capacity is that of gradient loading. The initial section of the column has a very high loading of liquid phase into which the sample can go into solution. Once the normal vapor and liquid phase concentrations are established, a more conventional liquid phase percentage is used. Thus a gradient exists from the inlet end of decreasing amount of stationary phase on the solid support. In fact, the detector end has a lower than conventional loading percentage. This gradient may be linear or nonlinear as desired. Locke and Meloan (33) obtained approximately equivalent efficiencies in linearly loaded and conventional columns, but the resolution in the gradient-loaded column was improved. For this improved resolution, however, the added complexity of column fabrication makes the net advantage questionable.

Using a Beckman multisection column of ⅝-in. diameter, Duty (34) studied normal and linear gradient-loaded columns for efficiency, resolution, and peak skew. This type of column consisting of 12 sections interconnected with narrow tubing is considerably easier to prepare with a desired gradient. Results similar to that of Locke were obtained generally. While the intrinsic efficiency was comparable for both column types, as the sample size increased the gradient loaded column decayed much more slowly and thus was the more effective column. As the end of the column near the detector had a reduced coating percentage, substrate bleed was less significant.

Preston (35) has commented that in general an increase in HETP results from gradient loading and little benefit can be derived. In fact, in one instance a reverse gradient produced better results. From the work advanced to date, while the technique appears intriguing, little practical benefit is obvious and column preparation is necessarily more complicated.

IV. Scale-Up Techniques

Suggestions for the potential of preparative GC date back to the origination of GC itself. In fact, many workers envisioned entire large chromatographic processes with throughput rates similar to present-day chemical plants. The limitation primarily preventing realization of these goals is the inherent loss in efficiency as the column capacity is scaled up. Only relatively recently has production-scale chromatography, discussed later in this chapter, become a reality.

Four methods have been proposed for increasing the capacity of the preparative column:

1. Increase of the column length.
2. Use of multiple parallel columns.
3. Cyclic automatic operation.
4. Increase of the column diameter.

The last-mentioned two techniques have in fact proven to be significantly more popular because of the practical constraints limiting other approaches. Although preparative GC is normally run in an overloaded condition, the degree of overload can likewise be minimized by cyclic operation and increased column diameter. These same principles are used to scale up to production level throughputs but in these cases ancillary means are used to hold the HETP to appropriate working levels, generally by addition of column internals.

A. Increased Length

Basically, the column capacity is proportional to the amount of stationary phase contained therein. In order to maintain the intrinsic efficiency associated with normal diameter columns and increase the mass of liquid phase, the column length must be increased. This fact has been experimentally confirmed by deWet and Pretorius (36). However, this length increase cannot be extended too extensively, as the pressure drop associated with the length increase becomes limiting. In order to overcome this increased pressure drop, a coarser-mesh solid support is usually employed. While the coarser mesh means an increased particle diameter and increased permeability, this also results in an increased resistance to gas phase mass transfer and a reasonable compromise must be reached. It has been previously shown that the HETP for preparative columns remains essentially constant with increasing length above a certain minimum length (6,8,13,37).

Another problem arising because of the increased length is prolonged retention time. Recalling the correlation between capacity, resolution, and speed, if narrow, long columns are used to increase capacity of a difficultly resolved pair, analysis time will be increased. Long residence times in the column generally lead to more band spreading and a more diffuse peak. This higher retention time can be compensated for in three ways. A lower liquid loading may be used, but this defeats the increased capacity goal. Higher gas flow rates help speed the analysis. While a u_{opt} does exist, with preparative size samples this effect is much less significant and only a small efficiency penalty is paid for large increases in flow rate. Finally, by operating the chromatograph at higher temperatures, retention time is reduced. For this reason preparative analyses are normally run at temperatures elevated over the corresponding analytical system. This must be held in control, for increasing temperature increases column bleed and this stationary phase bleed causes contamination of the collected fractions. For this reason the number of suitable liquid phases for preparative work is quite limited.

From the preceding discussion it can be readily concluded that increased capacity via increased length is obtained only with numerous other operational compromises. These compromises generally limit the utilization of this method to mixtures that are very difficult to separate, i.e., are of low α, and cannot be handled in any other way. The primary limitation is the increased pressure drop which imposes stricter requirements on the instrumentation.

B. Multiple Columns

Utilizing several parallel columns for increased sample capacity has the advantage of allowing each individual column to have a reasonable diameter and length. The total amount of liquid phase remains constant, but the sample charge to any single column more closely resembles the analytical system. On the surface it appears that this approach overcomes all the limitations of increased length or increased diameter. However, the system still has a considerable pressure drop as the flow must traverse an equivalent quantity of packed column.

The first commercial preparative gas chromatograph embodied this technique with eight parallel columns of 5/8-in. diameter (38). The system employed a recirculating gas system and the columns were manifolded together in a thermally controlled environment. A comparison of efficiency for single columns as well as the eight parallel columns is shown in Fig. 3.7.

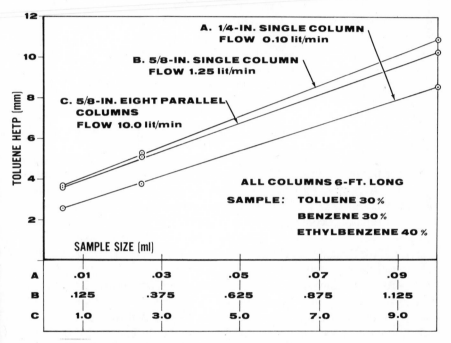

Fig. 3.7. Comparative efficiency of single and parallel columns (38).

For an equivalent column diameter the HETP on the column bundle is essentially the same as on the single column. The sample capacity, however, is increased by a factor of 8. The maximum sample capacity of the system was about 20 ml. In a later study of this parallel column array, a minimum HETP of 2 mm was achieved for very small samples (39).

The general method of multiple columns has not gained wide acceptance because of the difficult and tedious effort involved in balancing the array of parallel columns. At the inlet to the column manifold, the flow is divided into discrete segments. The flow resistance in each of the segments must be identical or the individual peaks from each column elute at slightly different times. This results in broadening of the total peak and a loss in resolution relative to the adjacent peaks. While this is a major factor to be considered, it is not sufficient to guarantee a balanced system. The relation for retention volume may be expressed as

$$V_R = V_G + KV_L \qquad (3.9)$$

As can be seen, two terms must be balanced on all columns in a parallel array in order to achieve synonymous retention. Both the residence time in the liquid phase as well as the gas phase transit time must be considered. Liquid phase balancing is accomplished by packing the same mass of coated support in each individual column segment. Once this is accomplished, flow resistance balancing by adjusting back pressure via small plugs or other equivalent means insures equal retention volumes for all columns in the manifold. While certainly feasible, this balancing operation generally requires a fairly high level of skill and has limited multiple-column applications.

C. Repetitive Cycle Operation

As the sample size increases, the column efficiency very rapidly degrades. In order to prevent this, the sample size may be kept small and the analysis run repetitively until the desired quantities of each pure material are collected. If the operator is required to inject each sample and monitor the analysis, the task rapidly becomes very burdening. It is here that automatically performed repetitive cyclic operation is mandatory. With the reduced sample sizes, the overloading previously referred to is minimized and more difficult separations can be achieved.

In order to automate a repetitive operation, the preparative gas chromatograph has evolved to a high state of sophistication. In a typical unit the sample is automatically discharged to the injector with or without preheat. This is accomplished by means of a pneumatically operated piston pump device or a unit that applies a constant pressure to a sample reservoir for a fixed time interval. Once the sample is injected, an automatic programming sequence occurs in which the column is held isothermally for an initial period while any low-boiling solvent materials elute. Then a normal programmed analysis or isothermal run occurs. As each desired peak elutes from the column and is detected, a means is provided for collection of the component and then repositioning of the collection mechanism to the next position which is either another component or waste. Following elution of all desired compounds and the end of the column thermal program, an isothermal time delay may be incorporated to permit any other high-boiling material to elute. The column temperature is then reset at the initial condition via cooldown if required, and the entire cycle is repeated as described. Such machines operate unattended and may be terminated after sufficient material is collected. Although somewhat time consuming in terms of requiring a waiting period to obtain the desired compounds, the technique requires a minimum of human effort and interaction.

Such chromatographs generally operate on a time or event base for collecting individual peaks. A number of precautions must be taken with either method. If a time base is employed for collection sequencing, then the absolute reproducibility of retention time must be extremely high. Any differential in cycle time or periodic aging of the column means shifted elution and false triggering of the collection mechanism. As any normal column has a finite bleed rate, the amount of liquid phase constantly decreases and the retention is reduced. Correction for this must be incorporated either by means of automatic program compensation or by sufficient resolution to allow broad "time windows" so as to prevent cross contamination. As most of the separations encountered practically are difficult and run at high temperatures, the column characteristics shift noticeably and time base control is difficult.

With event base control the signal generated by peak detection is used to advance the collection system. Here shifts in retention attributable to varying column characteristics or operating variables do not cause false movement of the collector. However, the question arises what level of signal generation should be used for collection sequencing, i.e., how is peak elution differentiated from spurious noise? Likewise, if insufficient resolution occurs between pairs of peaks, what valley level is adequate to prevent joint collection in the same trap and still not omit obtaining compounds with peaks of small magnitude? This problem may be resolved using more complex means of peak detection logic such as is incorporated in peak-integrating devices. In fact, although the method must be used with caution, event base control is the technique primarily used in the apparatus for repetitive cyclic operation. With this approach smaller samples are employed, overloading is avoided, and the intrinsic column efficiency is maintained while still obtaining sufficient amounts of pure materials, although at the expense of time.

D. Larger Cross Section

The last method discussed here and the current most popular means for increasing capacity is to increase diameter. Often increased cross section is used in conjunction with repetitive cyclic operation, especially for production-level operations. As the columns need not be as long, the pressure differential is generally moderate and easily accommodated in present-day instrumentation. Larger column cross section represents the only way of truly increasing sample size to any appreciable level and has therefore been most thoroughly investigated. The principal drawback of this approach is the characteristic decrease in intrinsic efficiency.

This efficiency loss evolves mainly from the difficulty in uniformly packing larger columns. As the packing material consists of a definite mesh range, particle segregation can occur. This segregation causes distinct profiles when a cross section is viewed. In regions where a less dense packing exists, the permeability increases and so does the velocity. For a tight packing with smaller particles, the sample resides longer in that zone and peak broadening occurs when the column is observed overall. This peak broadening results in lower efficiency and resolution.

In order to overcome this profile generation, a number of techniques have been developed. One approach is to pack the column very carefully with special tamping, vibration, and filling. By adding the packing in very small batches to the column and following rigid procedures with respect to axes of vibration and modes, these methods do produce columns of high efficiency but unfortunately the reproducibility is not overly good. Likewise, the method functions well only in the hands of the originator. Another means of overcoming cross-sectional gradients is to remix the flow at regular intervals. Any profile once initiated is broken up and redistributed. Remixing is commonly accomplished by placing obstructions to the flow within the column such that flow is diverted in a radial as well as a longitudinal direction. This method is used exclusively in production-scale columns and requires less attention to be paid to the actual packing procedure. For columns of medium cross section, insertion of flow diversion devices becomes increasingly difficult. A third way of achieving uniform flow without special devices involves packing fluidization. The column is filled and then a gas flow is passed through against gravity until the packing just becomes mobile. After a gentle agitation under this condition, the bed remains in a low-density configuration with particle size well mixed radially. Although such a bed is somewhat unstable, the resultant flow is quite gradient free.

The preceding scale-up techniques are all burdened with distinct limitations. Such is the current state of preparative column technology. Rather lengthy columns of conventional diameter or large-cross-section columns of moderate length can be used to increase the sample capacity. Often both are used in automated chromatographs which operate repetitively unattended.

V. Long-Narrow/Short-Wide Comparison

The age-old question remains today still mainly unanswered, which column is better for this separation, a long, narrow one or a short, wide one? Currently, the realization is growing that rather than the two column types being competitive for various tasks they are largely complementary and a careful review of separation requirements should be made before reaching a choice. Both columns produce the desired quantity of pure material. With increased diameter the sample size can be increased per injection while collecting the components of interest. By using repeated injections automatically, the long column of standard diameter separates the same total amount of a mixture.

The basic condition defining the column needed is the separation factor α. If a sample has components that are difficult to resolve and α is less than 1.15 then, practically speaking, the separation requires the higher efficiency of long, narrow columns and sample capacity per injection must be decreased. However, if the mixture is readily separated and α is greater than 1.15, then the throughput per unit time can often be increased considerably using a larger-diameter column. There are a multitude of mixtures, however, with a separability such that the resolution may be accomplished on either type of column. It is the purpose here to discuss the relative advantages and drawbacks of each and to contrast the performances.

With long, narrow columns and with preparative sample sizes, Verzele (13) has shown that little change in column efficiency occurs with changes in carrier gas type, liquid phase loading, and support mesh range. Thus in considering economics an inexpensive carrier gas and solid support can be used with low amounts of liquid phase. The coarse mesh range solid support is not only less costly but also produces a lower pressure differential across the column. As narrow columns require a much lower flow rate, not only is a slight gain in column efficiency obtained but also the trapping efficiency in the collector is enhanced. The lower flow rate permits initiation of condensation to occur more easily and minimizes losses resulting from the sample being entrained or swept away. The importance of temperature programming in gas chromatography has been adequately demonstrated, and this is true for preparative work also. It increases the column-loading capacity, decreases the analysis time, and can often increase the separation factor. With long, narrow columns, a gradient-free temperature-programmed analysis is readily accomplished. As the column radius is generally quite small, heating of the column throughout is rapid and uniform. This increases the reproducibility considerably between runs.

With short, wide columns, many of the features of long, narrow columns apply in a converse sense. Because of the increased diameter, much larger throughput rates are attainable. Commonly, with 4-in.-diameter columns sample sizes in the range 50–100 ml are injected. Under favorable conditions with large separation factors, samples of 200–300 ml can be accommodated. Again, the critical importance of ease of separability cannot be overly emphasized. Long columns are synonymous with long analysis times. Often this also means low throughput rates unless the mixture is composed of few components and the technique of overlapping injections can be used. Short, wide columns have short analysis times. This characteristic when coupled with the larger sample capacity provides for much higher throughput rates approaching a production-scale level. The economics of large-diameter columns often plays a large part in the selection process. Roughly speaking, as the same solid support, liquid phase, and carrier gas are used with any column, the unit prices are equivalent. However, much more of these materials must be used in the operation of a large-diameter column. The key word here is unit price. While many more units of supplies are needed with a short, wide column, many more units of the product material are also produced. Thus the unit cost of the pure component does not necessarily increase in proportion to the supplies expended. Although short, wide columns have significantly higher flow rates, the period of time during which the flow is applied is much shorter so the net consumption of carrier gas is not largely different. Excellent cost comparison data between long, narrow and short, wide columns have been provided, and the decision on column choice certainly cannot be made solely with this as a criterion (40,41).

It has been stated that the column choice is made primarily on the ease of separation of the sample. As an aid to this choice, Table 3.2 has been prepared (40). For the difficult separation with a low α value, a long, narrow column is advantageous, while for higher α-value mixtures a short, wide column is preferred. The α value of 1.15 suggested here is arbitrary and is based on practically achievable HETP values rather than feasible efficiencies. With wide-boiling range mixtures consisting of a number of components, the ability to program temperature is distinctly advantageous, especially at slow programming rates. If a given pair within this mixture is narrowly resolved, the temperature-programmed, long, narrow column provides the only means of accomplishing the separation. Careful perusal of Table 3.2 indicates that while often the choice of column is not clear-cut, the two types allow complementary utilization to effect the needed resolution. No one column is generally best.

VI. Solid Supports

Generally, when considering preparative separations, significantly less attention is paid to the solid support than in the equivalent analytical case. This is primarily because of cost considerations. Another reason is the level of the liquid loading on the solid support. Sample capacity has been shown to be directly proportional to the amount of liquid phase present in the column. Thus the average preparative column has a significantly higher percentage of liquid phase, typically 20–25 % liquid phase on Chromosorb W.

TABLE 3.2

Preparative Column Selection (40)

Separation conditions	Sample component	Column type	Collection rate[a]	Component resolution[a]	Sample capacity[a]
Easy	Major	Long, narrow	Fair (about 5 ml/hr)	Excellent	Good
		Short, wide	Excellent (about 50 ml/hr)	Excellent	Excellent
	Trace	Long, narrow	Too slow	Excellent	Too low
		Short, wide	Excellent	Excellent	Excellent
Difficult	Major	Long, narrow	Fair	Excellent	Good
		Short, wide	Fair	Poor unless multiple pass used	Good
	Trace	Long, narrow	Fair	Excellent	Poor
		Short, wide	Fair	Good with multiple pass	Excellent
Wide-boiling range mixture		Long, narrow	Fair (with programming)	Good	Fair
		Short, wide	Slow	Poor	Good

[a]"Fair" to "poor" may indicate the only reasonable solution to some sample problems.

The effect of this increased liquid loading is to better cover completely all the pores and interstices, which decreases potential sample adsorption on the solid support.

A. Classification

In analytical-scale work a large number of solid supports are used including diatomaceous earths, glass beads, metal shapes, commercial detergents, ceramic materials, and granulated plastics. In general, the unit cost of these materials is quite high but the small quantities employed do not practically limit the choice. In preparative work the column diameter is larger and much more solid support is required. Thus only the first two materials listed have found acceptance. In fact, glass beads have been used only in $\frac{3}{8}$-in. columns.

1. Diatomaceous Earths

Diatomaceous earth supports are prepared from filter aids mined from deposits of marine diatomites primarily by the Johns-Manville Corporation. These support materials probably represent in excess of 90% of all the solid supports used today. A family of diatomaceous earths trademarked as Chromosorb are prepared by selective calcination and treatment of the diatomite. For preparative work we restrict the discussion to consideration of Chromosorbs P, W, G, and A. No attempt is made at a comprehensive review of support preparation and treatment, this being adequately discussed by Ottenstein (42) and Horváth (2). In Table 3.3 a few of the main properties of each of these supports are presented.

TABLE 3.3

Chromosorb Support Properties

	P	W	G	A
Color	Pink	White	Oyster white	Pink
pH	6.5	8.5	8.5	7.1
Free fall density (g/cm³)	0.38	0.18	0.47	0.40
Packed density (g/cm³)	0.47	0.24	0.58	0.48
Surface area (m²/g)	4.0	1.0	0.5	2.7
Surface area (m²/cm³)	1.9	0.3	0.3	1.3

Chromosorb P is calcined from firebrick and is a relatively hard support. It is generally quite adsorptive and thus only useful with a high liquid loading, as found in preparative GC. Its primary advantage is the low cost, being the least expensive support. Chromosorb W is a flux-calcined diatomite, which accounts for its white color. The material is much less adsorbent than Chromosorb P but is very friable. If strenuously vibrated or tamped, even with high liquid loadings, bare support will exist. Chromosorb G is very similar to Chromosorb W in adsorptive properties, being somewhat more inert. This material is the hardest chromatographic support. As Chromosorb G has a very low surface area, liquid loadings are restricted to about 5–10%. Chromosorb A was developed specifically for preparative use. It has the liquid capacity of Chromosorb P as well as the hardness, but the adsorptive properties are significantly reduced. Naturally, with a high surface area it is not as inert as Chromosorb W or G, but its properties represent a useful compromise between Chromosorb P and W.

The final choice of support often rests with the sensitivity of the sample. If the sample components are prone to decomposition, inertness becomes the prime factor and some liquid phase capacity must be sacrificed. Verzele (43) found that with long narrow columns sample decomposition occurred on the catalytically active hot support. The studies were made on a 75-m column and because of the long residence time in the column ample opportunity existed for sample–support interaction. Likewise, the support that proves best for analytical work may not fare as well for preparative separations. Gas-Chrom P (Applied Science), Chromosorb W, and Anakrom ABS (Analabs) were evaluated with both analytical and preparative size samples using an SE-30 liquid phase. For analytical samples ($< 10 \mu l$) the Gas-Chrom P column was more efficient than Chromosorb W or Anakrom ABS (13). However, above 50 μl Chromosorb W proved to be the superior support. In this case the differences in apparently similar supports had affected the capacity for preparative separations. Based on the above considerations, for any compounds that may be subject to degradation, Chromosorb W or Chromosorb G is to be preferred, the liquid phase capacity of the latter being considerably lower.

2. Other Materials

As previously mentioned, because of higher costs few supports other than diatomaceous earths have been used in preparative work. Two specific examples are worth noting, namely, glass beads and protruded metal.

In large-diameter columns nearly all separations are performed isothermally. Attempts at temperature programming these wide columns usually give unreproducible results because of the gradients imposed across the column. While diatomaceous earths are good support materials, they are also good insulators. Thus heat transfer from the wall to the column center line is very poor with large gradients. In order to overcome this problem, Mitzner and Jones (44) have evaluated protruded stainless steel packings. (The metal has a much higher thermal conductivity than diatomaceous earth and thus gradients are markedly reduced during temperature programming. Protrusions that were ¼-in. in size and in a U shape were compared with Chromosorb W in a ¾-in.-diameter column. The protruded packing showed twice the efficiency at small sample sizes, but the difference was quite small with larger samples.

When sample decomposition is severe, less active surfaces must be employed. Verzele (24) investigated both glass beads and glass wool in comparison to Chromosorb G. The glass beads were 0.8 mm in diameter, while the glass wool was in strands about 5 mm long. The packing of glass columns with either of these materials was extremely difficult and time-consuming. Both materials are much less reactive than Chromosorb G. Glass wool in a $\frac{3}{8}$-in.-diameter SE-30 column was comparable to Chromosorb G in efficiency variation with sample size. Glass beads produced a low-efficiency column significantly worse than the other two, except at very large sample sizes. Unfortunately, the flow resistance of a glass wool column is prohibitively high with extremely long analysis times. Thus practical work is restricted to the utilization of glass beads for very sensitive samples.

B. Mesh Size and Range

1. Pressure Drop

In the theoretical development, the expanded van Deemter equation indicated that the highest column efficiency was obtained with the minimum possible particle diameter. In principle then, the particle diameter should be very small or the mesh size very large. Practically, this particle size optimization is limited by the flow resistance of the finished column. While the pressure differential across the column is inversely proportional to the square of the particle diameter, the HETP decreases only proportionally to particle size. Thus any increase in efficiency gained by reduced particle size is obtained only at the expense of a greatly increased flow resistance.

For most high-efficiency analytical columns, the usual limiting particle size is 120–140 mesh or 105–125 μ. Pressure drops associated with such columns which are 5–10 ft in length are usually less than a few atmospheres. However, column pressure differential is also directly dependent on column length. In preparative work with a typical $\frac{3}{8}$-in.-diameter column, the length may be 100–200 ft. In this case, with the small particle sizes employed for high efficiency, pressure drops approaching hundreds of atmospheres would result. Clearly such columns, although probably very efficient, could not be easily handled in standard commercially available apparatus. In addition, with such dense columns the flow rates would be reduced and the analysis time greatly increased. Such was the case with the column packed with glass wool just described. For the long, narrow columns required for difficult separations, in order to keep the pressure differential within reason, recourse is made to coarser supports. Most commonly used for preparative work is 30–60 mesh which is 250–590 μ, or for very long columns 10–30 mesh or 590–2000 μ may be employed. In comparing 10–30 mesh with 120–140 mesh, it can be expected that nearly a 100-fold decrease in pressure differential will occur. Thus by using these coarse mesh supports, the pressure drop can be maintained within reasonable bounds when operating with long, narrow columns.

2. Efficiency

Having now obtained a reasonable working pressure differential by increasing the particle diameter the question now arises what price has been paid in efficiency? In an analytical system a large rise in HETP would have resulted. However, in preparative work the penalty paid is relatively slight. This is because of the much greater preparative sample sizes commonly used. The effect of mesh size and range has been studied by many investigators (8,45–47). The conclusion reached is that as the sample size increases the mesh size is of little importance and all columns produce about equal efficiency. This is demonstrated in Table 3.4 for Chromosorb A(46).

TABLE 3.4

Effect of Mesh Size on Efficiency (46)

Isopropyl benzene sample size, μl	200 ml/min			500 ml/min
	30–40	20–30	10–20	10–20
25	12,416	7,120	5,024	4,736
250	2,176	1,944	2,304	2,048
1250	496	416	560	496
2500	240	256	229	208
5000	172	170	157	115

The data was obtained on a $\frac{3}{8}$-in.-diameter column with 10% SE–30. Hydrogen was used for carrier gas. It is very evident that at low sample sizes the small particle diameter produces more efficient results, but as the sample amount increases the difference becomes negligible. Likewise, the effect of increasing flow above u_{opt} is clearly to decrease the column efficiency. Verzele (48) has also demonstrated this behavior for Chromosorb W and Sterchamol as shown in Fig. 3.8. Sterchamol is a support manufactured in Germany which is very similar to Chromosorb P. Not only does the efficiency become invariant of support mesh size, but it is also seen that Chromosorb W is a more efficient support than Sterchamol both for analytical and preparative purposes. Further studies by Spencer and Kucharski (45) have also confirmed Chromosorb G to be more efficient than Chromosorb P at an equivalent mesh size.

While much has been said concerning mesh size, little consideration has been devoted to mesh range. Giddings (10), in his theoretical considerations, advocated a narrow mesh range for minimum HETP. The advantage of this,

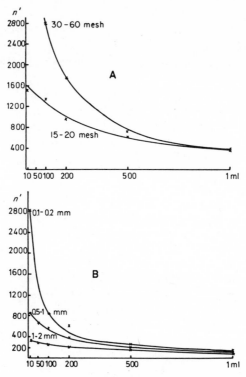

Fig. 3.8. Efficiency variation as a function of support mesh size and sample amount (4). *A*, Chromosorb W, 30–60 mesh and Chromosorb W–type support, 15–20 mesh; *B*, Sterch-amol support.

especially for short, wide columns, becomes apparent by a consideration of the reasons for efficiency loss in large-diameter columns briefly discussed previously. As wide columns are filled, particle segregation takes place, the larger particles being located near the column wall. In any mesh range a particle size ratio of at least 1.2 exists. For the common 30–60-mesh range, the size ratio is 2.4. Based on this particle segregation, the narrower the mesh range the less serious the effect will be. Ideally, if the particle size ratio were unity no segregation could occur and no efficiency loss would result. Thus in preparative work a narrow mesh range coarse support is desired in order to have high efficiencies and low pressure drops.

C. Surface Treatment

A large variety of techniques has been proposed to minimize sample-support interaction. These include calcination, acid and/or base washing, silanization, and precoating with deactivation agents. These are discussed in detail in standard references (2,42). The best method of masking support adsorptive surfaces is by ample coating with the stationary phase. As this is also a desired goal of preparative GC in order to increase sample capacity, a natural deactivation occurs. However, there are cases in which sample degradation occurs, forcing the use of special inert supports such as glass beads. Attempts in these instances have been made to try acid washing and silanization techniques but with few beneficial results. Verzele (24) in studying hop oil constituents with long analysis times, obtained some improvement by silanizing Chromosorb P but no marked reduction in decomposition. Dixmier et al. (20) acid-washed and salanized Chromosorb P but gained noticeable improvement only by coating the support with potassium hydroxide. With the rather slight improvement gained by support treatment and the significantly higher cost associated with such pretreated supports in general, preparative work is carried out with standard supports.

D. Economics

Any attempt to state absolute economic considerations naturally results in data that are obsolete at the time of publication. Therefore an attempt is made to present relative values so as to indicate advantages in utilizing certain materials. While most preparative studies have been made on diatomaceous earth supports for large-diameter columns (multiple inches), a less expensive material would be very desirable. One very economical substitute is common sand. The material is composed primarily of silica and is readily available. Of course, extensive size separation would be required to prepare narrow mesh range fractions, but the raw material is inexpensive. In many areas as a result of natural tidal action and water drainage a particle classification can be ready made. A common complaint with normal sieving methods is that only two-dimensional classification is possible. Natural water action can provide the wet screening advocated in these cases. One particular example employing sand as a solid support was given by Pretorius (49). The study was conducted on a 3-in.-diameter column with less than 1% liquid phase loading. Again the particle size was not a critical parameter and at the maximum a 30% loss in efficiency was experienced with a poor choice of particle diameter.

Extensive comparison between various solid supports for performance and cost were conducted by Carel and Perkins (41) and Spencer and Kucharski (45). For larger sample sizes raw Celite 545 (Johns-Manville) produced efficiencies equivalent to or better than that attainable with Chromosorb W. The raw Celite 545 of range 35–120 mesh costs about 500 times less than the Chromosorb W (41). The comparative results from studies made on a 4-in.-diameter column are presented in Table 3.5 (45). The first conclusion reached is that Celite 545 is the best support when balancing cost and efficiency.

TABLE 3.5
Column Support Characteristics (45)

Material	Mesh size	HETP, mm	Relative cost
Chromosorb G	60–80	0.8	11.00
Chromosorb P	60–80	0.8	8.00
Chromosorb P	10–60	1.7	1.00
Chromosorb W	60–80	3.0	10.00
Chromosorb W	10–60	3.0	1.10
Celite 545	~200	1.5	0.04

However, because of the small particle diameter a very large pressure differential existed across the column. Likewise, preliminary indications are that
the material prossesses unduly high catalytic activity. However, if nonpolar
materials are to be separated with heavy liquid loadings, then Celite 545
represents a good compromise. Otherwise, the broad mesh range Chromosorb P or W represents the next best alternative. Using the fine mesh range
supports gained increased efficiency but at greatly inflated costs.

Finally, a brief comparison is made of treated and untreated supports to
indicate why surface-treated supports are not often used. The values in
Table 3.6 are for 30–60-mesh supports in bulk quantities. The important
point to note is that it costs 30–60% more to silanize the support and 60–120%
more to obtain acid-washed support than the untreated material. As the
improvement factor gained by utilizing these treated supports was slight or
non-existent and preparative liquid loadings are high, the justification for
surface treatment of the support cannot be made.

TABLE 3.6
Relative Costs of Treated Supports

Material	Untreated	Acid washed	Silanized
Chromosorb P	1.0	2.2	1.6
Chromosorb W	1.2	2.2	1.8
Chromosorb G	1.7	2.7	2.2
Chromosorb A	1.7	2.7	2.2

VII. The Stationary Phase

The stationary phase is coated onto the solid support. This stationary phase
may be a liquid or a solid but is most commonly a liquid. It is in this liquid
that the actual chemical partitioning and separation take place with very

rapid equilibrium being established on transfer of the sample from the vapor phase. The choice of liquid determines the separating ability of the column for a given mixture, which depends on the relative affinity of the stationary phase for each specie in the mixture. This relative affinity is quantitatively embodied in the separation factor.

$$\alpha = V_{R_2}/V_{R_1} = K_2/K_1 \qquad (3.10)$$

The partition coefficient may in turn be related by solution thermodynamics to the activity coefficient of the species in the liquid phase and its vapor pressure. The separation factor of eq. 3.10 can then be expressed in terms of these new variables as

$$\alpha = \frac{\gamma_1 p_1^0}{\gamma_2 p_2^0} \qquad (3.11)$$

The activity coefficient is a measure of the deviation from ideal behavior of the resultant solution. If the components in the sample all form nearly ideal solutions, the activity coefficients will be approximately unity and a Raoult's law or boiling point separation will result. Such is the case for a mixture of saturated hydrocarbons on an SE-30 column. Many other types of sample–liquid phase interactions are feasible, but a general treatment of this area is beyond the scope of this work. The interested reader may refer to the text of McNair and Bonelli (50) for a practical working discussion of this subject. Suffice it to say at this point that the established rule of "likes dissolve likes" is a good practical relation to guide in the choice of stationary phase. Thus if it is desired to separate a mixture of polar compounds, a liquid phase of high polarity should normally be sought. A similar analogy exists for nonpolar samples. This general approach should be adequate for the majority of separations encountered in preparative GC.

A good liquid phase should have a number of desirable properties if it is to function properly in a chromatographic column. As a solvent for the sample mixture, adequate solubility for each component must be available. Likewise, to effect sufficient resolution good differential solubility must also be present. In order that the sample see the same stationary phase at all times, the liquid must be thermally stable at the operating conditions. It should also be chemically inert, as sample–liquid phase reactions are equally as deleterious as sample–support reactions. Finally, a low viscosity or high diffusivity is desired to reduce mass transfer resistance in the liquid phase. Theory also implies for maximum efficiency that a thin liquid film or low loading of the stationary phase be used. This, however, is not compatible with the general requirements of preparative work and as such must be compromised for increased sample capacity.

A. Thermal Stability

Thermal stability has two implications, the most obvious being the upper thermal limit of the material. However, concern should also be given for the lower thermal limit, i.e., the minimum temperature at which useful efficient separations can be achieved. Generally, for most liquid phases this corresponds to the solidification point of the material. However, with some of the polymeric materials used as liquid phases today, a lower temperature is reached below which the viscosity increases so rapidly that liquid phase mass transfer resistance becomes predominant and the HETP increases drastically. For the common preparative liquid phases, the lower operating temperature is that at which transition from solid to liquid phase occurs.

The upper temperature limit in gross terms is defined as that point at which the stationary phase ceases to be stationary. Bleed from the column is inadvertent, as any material has a finite vapor pressure. The object is to operate the column at a condition such that the amount of stationary phase present decreases only infinitesimally during any finite period. If the rate of loss of liquid phase is high, a number of undesired consequences result. The reproducibility of retention time from one analysis to another will be poor, and automation on a time base as previously described will be very difficult. The larger resultant problem is that of contamination of collected fractions with the stationary phase. An advantage of GC is the high purity of the collected products, and this can be very easily destroyed with trace amounts of liquid substrate. In order to avoid this stationary phase bleed and fraction contamination, preparative phases are normally run about 50°C below the recommended temperature limit of the particular material. In this manner the liquid substrate vapor pressure is substantially reduced and the stability of the column remains intact.

In addition to bleed from the column caused by elevated operating temperatures, the higher thermal conditions can also promote reactions with the liquid phase. The main type of reaction occurring is oxidation, which may take two forms (23,24). Some samples are in a reduced state when injected into the column and are very susceptible to oxidation by the liquid phase, especially with more polar substrates. This reaction is naturally enhanced by increasing temperatures. The result is the synthesis of two peaks for one component on the chromatogram which greatly complicates collection and identification of the material. A second more subtle type of reaction is the gradual oxidation of the liquid phase itself by trace amounts of air which enter the column by injection or back diffusion. The effect of this is to rather slowly decrease the life of the column and to change its characteristics. The addition of a very slight amount of foreign material in a column can cause significant shifts in the elution order of sample components. In general, all analyses should be performed at as low a temperature as is feasible, but unfortunately this usually coincides with the upper thermal limit of the stationary phase.

B. Common Partitioning Agents

Two types of partitioning media can exist, namely, a liquid phase or a solid phase. In analytical work separations on solid adsorbents and other materials are conducted frequently, but in preparative GC very little mention is made of solid partitioning agents. However, a third option has arisen, that of a liquid–solid phase. This is the porous polymer material which has gained wide acceptance in analytical work and is starting to be employed for preparative separations also.

1. Liquid Phases

The scope, number, and variety of potential liquid phases is immense. In analytical work it is this diversity of choice that has promoted the growth of GC. In preparative separations the amount of material required to fabricate a column and the high unit cost of support and stationary phase prohibit the preparation of large numbers of columns for occasional use. As an example a 5 ft by $\frac{1}{8}$-in. column packed with Chromosorb P requires only 5 g of material, while the same support used in a 200 ft by $\frac{3}{8}$-in. column needs about 1600 g for complete filling. Another interesting point is that upon careful review of the vast array of liquid phases many are very similar in terms of chemical structure, polarity, thermal range of usability, and other important properties. In fact, a relatively few broad spectrum liquid phases can suffice for the majority of the preparative separations. However, if in a particular situation the separation factor of the most difficultly resolved component pair falls below 1.10, alternate liquid phases should be investigated on an analytical scale as the number of plates required becomes very high and no practical column will achieve the separation.

Basically, two column types are mandatory, namely, those with a polar and a nonpolar liquid phase. Very often the sample encountered consists of a group of components of a similar polarity to the exclusion of the other class. The most popular phases are SE-30, a methyl silicone gum rubber which is nonpolar and gives boiling point separations, and Carbowax 20M, a high polymer of ethylene glycol which is quite polar and used for separation of alcohols, ketones, and such materials. Although the polarity of the mixture determines the liquid phase using the "likes dissolves likes" rule, examples do exist of polar compounds separated on nonpolar phases. Versele (43) has separated ketones, alcohols, and esters as well as aromatic hydrocarbons on SE-30. However, with small sample sizes the peak shapes were quite distorted, indicating some form of sample–liquid phase reaction.

In an extensive review of the literature, Dimick (31) has compiled the most frequently used preparative liquid phases and these are reproduced in Table 3.7. By employing one of the named liquid phases, nearly any preparative mixture can be separated and only seven columns need be prepared. In this manner the cost of preparative GC can be minimized.

TABLE 3.7

Common Preparative Liquid Phases

Stationary phase	Upper thermal limit, °C	Major application
Carbowax 20M	200	Polar materials, olefinic compounds
SE-30	300	General hydrocarbons
DEGS	150	Fatty acid esters
QF 1	200	Highly substituted unsaturated ring structures
Apiezon L	250	Aromatic mixtures, terpenes
Versamid 900	250	Nitrogen-containing compounds
FFAP	225	Aromatic mixtures, free fatty acids

2. Porous Polymers

Porous polyaromatic polymers have been widely used recently in analytical separations because of the rapid elution of polar materials with very symmetrical peak shapes. The polymer beads are primarily cross-linkages between styrene, ethylvinylbenzene and divinylbenzene. Some of the more recent materials are copolymers with small additions of very polar agents. These polymers are characterized by a large surface area and a fine pore structure. The various types of porous polymers are usable to about 250°C, being limited by decomposition rather than volatilization. As the material is essentially bleed free, no fraction contamination is expected of the collected components.

The purification of propylene carbonate from an aqueous solution was accomplished by Booker (51) using a Porapak Q column. As the water elutes prior to the propylene carbonate, contamination problems were eliminated. Using a 5 ft by ⅜-in. column, 4-ml injections were resolved in 53 min. The sample capacity for a column of these dimensions is quite high, implying that the porous polymers may find a broad spectrum of uses in preparative separations.

C. Support Loading

Current trends in analytical GC are toward very lightly loaded columns giving good resolution and rapid analyses. Of course with fractional microliter samples this compromise can easily be made. However, the goal of preparative work is the generation of pure materials and therefore either speed or resolution must suffer. The sample capacity of the column increases as the quantity of liquid phase in the column increases. However, other factors also increase, such as analysis time and liquid phase mass transfer resistance, and a compromise must be reached. Practically any solid support has a finite loading capacity dependent on the wettability by the particular liquid phase and by the surface area of the support. If the surface tension is quite high between support and substrate, then little liquid phase can be applied to the support surface as none of the pore structure is available to the liquid. Supports with larger surface area have a much greater stationary phase capacity. As an example, Chromosorb W can be loaded to 25–30%, about 5–10% liquid phase is feasible on Chromosorb G, but only a 2–3% coating of glass beads is possible. Attempts to increase the amount of liquid phase per unit mass of support cause flow of the surface liquid to particle contact points and a liquid buildup or puddling at these points.

While increasing the percentage liquid loading increases sample capacity, eventually resolution is impaired to the point where sample size must be reduced or pure components do not result. Likewise, as the percentage coating of the stationary phase increases, so does the residence time of the sample components. Thus extremely long runs result and the actual throughput rate declines. Verzele (13) has investigated the effect of various coating percentages on the column efficiency. The plate equation indicates that thin liquid layers are desirable to reduce the liquid phase contribution to the plate height. Often, however, gas phase mass transfer is the controlling factor. Figure 3.9 shows that for small sample sizes an optimum loading exists above which column efficiency decreases. With preparative samples, however, little effect is noticed as the liquid phase percentage is varied from 10 to 35%. Since the efficiency does not increase appreciably at higher liquid loadings, the maximum allowable sample size does not increase either and so in order to reduce the analysis time to a reasonable value, the percentage of liquid phase should be reduced to 10–15% on Chromosorb W–type supports. Similar considerations hold true for the other types of support materials.

D. Stripping

The process of stripping the liquid phase from the support does not occur to any significant extent in normally operated analytical or preparative

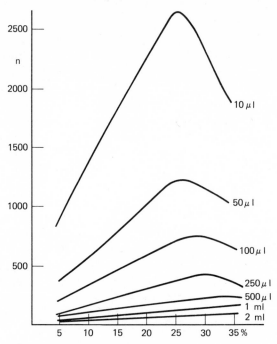

Fig. 3.9. Effect of liquid coating percentage on efficiency as a function of sample size (13).

columns. It does happen in columns that are overloaded, and since most preparative separations are run in this manner, stripping is much more prevalent than realized. The fact that stripping does take place often goes unnoticed because of the relatively poor efficiency of many preparative columns. When large samples are injected into a column, the feed volume capacity of the column is significantly exceeded and the liquid phase concentration of the sample components is quite high. In these cases not only does the liquid phase exhibit solubility for the sample, but the sample also has a marked solubility for the liquid phase. As the sample components proceed through the column, the band broadens and the local component concentration decreases. When this occurs, the amount of liquid phase in solution with the sample also decreases and the liquid phase redeposits back onto the support. All of this process takes place within the first few feet of the column. The effect is to strip the support bare in the column inlet and produce a region following this in which the percent coating is elevated. This higher local loading increases the band broadening resulting from longer residence in the liquid phase and thus decreases the column efficiency. The now bare support in the column inlet acts as a holding bed from which the samples percolate onto the coated section of the column where partitioning occurs.

Carel and Perkins (41) observed stripping in their work with 4-in.-diameter columns. They emptied the inlet section of the column and refilled it after thorough remixing of the packing. They determined that the support did not need recoating, as a natural redistribution took place in the column following the remixing process. Suspecting that stripping was occurring in his overloaded ⅜-in.-diameter columns of SE-30, Verzele (46) undertook a set of careful experiments to ascertain if and to what degree it was taking place. The intrinsic efficiency of the column was determined using a 50-μl probe sample of aromatic hydrocarbons. Then 100 injections of 5 ml each were made of the same sample mixture with intermittent efficiency determinations. Allowing for experimental fluctuations, the number of plates the column possessed was not decreased because of these repetitive large injections. Following this, the column was then segmented and the amount of stationary phase in each section determined. The original distribution had not changed to any extent. Verzele did point out that for a more soluble stationary phase stripping could be a larger factor. Later work with more efficient columns clearly demonstrated that stripping could reduce the apparent efficiency to one-third the original value.

A similar study to that just described was conducted by Mikkelsen et al. (52). Analytical (⅛-in.) as well as preparative (⅜- and 4-in.) columns were studied but with results quite different from the previous work. The columns were all Carbowax 20M, and methyl octanoate was used as a sample. Up to a given critical sample size, no stripping effects could be ascertained as judged by the residual efficiency with probe samples. These values were roughly 1 ml for the ⅜-in.-diameter column and 160 ml for the 4-in.-diameter column. However, above these sample sizes severe column damage resulted. By using injections five times the critical size, a 30% decrease in efficiency was observed for the 4-in. column while the ⅜-in.-column efficiency dropped by 70%. Repeated injections made at less than the critical sample size showed no loss in efficiency. By way of confirming the efficiency measurements, the segmented columns showed stripping of the liquid phase in the early sections followed by a buildup above the original coating level and then a tapering back to the standard percentage coating. The same tests were also run using dodecane samples with similar results. It was concluded that liquid phase buildup was most detrimental to the column efficiency because of increased diffusion and greater mass transfer resistance, but that catalytic effects on the bare support could also be harmful.

With these two conflicting results in hand, attempts were made in our laboratories to clarify the situation and define the conditions under which stripping is a problem (53). Both SE-30 and Carbowax 20M were studied, using *cis*- and *trans*-decalin as well as methyl hexanoate as samples, by the methods outlined above. Early work with inefficient ¼-in. columns confirmed the results of Verzele and showed no damage to column performance via stripping. This was true for polar and nonpolar samples with heavy and light liquid phase loadings. Although no efficiency losses were detected after injection of 0.6-ml samples, extraction of the column sections demonstrated a liquid phase redistribution with the first foot of the column. Following this, new ¼-in. columns with an HETP of 0.6 mm were prepared from both SE-30 and Carbowax 20M. With decalin samples of 1.6 ml, no efficiency decay was detected on the Carbowax 20M. However, 2.5-ml samples of methyl hexanoate caused an 85% decrease in efficiency. In this case the liquid phase stripping and piling extended over the initial 2 ft of the column. Finally, with the SE-30 stationary phase, both polar and nonpolar samples caused stripping but the nonpolar sample was more damaging. With 1.6-ml samples the residual efficiencies were 70 and 45% decayed from the original values, respectively, for decalin and methyl hexanoate. Extractive studies on the segmented columns again showed that damage from liquid phase redistribution was confined to the introductory segments of the column.

From this study the following facts on stripping can be concluded. With overloaded columns stripping is definitely a problem but is detectable only with efficient columns. No clear-cut critical sample size could be detected for inception of damage, the efficiency drop increasing as the sample size increased above analytical levels. However, if normal sample sizes are utilized with ¼- or ⅜-in. columns avoiding overloading, then stripping should not be a problem. The extent of damage to the column not only depends on the magnitude of the sample size but also on the relative solubility of the liquid phase in the sample. Likewise, the nature of the liquid phase and the percentage loading can markedly affect the degree of damage resulting from stripping. The efficiency loss appears to be attributable to the piling of the liquid phase or irregular redistribution rather than the bare support reaction with the sample. Fortunately, this irregular redistribution is confined to the earlier section of the column and upon decay of the performance may easily be corrected. For expensive columns an inlet leader section of column will serve this purpose and provide simple replacement and rejuvenation of performance.

VIII. Column Fabrication

Many of the techniques and procedures presented in this section are completely analogous to those used in the preparation of analytical columns. Just because a column is used for preparative purposes does not mean that it has to be different. As might be expected, the same procedures and details that yielded efficient analytical columns can produce efficient preparative columns. There is no magical way to make high-quality columns, nor can the method be conveyed in a text such as this. Column fabrication is much more an art than a science and that is why each technique has its own proponents in whose hands the method is eminently successful while for others it generates mediocre efficiencies. There is no substitute for quality and tender loving care. The best way to prepare columns is with practice and experience.

A. Tubing Material

Preparative chromatographs commercially available accept columns with outside diameters of $\frac{3}{8}$, $\frac{1}{2}$, $\frac{5}{8}$, $\frac{3}{4}$, 2, $2\frac{1}{2}$ and 4 in. For laboratory-scale work the tubings most commonly employed are stainless steel, aluminum brass, copper, and glass. For many preparative applications the cost of stainless steel prohibits the fabrication of many columns. With 4-in. columns as long as 80 in. and $\frac{3}{8}$-in. columns as long as 300 ft, the column tubing alone costs $400 to $900. Needless to say, these columns are emptied and reused. Because of this high cost, aluminum is the preferred material although it is not as inert. Also aluminum, being so ductile, allows simple shaping and coiling as needed. Utilizing aluminum brings the column tubing cost down to about $100. The use of brass or copper is more expensive than aluminum, affords no advantage chromatographically, and increases the potential for catalytic reaction with the sample. For maximum inertness glass should be used, but long, coiled $\frac{3}{8}$-in. columns are quite fragile and susceptible to breakage from fatigue. With 4-in. columns glass pipe is extremely strong and can be readily flanged together in a flow system.

Any tubing for best use should be washed with a suitable solvent and dried with a clean gas stream prior to filling with packing. Lubricants and drawing compounds used at the tubing mill can often produce extraneous peaks and also react with the sample. Tubing is best filled straight, with any coiling done subsequently. This avoids or minimizes the possibility of plugs or packing voids and generally allows a more dense and thus more efficient packing job. To prevent the racetrack effect of varying velocity on the inner and outer column walls in the coil, a coil diameter at least 10 times the column diameter should be used. When connecting preparative columns the tube fittings are often also packed to prevent component remixing and dead volume.

B. Support Coating

Two techniques are currently used to coat the liquid phase on the solid support, namely, batch coating and in situ coating. The former is much more prevalent, as much larger quantities of packing material can be prepared and the ratio of liquid phase to solid support is easily determined. When preparing coated supports, the amounts of material are always expressed as weight percents of the total, for example, 20 % SE-30 on Chromosorb W means that 20 g of SE-30 are coated on 80 g of Chromosorb W producing 100 g of packing. In order to spread the liquid phase evenly on the support, the liquid must be dissolved in a suitable solvent of very low viscosity. The most commonly used solvents are methylene chloride, acetone, methanol, and toluene. These solvents all possess high volatility to facilitate rapid removal after coating.

The batch-coating process, as the name implies, produces batches of coating rather than a continuous operation. In preparative work it is wise to prepare large batches at the same time so that groups of columns will have more similar characteristics. After carefully weighing both liquid phase and support, the liquid is dissolved in the solvent to a dilute solution. Care should be taken to use only high-quality solvents, as any heavy trace impurities become an integral and nonreproducible part of the stationary phase. Very small amounts of impurities can markedly effect the elution order of the sample components. The dissolved liquid is then poured over the support, carefully stirring the slurry until all the solid support is well mixed and coated. The material is then transferred to a rotary vacuum evaporator, which is turned at low speeds while the volatile solvent is removed. Following this drying when the coated support is free flowing, the packing is pan dried in an oven to remove the last traces of solvent to produce a low-bleed column which minimizes the required conditioning. All containers used should be clean and the finished packing should be stored capped in order to prevent contamination.

By using the batch-coating method followed by column filling, the possibility exists of crushing the packing and exposing the bare support in the column. In order to prevent this, the support may be coated in situ. The column is first filled with the bare support and then the liquid phase dissolved in the solvent is forced through the column via pressure applied from a gas source. By continuing this gas flow, all the excess solution as well as the volatile solvent is removed from the column. The principal drawback to this technique is the lack of accurate knowledge of the amount of stationary phase deposited on the support. Bayer and co-workers (12) have used this process for coating $\frac{1}{2}$-in.-diameter columns, and the packing density and minimum HETP were equivalent to those columns prepared from standard batch coatings. Bayer (28) also discovered that, for columns in excess of $\frac{1}{2}$-in. diameter, by coating the tubing walls with stationary phase as well as the solid support a 15% increase in efficiency could be gained.

C. Packing Methods

A fairly large number of techniques are considered here for packing preparative columns. As the filling technique so critically influences the final efficiency of the column, much care is put into this phase of the preparation.

In order to handle the broad range of preparative samples, the length and diameter of the column can vary over much larger ranges than those covered in analytical columns, which complicates the task. Basically, all columns are filled by pouring the coated support into the tubing while tapping the column to prevent consolidation and plugging of the packing. After this process the column is often vibrated in various modes to settle the packing and increase the packed density to yield a higher efficiency. Then a slight additional pouring and tapping takes place until no more material can be added to the column. The packing is then complete and small gauze plugs are placed in the ends to maintain the packing within the column. For columns in the 1- to 4-in.-diameter range retaining plates are used to confine the packing. While some of the subsequent material may appear conflicting, it must be recalled that seldom does the same method work as well for two different individuals and for each column type a preferred technique usually does evolve.

The long, narrow columns, being analogous to the analytical case, are packed in general as described above with few problems. Generally, columns at most 50 ft in length are packed in single sections and then coupled together for greater lengths. While the efficiency of larger-diameter columns increases significantly with increased packing density, for the long, narrow type very marginal improvement is realized (13). Columns 200 ft long have been produced with intrinsic efficiencies of HETP = 1 mm. In columns as long as 200 ft, significant pressure differentials exist between the inlet and outlet. During packing, the end that is first capped for filling is generally more dense because of compaction than the end where material is being added. Verzele (24) recommends that this dense end be attached to the injector where the available pressure is greatest and this will result in a faster analysis time.

Short, wide columns present the major difficulty with respect to packing, as attempts are made to maintain the intrinsic efficiency of analytical columns while increasing the diameter. In the larger-diameter columns both radial and longitudinal diffusion take place and these factors contribute to band broadening. The major limitation, however, is the spreading caused by the particle size segregation across the column. As the packing is poured into the column, the larger particles tend toward the wall with the fines congregating near the center line. Exactly why this occurs is not well known but the effect most certainly is real. Efforts have been directed toward the elimination of this segregation in order to improve efficiency.

Early investigations showed that efficiencies were high as long as column diameters remained between ½ and 1 in. (38). Above this value the number of plates decreased rapidly. Huyten et al. (8) then investigated packing techniques for 3- and 10-in.-diameter columns. They found a definite particle segregation to exist in the columns, which was strongly dependent on packing method. A coarse, narrow mesh range support produced the best packing. The more dense packings provided increased efficiency although the beneficial effects of radial diffusion were lost. Pouring, pouring with tapping, tapping, vibration, and tapping with vibration were all studied, the latter technique generating a 3-in.-diameter column with an HETP of 2.5 mm. The rate at which the packing was added to the column was determined to have a pronounced effect, and an HETP of 2.0 mm was achieved with tapping while adding material at the rate of 20 g per min. Bayer (54) also found that tapping and slow filling produced the best columns. By using a 1½-in.-diameter column, an improved HETP of 1.8 mm was achieved after the inner column wall was impregnated with sintered firebrick onto which was coated the stationary phase. This wall treatment aided in counteracting the particle segregation effect.

In attempting to better study this particle segregation, Pypker (55) mixed green 20–40-mesh Celite with red 70–120-mesh Celite in a 1:1 ratio and visually observed the particle size distribution for different modes of packing. The results obtained are shown in Fig. 3.10. The normal vibration in the funnel necessary to make the packing flow into the column causes particle segregation. By filling slowly with tapping, a nonuniform packing was ob-

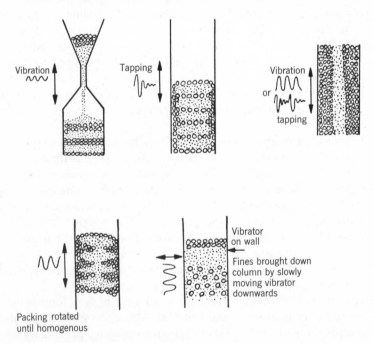

Fig. 3.10. Particle segregation resulting from various packing techniques (55).

tained with the fines in the center. By vibration at 50 cps with tapping, the coarse material is at the wall and the fines in the center. Other means of vibration with rotation generate very similar particle segregation with the fines partially sifted through the coarse material. The difficulty in obtaining a homogeneous packing with any combination of pouring, tapping, and vibration is clearly demonstrated with this study.

Frisone (56) studied 2-in.-diameter columns filled by a variety of packing techniques. Initially, packing was poured in with vibration and a vacuum applied to the column. As this method proved unsuitable, a technique was tried involving addition of 250 ml of packing followed by tamping with a special tool and application of a vacuum. Subsequent additions of 250 ml of packing were repeated until the column was full. This method also was unsatisfactory. A third approach was to slurry-pack the column as in liquid chromatography with a hexane solvent. This attempt also failed to produce an efficient column. Finally, acceptable columns were fabricated by incorporating flow homogenizers.

With each of the above techniques, operator interaction was quite severe and reproducibility of column packing technique was poor. Bayer and associates (12) developed a method incorporating mechanical column tapping with a rocking table so as to improve reproducibility. The column is supported by a stand-off from the rocking table so as to create standing waves as the column travels a 1-mm distance before impact. The packing is poured into the column and then subjected to mechanical tapping at the rate of 50 cps for about 1 hr. Then the packing is pressurized with gas for a brief period, followed by an additional 10 min of mechanical tapping. For $\frac{1}{2}$-in.-diameter columns the HETP was 0.9 mm, with a reproducibility of $\pm 3\%$.

Four different packing methods were investigated by Higgins and Smith (9). They attempted pouring and tapping, pouring through an orifice followed by vibration and tapping with two different orifice diameters, and mountain packing, in which the packing leaving the funnel trickled down a glass tube and through an orifice centered above the column. In the latter technique the orifice was constantly raised to be just above the level of the packing in the column. By so doing the packing formed a small cone on top of the settled bed and particles actually ran down the sides of this cone. Columns of 1-in. diameter were studied with the surprising result that the mountain packing produced an HETP of 1.0 mm which was better than the standard pour-and-tap packing technique. However, as the pour-and-tap method produced an HETP of 1.1 mm, the additional effort to pack columns by the mountain packing method is not justifiable. The results do demonstrate that packing techniques especially designed to aggravate particle segregation do not necessarily produce inefficient columns.

A comparison of the pouring slowly and tapping approach with packing tamping using special plungers was performed by Hupe et al. (57). With 4-in.-diameter columns, by using normal pouring and tapping, the HETP was 6 mm. Careful study of the resultant profiles in the column indicated that the maximum packing density was at the column center line. By using a conical plunger, the peripheral packing density was increased and the resultant HETP of 2.7 mm was obtained. The cross-sectional profile once established was not counteracted to any appreciable extent by radial diffusion.

Albrecht and Verzele (58) have recently developed an improved packing technique for larger-diameter columns described as the "STP" method. After a careful and detailed study of the effect of vibration modes on the column profile, it was concluded that pouring, tapping, and vibration are the cause of asymmetry and do not aid in reducing it. It would be preferable to impart a radially directed shaking to the column. Therefore, the column was rotated while simultaneously being shaken at a rapid frequency. During the entire packing process, the column was pressurized. Packing is added successively in small aliquots and the shake-turn-pressurize procedure is repeated. Packing a 3-in.-diameter column by this method yielded an HETP of 1.2 mm, by far the best value yet attained.

The preceding discussion on packing methods points up the critical importance of column packing in preparative GC. Long, narrow columns similar to their analytical counterparts are reproducibly packed with high efficiency. However, short, wide columns present immense difficulty in preparing uniform reproducible high-efficiency packings. Each of the techniques discussed worked admirably well in the hands of its developer. However, the results when compared are often somewhat conflicting. It does appear that larger-diameter columns can be prepared with plate heights on the order of 2–3 mm. Methods that are mechanically performed such as that of Bayer et al. and the new approach of Albrecht and Verzele lend hope that reproducible packings can be obtained. If preparative GC is to be successfully scaled up to production levels, such improved packing methods are definitely needed.

D. Fluidization

The effect of particle segregation during packing on the efficiency of the column was extensively treated in the preceding discussion and was seen to be extremely detrimental. A technique has been proposed by Guillemin (59–62) for negating particle segregation and producing uniformly distributed packing beds. The method consists of fluidizing the column packing after filling, which causes a rearrangement of the particles to a homogeneous distribution across a cross section. By fluidization the coated particles that are originally in a close, packed triangular structure are transformed to an open tetrahedral packing. Thus the pressure drop is reduced and less resistance to flow is presented by the column. Fluidization is accomplished by the following process. Once the column is filled, an extension equivalent in diameter is connected above the column. A suitable gas flow is then percolated through the bed opposite the direction of gravity. As the flow is slowly increased, a point is reached where the packing is no longer a compact bed but the particles are each moving independently without contact. In this fluidized state cross-mixing occurs among particles and this agitation is continued for about 5 min. The gas flow is then gradually decreased until the packing settles back down to a contact position. The column is then ready for use.

Employing such a technique, Guillemin obtained HETP values as low as 1.6 mm for a 6-in.-diameter column. This remarkable efficiency is somewhat tenuous as the extremely porous packing bed created by the fluidization process is unstable and very susceptible to collapse as a result of vibration or mechanical shock. This aspect has been investigated with the result that 15–30% of the efficiency originally obtained is lost, presumably by collapse of the packing bed into a more dense configuration.

Intrigued by the potential of this technique, we began an investigation of the effects of fluidization. Glass columns of 4-in. diameter were employed in order to visually observe the particle agitation and mixing. Early results produced an HETP of 3 mm which decayed linearly as the sample size was increased to 8 ml. While Guillemin used 20% DC200 on Chromosorb P, our studies with 5% Carbowax 600 on Chromosorb G yielded equivalent performance improving to an HETP of 1.6 mm. The effect of particle size was investigated, and 80–100-mesh support was found to be preferable to 60–70-mesh material even though the particle size ratio in both cases was 1.2. The results obtained were reproducible both with a given fluidization and between repeated fluidizations on the same column. An interesting observation was that an efficiency improvement of about a factor of 4 was obtained between the poured-only packing and the fluidized column. However, when the column was deliberately subjected to mechanical shock so as to cause collapse, the efficiency only decreased by 25–30% from the fluidized value which still was significantly better than the poured-only column. This is most probably attributable to the primarily longitudinal collapse of the bed when shocked and as such little radial cross-mixing occurs. Thus the benefits gained by fluidization are largely retained with respect to the reduction in particle segregation.

E. Column Shape

1. Circular Cross Section

The vast majority of preparative columns are of standard circular cross section. As tubing normally comes in this shape it is also the cheapest to produce. However, in an effort to increase the efficiency of large-diameter columns, alternate forms are used to avoid particle segregation. Another reason for deviating from the standard configuration in larger columns is temperature-programmed runs. As the column packing is an excellent in- sulator, various means are employed for increasing heat transfer from the wall to the center line. Practically speaking, only the long, narrow columns can be reproducibly temperature programmed with gradient-free operation.

In preparing very long, narrow columns, care must be exercised in coupling the individual sections together. If changes in diameter occur, band-broaden- ing may result with a loss in efficiency. One technique previously mentioned is to pack the connections with coated support so as to approximate a con- tinuance of the column itself. With large-diameter columns the reduced diameter couplings can be beneficial in helping to offset profiles caused by the particle segregation. Remixing and redistribution occurs in the connec- tions, but of course some degradation in resolution may result.

2. Multidiameter

The sample often overloads the column at the inlet end where the concen- tration is the highest. This effect can be partially compensated by the use of columns with sections of varying diameter. The larger-diameter section is connected to the injector, and after a suitable length transition is made to a narrower diameter of higher efficiency. This type of column was used by Walker (63), coupling $\frac{3}{8}$-in. diameter with $\frac{1}{4}$-in. diameter. The optimum carrier gas flow rate was shown to vary with length as well as diameter. Thus by carefully matching lengths and diameters, two columns can be paired that have the same optimum gas flow rate. When this was done, how- ever, the new optimum for the combined column had shifted, generally to a higher flow rate. With such column combinations much improved resolution was obtained for trace components with larger samples.

3. Ovalization

When long, narrow columns of $\frac{3}{4}$-in. diameter are coiled for compact placement in the chromatographic oven, a racetrack effect often occurs. In the coiled configuration the gas velocity in the column on the inside of the coil is higher than on the outside of the coil. This velocity gradient causes an efficiency decrease in the column separating ability. Likewise, with the $\frac{3}{4}$-in.-diameter columns thermal gradients can exist across the diameter during temperature-programmed runs. Both these effects can be reduced by ovalizing the column. This technique was proposed by Taft (64), who compared ovalized columns with a diameter ratio of 1.5:1 with standard columns. The ¾-in.-diameter ovalized column gave much sharper separations than an equivalent circular cross-section column during temperature-programmed and isothermal runs. Verzele has also used ovalized columns with a diameter ratio of 2:1. Equivalent separations and sample capacity were obtained for a 15-m column with a cross section of 1.2 × 2.4 cm and a 75-m column with a 0.9-cm diameter (24). The key factor here is equivalent column volumes, being quite independent of particular lengths and diameters.

4. Annular

Annular or biannular columns represent the best solution to the problems of column profiles and thermal gradients. Giddings and Fuller (65) studied the problem of particle segregation and concluded that one means of overcoming this source of column degradation was the use of the annular space between two concentric cylinders that have angular symmetry. Giddings (66) also demonstrated that such an annular column would have a sample capacity equivalent to a standard column when the product of the cylinder diameter and the annulus thickness was equal to the square of the radius of the conventional column.

Such annular columns have been used experimentally with very good results. Kishimoto and Yasumori (67) compared annular and conventional columns of an equivalent diameter and obtained a 30% increase in efficiency with the annular column. The ¾-in.-Biwall column (Nester Faust) was compared with a normal ¾-in.-diameter column for resolution and peak skew by Mitzner and Jones (44). At low sample sizes the Biwall column was clearly superior, while for 1-ml-size samples the two were approximately equivalent. The Biwall column contains packing between two annular cylinders with circulation of the heating medium through the center as well as about the exterior of the column. This permits heat penetration through both metal walls, allowing temperature programming to be used without gradients.

5. Finned

Another means of alleviating the thermal gradients in large-diameter columns is by the placement of fins internal to the column such that they extend radially inward from the wall to the center line. Not only should such an arrangement increase heat transfer to the column center, but also the creation of smaller partially enclosed cross sections should aid in reducing column profiles. An important feature of these somewhat independent cross sections is the fact that the periphery of the element adjacent to the center line has no physical boundary and as such free diffusional interchange between elements along the column axis is possible. The design aspects of such a column have been presented by Wright (68), who proposes an efficiency gain by a factor of 4 over conventional columns of the same diameter. Unfortunately, no data are presented to substantiate the hypothesis.

Reiser (47) has also studied columns with internal fins having 1½- and 3-in. diameters. The finned columns were prepared by welding together eight lengths of angle having about 90° bends. Packing with gentle tapping followed by refilling the empty space thus created gave the best column resolution. The finned columns showed a 30–70% increase in efficiency over the conventional columns depending on the throughput rate. This gain in efficiency is ascribed to the improved dissipation of heat attributable to localized solution effects and reduced column profiles as described above. With fast programming rates thermal gradients still existed between the circulating air temperature and that at the column center line, but programming at slow rates was accomplished successfully. Columns prepared of aluminum and stainless steel were studied, the aluminum column providing about 20% higher efficiency than the stainless steel unit. The minimum HETP for the 3-in.-diameter column was 3.0 mm, while for the 1½-in.-diameter column the value was 1.8 mm.

6. Shamrock

Very similar in concept to the internally finned column but fabricated in a different manner is the four-leaved shamrock column. The goal of this column type is the reduction of thermal gradients during programming and particle segregation effects. As the name implies, the column has the cross section of a four-leaved shamrock and as such consists of five cylindrical columns in parallel with radial diffusion unobstructed between them. The total cross-sectional area of the columns as fabricated is equivalent to a standard 1¼-in.-diameter column (69). While a distinct thermal gradient existed between the interior of the column and the circulating air, as long as the programming rate was below 15°C per min the column itself was gradient free. The efficiency of such columns dropped much more slowly than the conventional type as the sample size was increased. Significantly better resolution of the *cis*- and *trans*-decalin pair was obtained as compared to the standard circular cross-section column.

7. Internal Rods

Rather than fabricating columns with special cross-sectional shapes in order to reduce the effect of particle segregation, a more economical approach is to insert solid rods in the conventional circular cross-section column. This modification of the standard column was investigated by Dixmier et al. (70).

Rods of $\frac{1}{8}$-in. diameter are placed parallel to each other about the column axis at regular intervals. In a $\frac{3}{4}$-in.-diameter column four rods were used, while for a $1\frac{1}{2}$-in.-diameter column 13 rods were needed. The largest column of this type fabricated was 4 in. in diameter and required 80 rods. The columns were packed using the mechanically tapped rocking-table technique previously discussed. For the $1\frac{1}{2}$-in.-diameter column, the HETP for the standard configuration was 3 mm, while the modified version with internal rods had an HETP of 2 mm. This improvement was retained up to sample sizes of 3 ml. Rods of copper, brass, and quartz were evaluated and all three types produced approximately equivalent results. Although the tests were performed under isothermal conditions based on the above fact, it would be anticipated that gradients would exist during temperature-programmed separations.

F. Entrance and Exit Cones

Flow profiles in large-diameter columns that are not flat. The primary reason for these profiles is the nonhomogeneous distribution of particle sizes across the column. The larger particles concentrated at the column wall present less resistance to flow than the smaller particles at the center line. However, this particle segregation is not the only cause of flow profiles. The method in which the sample is introduced to the column also plays a very significant part in establishing these profiles. In fact, an inlet profile once established generally propagates through the entire length of the column. As interconnecting tubing between the inlet and the column is of relatively small diameter, the sample, when it arrives at the column is distributed over a small segment of the available area. Spreading to the full column cross section is accomplished by radial diffusion and random flow motion through the tortuous paths provided by the packing bed.

In order to distribute the sample better over the entire column, entrance cones should be employed. These conical adaptors provide a smooth transition from the narrow connecting tubing to the column and allow the sample to be distributed over the entire area before encountering the stationary phase. Likewise, at the exit of the column inclusion of a conical adaptor provides a flow volume in which any profile effects can be counteracted before collection. By far the entrance cone is much more critical.

The utilization of entrance and exit cones has generally led to improved column performance. Giddings has theoretically formulated the contribution of cones to the plate height and proposed an entire column composed of alternately inverted cones as a method of overcoming normal large-diameter column flow profiles (6). Whether or not these cones should be filled with column packing material seems to vary between investigators.

For 3-in.-diameter columns the conical angle had no effect, but filling the entire exit cone produced a 40% increase in efficiency in the work of Huyten et al. (8). However, the degree of filling of the inlet cone had no effect. For a 10-in.-diameter column an inlet cone partially filled produced an HETP of 2.2 mm, which was better than the totally filled or empty cone.

With 1-in.-diameter columns Hargrove and Sawyer (11) investigated standard right-angle inlets and conical inlets. While the minimum HETP with the right-angle inlet was 2 mm, the conical inlet improved this value to 1.7 mm or 15%. No study of packed cones was made for comparison. Rose and associates (32) also studied 1-in.-diameter columns with open and packed cones. The packed cones produced definitely superior performance, but an even greater improvement was gained when the entrance cone was heated to about 60°C above the column temperature. Without an inlet cone the maximum throughput rate was 4.0 ml/hr and with the heated filled cone this rate increased to 12.0 ml/hr. In an attempt to further improve the efficiency, Dixmier et al. (20) coupled conical column ends with fritted metal discs to reduce the sample inlet profile. By using this technique on 1½-in.-diameter columns, the plate height decreased to 1.6 mm from the value of 2.7 mm obtained without special column interconnections.

G. Flow Distributors

With the sample inlet profile minimized as much as possible by entrance cones, the flow profile attributable to particle segregation must still be coped with in large-diameter columns. The various attempts at reducing this segregation by specialized packing techniques were generally seen to be unsuccessful except in a few unique instances. Likewise, the reproducibility of these packing techniques was normally poor. As particle segregation causes contoured flow profiles which induce band broadening, if a means were generated of periodically redistributing the flow then columns could be scaled up without such large efficiency losses. It is this approach that has made possible the production-scale preparative GC. However, these techniques are also employed in the laboratory-scale preparative separations and are briefly described here.

1. Homogenizers

The most popular device to date for redistributing the flow is the homogenizer developed by Carel and Perkins (41). The device consists of three plates coupled together which are placed in the column at periodic intervals after the flow profile has become clearly developed (71). The first plate is a sintered metal disc with 20-μ holes. After the flow is homogenized by passage through this disc, it is then directed through the second plate. This plate has one central hole and as the flow passes through total remixing takes place with the eradication of the flow profile. Following this, a third plate identical to the first is used to redistribute the mixed flow across the entire column cross section. The spacing interval between homogenizers is not extremely critical but as incorporated into current 4-in.-diameter columns occurs every 20 in. With this type of column, plate heights of 2 mm were commonly achieved.

Verzele (72) studied the effect of flow homogenization using a stack of four rings. The first ring had a series of holes spaced on a given radius which moved the flow toward the center. The second ring incorporated a single central hole and provided the flow remixing and profile removal as in the previous design. A third ring consisting of many evenly spaced holes began flow redistribution. Finally, a fourth ring of fine mesh screen completed flow dispersion across the entire column. With this arrangement the best HETP value obtained was 3.3 mm in a 3-in.-diameter column.

Even with the flow homogenizers, careful packing technique and proper choice of solid support can bring a significant improvement in efficiency. Spencer and Kucharski (45) found that merely pouring in the packing was inadequate but rather the technique of adding the packing in small 200-ml aliquots coupled with tapping either on a vibration table or the floor was required for maximum efficiency. Depending on the choice of solid support, plate heights of 0.8–1.7 mm were obtained on 4-in.-diameter columns. Narrow mesh range supports gave the lower values but at the expense of increased pressure differential. Chromosorb W was an extremely poor support, producing very high plate heights on the order of 3 mm.

2. Baffles

Baffles as a means of flow distribution have been exclusively utilized in production-scale columns and are mentioned here for completeness. Early developments in chemical engineering technology employed disc and dough-nut baffles to provide good liquid contact and mixing in liquid-liquid absorption and extraction columns. This approach has been applied by Baddour (73) to GC columns of 4-in.-diameter and greater. At regular intervals along the length of the column, alternate discs with diameters less than that of the column and doughnut forms having a hole in the center are placed. Normal column packing material is placed between the baffles. Upon passing through the column the flow encounters a disc and is directed toward the wall, after which it strikes the doughnut and is redirected toward the center. In this manner actual radial flow prevents channeling which would normally occur because of particle segregation. The packing between baffles is merely poured into place, and so filling of a column proceeds rapidly. Virtually no efficiency data have been presented on the 4-in.-diameter column, primary attention being devoted to columns in excess of 1 ft in diameter.

3. Washers

For columns employed on a laboratory scale, the concept of baffles has been incorporated using only the doughnut form with the central hole. In smaller columns these suffice to create a flow pattern which expands and contracts relative to the wall as a washer is encountered. Two forms of washers have been studied to overcome column flow profiles.

Amy and co-workers (74) reported the use of washers spaced regularly along the column with 2-mm orifices in the center. With large preparative sample sizes of 5 ml, the column without washers had a plate height twice as large as that of the same column with washers. Bayer et al. (12) also report on using washers for improved efficiency in large-diameter columns. By inserting rings every 4 in. along the column length, a plate height of less than 2 mm was achieved in a 4-in.-diameter column.

In the above studies solid washers were placed within the column which physically diverted the flow, causing remixing and elimination of the established profile. As larger-diameter particles tend to segregate near the column wall, less flow resistance occurs here and the velocity is higher. Therefore, along the wall the sample component front is leading the portion of the band at the column center line. In an effort to slow down this peripheral band front, additional stationary phase may be placed in the region near the wall. Frisone (56) accomplished this by the use of chemical washers. Washers with decreasing orifice size in the direction of flow were constructed from filter paper and were saturated with the partitioning agent. These washers provided longer retention of the sample near the wall. With 2-in.-diameter columns and chemical washers spaced at one foot intervals, the efficiency as measured at the column center line using a sampling probe was equivalent to that determined at the wall. Verzele (75) attempted to utilize restricting devices in a similar manner but obtained no beneficial effect. Flat, conical, and semiconical shapes with identical and gradually diminishing openings were studied at various spacings along the length. However, in none of these cases was the efficiency of the column improved over that obtained with the unaltered configuration. The real benefits of such chemical washers are therefore at this point somewhat uncertain.

H. Conditioning

After the column is packed and shaped in its final configuration in the chromatographic oven, some thermal conditioning is required. Even with careful packing preparation including rotary vacuum evaporation followed by open pan drying of the free flowing support, traces of solvent and other materials still exist which are manifest as bleed and a drifting signal output with time. Likewise, as many stationary phases are not pure materials but a blend of substances or a range of polymeric composition, lower-molecular-weight fractions or more volatile components exhibit a sizable vapor pressure. These impurities and other substances can be removed by subjecting the column to a thermal condition somewhat above the intended operating temperature.

Normally, to prevent contamination of the detector the column outlet is left free to the environment. In order to avoid possible reaction with the liquid phase by air or other substances back-diffusing into the column, a low positive flow of carrier gas is maintained during the conditioning. A temperature about 25°C above the operating condition is generally chosen as the conditioning point. If programmed runs are to be attempted, the conditioning temperature is about 25°C above the upper limit of the program. Conditioning is generally done overnight, i.e., for a period of 8–16 hr. With the heavy liquid loadings employed in preparative GC, this time may be excessive and an alternate approach is to condition for a time until the signal output can be electronically balanced onto the readout device and remains relatively drift free for a period of hours. In temperature-programmed runs this conditioning is very important because if it is not done the signal will constantly drift as the analysis temperature is increased, which can result in severe problems relative to collection of the components.

IX. Performance and Behavior

GC column characteristics are very critically influenced by the flow rate of carrier gas and the temperature of the column. For large-diameter columns used in preparative separations, the packing technique and resulting bed geometry have a significant effect on the velocity profiles and thermal profiles that are established. Another factor that markedly determines column behavior is the sample introduction technique. A poorly injected sample can penalize the apparent efficiency of a column drastically. Inlet profiles once initiated in the column are propagated throughout the entire column. Although these profile effects exist in analytical columns and long, narrow preparative columns, they do so to a minor extent and can therefore normally be neglected. However, in short, wide columns consideration must be given to the effect of profiles on column performance and behavior.

A. Velocity and Concentration Profiles

When packing the larger-diameter columns, a very distinct particle segregation exists with the larger particles in the mesh range concentrated near the wall of the column and the fine particles near the center line. Because of this, a much more dense central packing exists as compared to that at the periphery. Attempts at gaining selective radial compaction using specialized tamping tools have in general not improved column performance as discussed in detail under packing methods. Even without this particle segregation, a packing density difference can exist at a given cross section which will cause velocity and concentration profiles. The effect of these profiles is to accelerate flow in areas of porous packing, generally at the walls, which spreads the component bands and produces poorer resolution and efficiency. Overcoming these profiles is accomplished by very careful packing techniques, fluidization, entrance and exit cones, and flow distributors. Certain of the column shapes previously described also aid in reducing the undesired profiles.

The profiles in a 3-in.-diameter column were studied by Huyten et al. (8). By doping the carrier gas with ammonia, the normal flow profile in the column was established, as the ammonia is not absorbed by the packing. A section of packing containing orthophosphoric acid is then added upstream from the normal bed but in contact with it and the ammonia flow continued until the distribution is established. The added column section is then cut into annular segments and titrated to determine the ammonia content. In nearly all cases the profiles were radially symmetric with the largest velocity at the wall. The distribution was independent of the column length and carrier gas velocity. With loose packings the profile was nearly flat in the center but as the density increased the shape became more nearly parabolic. Likewise, as the density of the packing increased the column efficiency also increased. This occurs in spite of the fact that radial diffusion, which normally tends to compensate the profile formation, is reduced in the more densely packed bed. When determining the plate height at the center line and at the wall, it was found that at the optimum gas velocity the wall plate height was about 30% higher than the center-line value. As the linear gas velocity increased, the center-line plate height remained constant while the value at the wall increased proportionately. This implies that the annular fractions should be collected separately with the outer fraction recycled.

Bayer and associates (12) studied profiles in larger columns, obtaining results very similar to the previous work. They attributed the profile formation to the wall effect caused by the porosity gradient across the column. In order to counteract this velocity profile, washers were installed in the column to redistribute the flow. The flow pattern in a 2-in.-diameter glass column was investigated by Frisone (56). The column was saturated with hexane which was then evaporated by means of a vacuum. Inverted parabolic flow distributions existed. Likewise, the column efficiency was much higher at the center line than at the wall. By inserting chemical washers saturated with stationary phase, the efficiency throughout the column cross section was nearly equivalent.

An ingenious experimental technique was employed by Giddings and Fuller (65) for quantitatively determining the degree of particle segregation. The 2-in.-diameter columns were packed by pouring the material through a capillary tube centered in the column. Following this, a hot gelatin solution was poured through the column and solidified. The packing core was then removed, cut into annular sections, and the mean particle size as a function of cross-sectional position determined. The results using 80–100-mesh Chromosorb W are given in Table 3.8.

TABLE 3.8

Radial Particle Size Distribution (65)

Column	Radius, cm	d_p, mm
I	0.0	0.154 ± 0.027
	0.5	0.157 ± 0.032
	1.0	0.154 ± 0.033
	1.5	0.166 ± 0.030
	2.0	0.172 ± 0.036
II	0.0	0.145 ± 0.035
	0.5	0.146 ± 0.037
	1.0	0.154 ± 0.028
	1.5	0.163 ± 0.029
	2.0	0.174 ± 0.032

Column I had a light tapping in addition to pouring, while no tapping was used with column II. As can be seen, the particle size is somewhat uniform over the central part of the column but increases considerably as the wall is approached. Tapping had little effect but did generate a greater amount of fines. The flow resistance associated with this particle size distribution is compatible with the results of Huyten and co-workers relative to the velocity profile measured.

Concentration profiles were studied in 1-in.- diameter columns by Rose et al. (76) for both absorbing and nonabsorbing samples. For the nonabsorbing species the concentration profile is in essence a velocity profile in which, as previously demonstrated, the flow is faster at the wall even with radial sample diffusion. With small sample sizes negligible concentration gradients existed for a variety of operating conditions. With overloading samples the radial concentration gradients were generally small though the method of column heating influenced this profile. Normal convective air heating did not generate concentration profiles. The degree of profile formation was very dependent on sample introduction technique and the gradient-free results were obtained only with the rotating spray injection method previously described. Rapid injection and vaporization gave extremely undesirable results.

Hupe and co-workers (57) studied a variety of packing techniques in 4-in.-diameter columns with rather different results compared to those just presented. With either fluidization or the pour-and-tap technique, adding the packing in small aliquots, a greater packing density was obtained at the wall rather than at the center line. This naturally resulted in a parabolic concentration profile but of the normal laminar flow distribution type rather than the inverted shape found by other investigators. Using different tamping tools of varying geometry, in order to homogenize the packing density in the column, succeeded in reducing the amplitude of the parabola but never eliminating it. Much of the resultant concentration gradients were the result of inlet distribution profiles propagated throughout the column. The normal laminar flow profile was also observed by Milli (77) in large-diameter columns. These profiles were overcome by placing disc baffles in the column. In this case again, the injection method was attributed as the cause of the profile initiation.

Concentration profiles in 3-in.-diameter columns were established by Albrecht and Verzele (58) using an indicator technique. Columns packed by pouring or pouring and tapping gave extremely asymmetric profiles. The profile shape depended on the direction from which the column was filled. Likewise, profile reversal often occurred during passage of the sample through the column. By using the STP packing technique described above, essentially flat concentration profiles were obtained. Some minor acceleration or retardation was observed at the walls. Radial diffusion did help reduce concentration gradients, but with the STP method it is not needed. Sample introduction method definitely influenced the inlet concentration profile.

B. Peak Skewness

One of the more disturbing facts about the operation of preparative systems is the asymmetric peak shape as seen on the recorder. Normally, with analytical separations, some peak tailing is perceived which is attributed to sample adsorption on the solid support. However, when employing preparative sample sizes and columns, the resultant peaks often exhibit both leading and tailing during the same separation. The peak skewness practically does not affect the ability to collect pure fractions providing that sufficient resolution is obtainable with the column. For some repetitive automatic devices that use an event-based actuation for cycle advancement, skewed peaks can cause erratic triggering and loss of synchronization. In the majority of cases, however, the dislike of asymmetry is based on the familiarity of observing a Gaussian output. Because of this, significant effort has been expended in counteracting this peak skewness and producing normal symmetrical peaks.

Leading peaks are caused primarily by overloading the column, with the resulting nonlinear partition isotherms for the sample components. In preparative cases in which solid support adsorption is minimal because of high liquid loadings, the tailing peaks are caused by thermal and pressure effects as the various species pass through the column. These effects have been studied both theoretically and experimentally by Scott (1,78,79). As a sample component is absorbed by the liquid phase, the heat of solution is evolved. Following the partitioning in the stationary phase, the component is desorbed with heat consumption. The result of this is that the peak front is at a higher temperature than the surrounding column, while the peak tail is often cooler than the adjacent packing. As the later peak section is cooler, it elutes slower and this gives rise to a tailing peak. Pressure effects that occur as a substance passes through the column cause a similar peak skewing but to a lesser degree.

During temperature-programmed runs over the boiling point range of a sample, the peaks that elute early are fairly symmetrical, but the larger-molecular-weight components, as they reside in the column longer at a lower temperature, exhibit overloading with leading peaks. Since the skewness attributable to thermal and pressure effects is enhanced by more rapid flow rates, if the temperature and flow are simultaneously increased during a run, the net result will be that the thermal effect skewing will counteract the overloading asymmetry and nearly Gaussian peaks will result. In effect, by increasing temperature and flow jointly, the peak is actually broadened on both the leading and tailing edges. In a practical case for a complex mixture, if the peaks are significantly broadened, a resultant loss in resolution will occur even though symmetrical peaks are obtained.

C. Thermal Profiles

The thermal effects caused by passage of the sample through the column are markedly increased as preparative sample sizes are used. Since the packing material is a poor conductor of heat, significant temperature gradients exist across a given column cross section. For a 1-in.-diameter column and a 5-ml sample, the temperature gradient can be as large as 25°C during the partitioning of a peak (71). In order to counteract these gradients and prevent the accompanying band broadening, resort must generally be made to long, narrow columns or to columns employing special geometry designed to improve heat transfer.

An extremely detailed study of thermal gradients in 1½-in.-diameter columns was made by Hupe and Bayer (80). The temperature distribution is theoretically formulated and solved in closed form. The distribution was then measured in a column packed with keiselguhr, a diatomaceous earth support. The gradients at a given cross section as determined for heating rates of 5, 10 and 15°C/min are depicted in Fig. 3.11. Thermocouples were placed in the

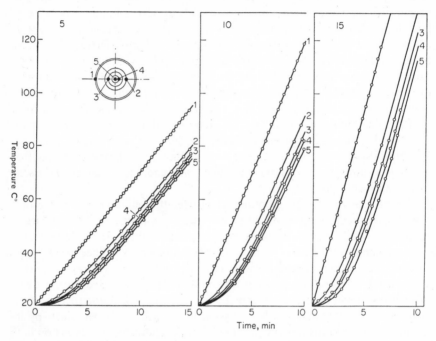

Fig. 3.11. Thermal profiles at different heating rates (80). 5°C/min; 10°C/min; 15°C/min.

bed at the respective radii of 0.0, 3.5, 7.0, 10.5, and 19 mm as shown in the insert. As can be seen, the higher the heating rate the larger the gradient from the center line to the wall. The experimental results correlated extremely well with the theoretical predictions. For supports with lower thermal conductivity, the gradients would be even larger. Varying the gas flow rate through the column by a factor of 4 had no effect on the profiles. The plate height in the column with gradients was about a factor of 8 higher than in an equivalent gradient-free column with high heating rates.

Peters and Euston (81) investigated the thermal profiles in 1-in.-diameter columns as a sample passed through. As the column operates neither adiabatically nor isothermally, the temperature at a point first increases and then decreases by 3–5°C before returning to the original packing bed temperature. These thermal excursions were maximum at the center line, decreasing as the wall was approached. With temperature-programmed operation the center-line temperature lagged that at the wall by about 1 min.

Radial and axial temperature gradients were determined in 1-in.-diameter columns by Rose et al. (76). Gradients existed both longitudinally and laterally but were very much dependent on the method of column heating. Maximum axial gradients for a 3-ft column were about 4–6°C. Radial temperature profiles decreased as the distance from the injector increased during passage of a sample. For large sample sizes temperature excursions at the center line were of the order of 35°C, while those at the wall were only 10°C in the inlet section of the column. The amplitude of this temperature transient was clearly affected by the method of sample introduction.

D. Temperature Programming

The advantages of temperature programming in GC are the reduced analysis times and the improved symmetry of the resulting peaks. In addition, for preparative work the sample capacity may be increased. With isothermal operation the band spreads considerably because of the finite injection time required for large samples. By using temperature programming and starting with low initial temperatures, this effect is considerably reduced because of the slower movement of the sample components (80).

As discussed in the preceding section, when using short, wide columns and temperature programming, rather large temperature gradients exist from the column center line to the wall. These gradients contribute to the band broadening and result in a decrease in efficiency. It is generally with complex mixtures that are difficult to resolve that temperature programming is most acutely needed. Likewise, this type of separation requires the maximum number of plates and so this efficiency drop in large-diameter columns is unacceptable. The use of annular or finned columns improves radial heat transfer because of the shorter path length through the insulating packing. However, if the column wall is not well heated or insulated, then the higher conductivity metal acts as a heat sink accentuating the gradient. Even with the use of such columns, rapid programming leads to gradients at a given cross section. With 1.5-in.-diameter finned columns, gradients existed at heating rates in excess of 1°C/min (47). Use of metallic powders or shapes as the solid supports may lead to more uniform column temperatures.

From a practical point of view, temperature programming with high resolution may only be accomplished utilizing long, narrow columns. With mixtures containing a large number of components, low α-values are unavoidable and high-efficiency temperature programming is indicated. As complex preparative separations characteristically require a long time, slower programming rates are normally used in the range of 0.5–2°C/min (25).

An example of such a separation is shown in Fig. 3.12. The column is 200 ft by ⅜-in. diameter, containing 5% SE-30 on 30–40-mesh Chromosorb G. Using nitrogen as the carrier gas, a 75-μl sample of C_{11} to C_{14} alkylbenzenes was separated by programming the temperature from 140 to 290°C at 0.5°C/min. The column had an HETP of 1.0 mm. The resolution obtained with this column for a 75-μl sample was nearly equivalent to that realized on a 200 ft by 0.02-in.-diameter capillary column of SF-96 utilizing a fraction of a microliter sample size. Of course the penalty paid for such resolution with a preparative sample size is analysis speed. While the capillary analysis was completed in less than 1 hr, the preparative separation required 7.5 hr. Preparative separations attempted without slow temperature programming on this sample suffered much resolution loss.

E. Column Heating

The effect of temperature on the behavior and performance of the column has been clearly shown to be very critical. The more uniform the environment of the column is maintained the more reproducible the results obtained will be. The most exacting means of thermal control would be accomplished by immersing the column in a fluid reservoir held at a very constant temperature. However, the practicality of such an arrangement for the large amounts of column involved in preparative work is somewhat questionable. Likewise, if the temperature is to be programmed, the rate would be extremely slow with a fluid bath. Therefore, techniques that respond more rapidly to thermal change and are not limited by size are used. Two particular approaches have been investigated, namely direct heating via resistance element and convective heating by means of recirculating air. The latter method has by far received the widest acceptance.

In their detailed study of thermal gradients in large-diameter columns, Hupe and Bayer (80) employed resistance heating. High currents with low voltages were applied directly to the column. Under isothermal conditions a uniform temperature was obtained. With this system no care is needed in arrangement of the column, and it may easily be removed from the system. Also, by transferring the heat directly to the wall, little thermal loss from the system occurs and the power required as well as the amount of insulation may be decreased. The axial thermal gradients with this method were less than 1°C.

Rose et al. (32,76) have utilized circulating air and resistance heating with Nichrome wire to maintain the temperature of the column. While not entirely comparable to the direct-heating method of Hupe, the results obtained are

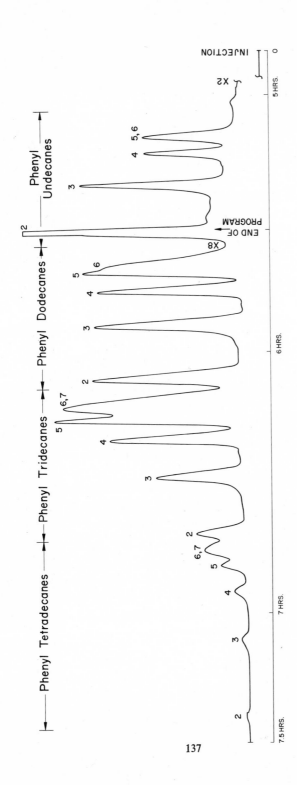

137

Fig. 3.12. Effect of slow temperature programming on the preparative separation of C_{11}–C_{14} alkylbenzenes. Sample: 75 μl. Column: 200 ft $\times$ ⅜-in.; 5% SE-30 on 30–40-mesh Chromosorb G. Program: 140°–290°C at 0.5°C/min.

diametric with respect to minimization of gradients. Axial gradients along the column were generally less than 2°C with forced air, while the heating-wire method produced linear differentials of 5°C and more. Both radial concentration and temperature profiles were improved using convective air thermostating. Even with overloading samples, no concentration gradients were evident when the column was heated with circulating air, but significant gradients existed when heated with Nichrome resistance wire. In the inlet section of the column radial temperature gradients resulting from passage of the sample components were about the same order with either heating method, but the amplitude of the temperature excursion using Nichrome resistance wire was about twice that experienced employing forced air convection. Based on these facts, the normal forced convection heating using recirculating air is recommended for more uniform control of the column temperature.

Symbols

A	Eddy diffusion term in van Deemter equation
b	Intercept in sample size equation
B	Molecular diffusion term in van Deemter equation
C	Preparative term in van Deemter equation
C_g	Gas phase mass transfer term in van Deemter equation
C_l	Liquid phase mass transfer term in van Deemter equation
d_f	Liquid phase film thickness
d_p	Particle diameter
D_g	Molecular diffusion coefficient for gas phase
D_l	Molecular diffusion coefficient for liquid phase
E	Preparative efficiency
h	Plate height
h_{prep}	Preparative plate height
HETP	Height equivalent to a theoretical plate (plate height)
HETP_{min}	Plate height at the minimum of the HETP curve
k	Partition ratio
K	Partition coefficient
K_i	Partition coefficient of the i-th specie
L	Column length
m	Slope in sample size equation
MASS	Maximum allowable sample size
n	Number of plates
p_i^0	Vapor pressure of the i-th specie
r	Column radius
R	Resolution
S	Sample size
$S(n)$	Sample size corresponding to a given number of plates
u	Carrier gas linear velocity
u_{opt}^E	Optimum carrier gas linear velocity for maximum E
u_{opt}^h	Optimum carrier gas linear velocity for minimum h

V_f Feed volume
V_G Volume of gas phase in column
V_L Volume of liquid phase in column
V_R Retention volume
V_{R_i} Retention volume of the i-th specie
x Mole fraction of larger sample component in pair
α Relative retention value
γ Labyrinth factor
γ_i Activity coefficient of the i-th specie
λ Packing irregularity factor

References

1. R. P. W. Scott, in *Gas Chromatography 1964*, A. Goldup, Ed., Institute of Petroleum, London, 1965, p. 25.
2. C. Horváth, in *The Practice of Gas Chromatography*, L. S. Ettre and A. Zlatkis, Eds., Interscience, New York, 1967, p. 129.
3. G. W. A. Rijnders, in *Advances in Chromatography*, Vol. 3, J. C. Giddings and R. A. Keller, Eds., Marcel Dekker, New York, 1966, p. 215.
4. M. Verzele, in *Progress in Separation and Purification*, Vol. 1, E. S. Perry, Ed., Interscience, New York, 1968, p. 83.
5. K. P. Hupe, *Chromatographia*, **1**, 462 (1968).
6. J. C. Giddings, *J. Gas Chromatog.*, **1**(1), 12 (1963).
7. J. C. Giddings, *J. Gas Chromatog.*, **1**(4), 38 (1963).
8. F. H. Huyten, W. van Beersum, and G. W. A. Rijnders, in *Gas Chromatography 1960*, R. P. W. Scott, Ed., Butterworths, London, 1960, p. 224.
9. G. M. C. Higgins and J. F. Smith, in *Gas Chromatography 1964*, A. Goldup, Ed., Institute of Petroleum, London, 1965, p. 94.
10. J. C. Giddings, *J. Gas Chromatog.*, **2**, 290 (1964).
11. G. L. Hargrove and D. T. Sawyer, *Anal. Chem.*, **38**, 1634 (1966).
12. E. Bayer, K. P. Hupe, and H. Mack, *Anal. Chem.*, **35**, 492 (1963).
13. M. Verzele, J. Bouche, A. DeBruyne, and M. Verstappe, *J. Chromatog.*, **18**, 253 (1965).
14. W. J. deWet and V. Pretorius, *Anal. Chem.*, **32**, 169 (1960).
15. S. M. Gordon and V. Pretorius, *J. Gas Chromatog.*, **2**, 196 (1964).
16. S. M. Gordon, G. J. Krige, and V. Pretorius, *J. Gas Chromatog.*, **3**, 87 (1965).
17. S. M. Gordon, G. J. Krige, and V. Pretorius, *J. Gas Chromatog.*, **2**, 246 (1964).
18. S. M. Gordon, G. J. Krige, and V. Pretorius, *J. Gas Chromatog.*, **2**, 285 (1964).
19. S. M. Gordon, G. J. Krige, and V. Pretorius, *J. Gas Chromatog.*, **2**, 241 (1964).
20. M. Dixmier, B. Roz, J. Taboureau, and G. Guiochon, Méthodes Physiques d'Analyse (GAMS), **5**, 250 (1969).
21. D. T. Sawyer and H. Purnell, *Anal. Chem.*, **36**, 457 (1964).
22. M. Verzele, *J. Chromatog.*, **15**, 482 (1964).
23. M. Verzele, *Planta Medica Suppl.*, 38 (1967).
24. M. Verzele, *J. Gas Chromatog.*, **4**, 180 (1966).
25. M. Verzele and M. Verstappe, *J. Chromatog.*, **26**, 485 (1967).
26. J. H. Beeson and J. Booker, paper presented at the Chicago Gas Chromatography Discussion Group Meeting, Chicago, Illinois, October 1967.
27. P. C. Haarhoff, P. C. Van Berge, and V. Pretorius, *Trans. Faraday Soc.*, **57**, 1838 (1961).
28. E. Bayer, *J. Chem. Educ.*, **41**, A755 (1964).
29. A. B. Carel, R. E. Clement, and G. Perkins, *J. Chromatog. Sci.*, **7**, 218 (1969).

30. A. Rose, D. J. Royer, and R. S. Henly, *Separation Sci.*, **2**, 257 (1967).
31. K. P. Dimick, *Gas Chromatography Preparative Separations*, Varian Aerograph, Walnut Creek, California, 1966.
32. A. Rose, D. J. Royer, and R. S. Henly, *Separation Sci.*, **2**, 211 (1967).
33. D. C. Locke and C. E. Meloan, *Anal. Chem.*, **36**, 2234 (1964).
34. R. C. Duty, *J. Gas Chromatog.*, **6**, 193 (1968).
35. S. T. Preston, Preparative Gas Chromatography Discussion Group, Fifth International Symposium on Advances in Chromatography, Las Vegas, Nevada, January 1969.
36. W. J. deWet and V. Pretorius, *Anal. Chem.*, **32**, 1396 (1960).
37. P. V. Peurifoy, J. L. Ogilvie, and I. Dvoretzky, *J. Chromatog.*, **5**, 418 (1961).
38. T. Johns, M. R. Burnell, and D. W. Carle, in *Gas Chromatography*, H. J. Noebels, R. F. Wall, and N. Brenner, Eds., Academic Press, New York, 1961, p. 207.
39. T. Johns, in *Gas Chromatography 1960*, R. P. W. Scott, Ed., Butterworths, London, 1960, p. 242.
40. *Facts and Methods*, **6**(5), F & M Scientific, Division of Hewlett-Packard, Avondale, Pennsylvania, 1965.
41. A. B. Carel and G. Perkins, *Anal. Chim. Acta*, **34**, 83 (1966).
42. D. M. Ottenstein, *J. Gas Chromatog.*, **1**(4), 11 (1963).
43. M. Verzele, *J. Gas Chromatog.*, **3**, 186 (1965).
44. B. M. Mitzner and W. V. Jones, *J. Gas Chromatog.*, **3**, 294 (1965).
45. S. F. Spencer and P. Kucharski, *Hewlett-Packard Tech. Paper* 37, Hewlett-Packard, Avondale, Pennsylvania, 1966.
46. M. Verzele and M. Verstappe, *J. Chromatog.*, **19**, 504 (1965).
47. R. W. Reiser, *J. Gas Chromatog.*, **4**, 390 (1966).
48. M. Verzele, *Proc. Chromatog. Symp. Belg. Soc. Pharm. Sci.*, Brussels, September 1964.
49. V. Pretorius, paper presented at the 2nd International Symposium on Advances in Gas Chromatography, Houston, Texas, March 1964.
50. H. M. McNair and E. J. Bonelli, *Basic Gas Chromatography*, Varian Aerograph, Walnut Creek, California, 1969.
51. J. Booker, *Previews and Reviews*, Varian Aerograph, Walnut Creek, California, October 1967.
52. L. Mikkelsen, F. J. Debbrecht, and A. J. Martin, *J. Gas Chromatog.*, **4**, 263 (1964).
53. J. H. Beeson, submitted for publication.
54. E. Bayer, in *Gas Chromatography 1960*, R. P. W. Scott, Ed., Butterworths, London, 1960, p. 236.
55. J. Pypker, in *Gas Chromatography 1960*, R. P. W. Scott, Ed., Butterworths, London, 1960, p. 240.
56. G. J. Frisone, *J. Chromatog.*, **6**, 97 (1961).
57. K. P. Hupe, U. Busch, and K. Winde, *J. Chromatog. Sci.*, **7**, 1 (1969).
58. J. Albrecht and M. Verzele, *J. Chromatog. Sci.*, **8**, 586 (1970).
59. C. L. Guillemin, *J. Chromatog.*, **12**, 163 (1963).
60. C. L. Guillemin, *J. Gas Chromatog.*, **4**, 104 (1966).
61. C. L. Guillemin and G. Wetroff, U. S. Patent No. 3,248,856 (May 3, 1966).
62. C. L. Guillemin, *J. Chromatog.*, **30**, 222 (1967).
63. J. Q. Walker, *Anal. Chem.*, **40**, 226 (1968).
64. E. M. Taft, *Aerograph Tech. Bull. W116*, Varian Aerograph, Walnut Creek, California, 1964.
65. J. C. Giddings and E. N. Fuller, *J. Chromatog.*, **7**, 255 (1962).
66. J. C. Giddings, *Anal. Chem.*, **34**, 37 (1962).
67. K. Kishimoto and Y. Yasumori, *Japan Analyst*, **12**, 125 (1963).

68. J. L. Wright, *J. Gas Chromatog.*, **1**(11), 10 (1963).
69. *Carlo Erba Short Notes*, Scientific Instruments Division, Carlo Erba, Milan, Italy, 2–66.
70. M. B. Dixmier, B. Roz, and G. Guiochon, *Anal. Chim. Acta*, **38**, 73 (1967).
71. F. J. Debbrecht, *Hewlett-Packard Tech. Paper* 28, Hewlett-Packard, Avondale, Pennsylvania, 1965.
72. M. Verzele, University of Ghent, Belgium, personal communication, 1967.
73. R. F. Baddour, U. S. Patent No. 3,250,058 (May 10, 1966).
74. J. W. Amy, L. Brand, and W. Baitinger, paper presented at the 12th Pittsburgh Conference on Analytical Chemistry and Applied Spectroscopy, Pittsburgh, Pennsylvania, February 1961.
75. M. Verzele, *J. Chromatog.*, **9**, 116 (1962).
76. A. Rose, D. J. Royer, and R. S. Henly, *Separation Sci.*, **2**, 229 (1967).
77. V. E. Milli, Estonian Academy of Science, U.S.S.R., personal communication, 1969.
78. R. P. W. Scott, *Anal. Chem.*, **35**, 481 (1963).
79. R. P. W. Scott, *Nature*, **198**, 782 (May 25, 1963).
80. K. P. Hupe and E. Bayer, in *Gas Chromatography 1964*, A. Goldup, Ed., Institute of Petroleum, London, 1965, p. 62.
81. J. Peters and C. B. Euston, *Anal. Chem.*, **37**, 657 (1965).

CHAPTER 4

Outlet System

K. -P. Hupe, *Hupe + Busch, Karlsruhe, Germany*

I. Introduction

Apart from the separation of the mixture, the main task of a preparative gas chromatograph is to carry the separated components to collection vessels where the compounds can be recovered from the carrier gas. Although the physical processes taking place can be described relatively easily, they still pose a series of technical problems which are as yet not solved completely. This is indicated by the multiplicity of different concepts embodied in various commercial instruments and by the almost boundless number of publications describing this problem.

II. Fraction Cutting

The first task is to lead the eluate from the column to one of several collection vessels. The process must be synchronized with the progress of the separation cycle so that upon the appearance of a certain peak the same vessel is always connected with the column. This part of the instrument must meet the following requirements.

1. Reliable operation without mechanical trouble or leaking. This requirement is best met by a system without mechanical parts and by using the correct materials with respect to temperature stability and compatibility with the eluted materials.

2. No premature condensation of compounds and stationary phase in cold spots. To do this the system must be able to be thermostated to at least 400°C.

Thermal decomposition of compounds at this point is much less likely than in the vaporizer, as they are greatly diluted with carrier gas (several percent) and flow through the narrow channels of the system for only a short time (several milliseconds). Both cause only a few molecular collisions with the wall. It is therefore not necessary, even for very sensitive materials, to use glass for this part of the instrument. When connections cannot be heated for constructional reasons, it is necessary to keep them straight and without change in diameter since turbulence at these points may cause condensation.

3. No remixing of separated compounds in processes (*1*) and (*2*) or as a result of diffusion. This causes the problems in the outlet system as mentioned above, as diffusion can be excluded completely only by using mechanical valves. This contradicts, however, requirements (*1*) and (*2*). An alternative solution is described later.

4. If possible, the changing of collection vessels should be carried out without changing pressure and flow, as this generally interferes with the detector signal which may cause incorrect vessel changes. Because changes in pressure can never be avoided completely, it is necessary in any case to disconnect the detector pneumatically from the rest of the system.

The extent to which these requirements must be met naturally depends on the properties and kind of substances to be chromatographed. Gases, for instance, are hardly likely to condense and do not demand special requirements for the temperature stability of the materials used. The essential fact to consider is the high diffusion velocity. The separation of a high-boiling substance, however, requires quite the reverse. Most systems described in the literature are designed for the requirements of a special application.

Figure 4.1 shows a rotary valve which is often used for fraction cutting.

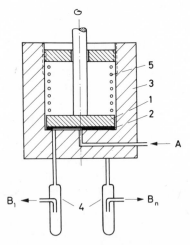

Fig. 4.1. Rotary valve.

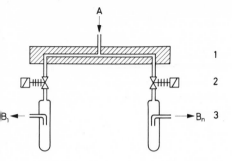

Fig. 4.2. Fraction collector. *1*, Outlet system; *2*, solenoid valve; *3*, collection vessel; *A*, Eluate inlet; *B*, carrier gas outlet.

The eluate from the column enters the fixed body *3*, at *A*, and may be directed to a collection vessel *4* via a series of vertical bore holes. The connection between *A* and *B* is formed by the radial groove in Teflon seal *2*. The Teflon seal is firmly connected to part *1*, which can be rotated by a motor. Eight collection vessels, for instance, require a torsion angle of 45° each. A further vessel for collecting the eluate between fractions is required between any two fraction-collecting vessels, so that only half the collection vessels present can be used for the main fractions. Parts *1* and *3*, which must have carefully manufactured sealing surfaces (1) are pressed together by a spring *5* to improve the seal. Cone-shaped surfaces improve the sealing properties (2).

With regard to the points of view expressed under (*1*) and (*2*), the problems of this arrangement lie in the seal between *1* and *2*, which always causes trouble at temperatures above 200°C. The same is true when the rotary valve is replaced by several solenoid valves (Fig. 4.2), which are placed either in a circle or in a straight line (3–6). Plastic sealing material such as Viton is unsuitable because of unsatisfactory temperature stability. Teflon is unsuitable for solenoid valves because of its tendency to flow. Nor has the combination ruby-steel proved satisfactory. The electrical part must, moreover, be far removed from the heat of the seal as it allows a maximum temperature of only 100°C. This difficulty can be overcome by activating the valve with compressed air (7). This, however, is by no means inexpensive.

The temperature stability requirements of the seal are lessened when it is placed at the inlet of the collection vessel, away from the hot zone. This requires the arrangement shown in Fig. 4.3. As the seals are further apart here, because of the size of the collection vessels, the latter must be moved against the fixed outlet. The collection vessels are placed in a rotating plate and are connected to the outlet in turn. The plate must make a lifting movement as well as turn. The traps are lowered during turning and are then lifted, which presses the seals together. Apart from the fact that this arrangement is costly, it does not meet requirement (*1*) for reliable operation. Furthermore, the carrier gas cannot be trapped or be carried away after leaving the collec-

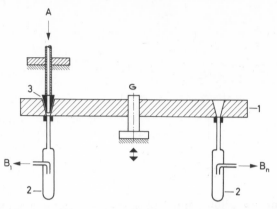

Fig. 4.3. Rotation fraction collector. *1*, Rotating plate; *2*, collection vessel; *3*, gasket; *A*, eluate inlet; *B*, carrier gas outlet.

tion vessels because of the rotary movement of the traps. This is absolutely necessary however, when separating poisons or strong-smelling substances. The problem is solved much more elegantly when the valves are put behind the collection vessels in the flow direction (8–11).

According to Fig. 4.4, the eluate reaches the collection vessels via a distribution system. A solenoid valve is positioned behind each vessel. The eluate always enters the vessel behind which the valve is open. To prevent materials from entering the wrong trap because of diffusion, a purge stream (ca. 5 ml/min) flows against the direction of the main eluate stream. The former flows through all channels that are not in use and in a direction opposite to possible diffusion. This arrangement has the advantage that no moving parts

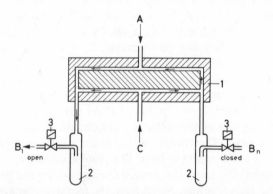

Fig. 4.4. Fraction collector with purging gas flow. *1*, Collection system; *2*, collection vessel; *3*, solenoid valve; *A*, eluate inlet; *B*, carrier gas outlet; *C*, purging gas stream.

are placed in the hot part of the system and that the seals of the solenoid valves are only in contact with cold carrier gas and the rest of the substances not trapped in the collection vessels. An additional adsorption tube can be included between collection vessel and solenoid valve; this completely removes residual substances from the solenoid valves. A rotary valve can be used instead of solenoid valves (12).

III. Collection Vessels

The collection vessels or cooling traps are the final destination of the substances on their passage through the gas chromatograph. A high collection yield is basically the essential requirement. Incomplete recovery of material is at least uneconomical as this extends the total separation time. The separation often involves mixtures that are the fruit of a lengthy synthesis and are therefore very valuable to the chemist, especially when only small quantities are available. The collection yield is the deciding factor in these cases when considering whether or not the method can be used for separation at all. From a preparative point of view, the best separation does not make sense when it is not successful in trapping the components of interest in a reliable manner.

The process that takes place during separation of substances from the carrier gas is a combined process of heat and mass transfer. Heat must be removed from the eluate to cool the solutes to below their saturation temperature, thus allowing them to condense, while demixing of condensate and carrier gas must take place on the walls of the vessel. It is therefore not a pure condensation process, but rather the deposition of a vapor from an inert gas as a dew. This process often leads to mist formation or the formation of small ice crystals (aerosols). Aerosols, however, often have a relatively large surface tension at their disposal and often carry an electrical charge. Both prevent agglomeration to larger drops of liquid. As the particles formed do not follow the vapor pressure gradient present, the condensation of the substance in this condition becomes much more difficult than when it is present in molecular form. Cooling the eluate stream must therefore take place without aerosol formation, a task not always easily accomplished. The problems connected with these processes are readily illustrated by theoretical considerations (13–15).

The carrier gas–substance mixture leaves the outlet system in a superheated form and, ideally, attains a form while flowing through the collection vessel which is determined by the temperature of the vessel wall and the corresponding saturation pressure of the substance in the mixture. The continuous line in Fig. 4.5 is the vapor pressure curve of the material to be condensed, $p'' = F(T'')$, which signifies complete saturation. To the left of

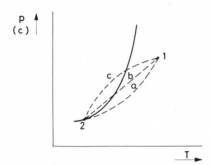

Fig. 4.5. Cooling process in $p(c)$ vs. T diagram.

the curve, the mixture is supersaturated, to the right, undersaturated (over-heated). *1* is the beginning, and *2* the final state of the condensation process. Misting occurs when the substance enters the supersaturated range during this process, i.e., when it is in a form in which its vapor pressure or concentration in the mixture is higher than that corresponding to the equilibrium. To take measures to prevent this, one must know which factors influence the transition from form *1* to form *2*; the function $p = f(T)$, or $c = f(T)$, which is the same, and therefore the partial quotient dc/dT must be known. dc/dT signifies, however, the ratio of the velocities of substance and heat transfer, which is determined essentially by the so-called Lewis number ϵ. By presupposing some simplifying assumptions, the following relation is valid.

$$\frac{dc}{dT} = \frac{1}{\epsilon} \frac{(c - c_w)}{(T - T_w)} \tag{4.1}$$

or

$$(c - c_w) = (T - T_w)^{1/\epsilon} \tag{4.2}$$

where c and T are concentration and temperature anywhere in the mixture, c_w and T_w the corresponding values on the wall of the vessel, presupposing that these correspond to the saturated form. Equation 4.2 then gives the transition from *1* to *2* for various values of:

$\epsilon < 1$: $c = f(T)$ corresponds to line *a* in Fig. 4.5 and lies in the super-heated range; no mist formation occurs.

$\epsilon = 1$: $c = f(T)$ corresponds to line *b* in Fig. 4.5. The supersaturated range is only slightly touched and mist formation is unlikely.

$\epsilon > 1$: $c = f(T)$ corresponds to line *c* in Fig. 4.5 and lies in the saturated range. Misting is very probable. The greater the value of ϵ, the more probable mist formation becomes.

The particular progress of the cooling curve in relation to the vapor pressure curve naturally depends on their shapes, and the statements made must therefore be taken in a qualitative way only.

The value of ϵ is essentially determined by the flow condition of the mixture and the properties of the constituents. For laminar flow:

$$\epsilon_{lam} = a/D \qquad (4.3)$$

where a = temperature conductivity coefficient [$a = (\lambda/\gamma)c_p$; λ = heat conductivity; γ = specific weight; c_p = specific heat] and D = diffusion coefficient. A simple correlation between the ratios of the molecular weights of substance m_1 and carrier gas m_2 can now be stated for the relation a/D based on empirical data. As a first approximation (13):

$$a/D = \sqrt{m_1/m_2} \qquad (4.4)$$

From this follows the important result that in the range of laminar flow the greater the ratio of the molecular weights of the solute and carrier gas, the greater ϵ. This is borne out by the experience that the tendency to mist formation is very great for high-molecular-weight substances and the reverse for smaller molecules. However, the smaller the molecular weight of the carrier gas, the greater ϵ. Nitrogen is therefore preferable to hydrogen for avoiding misting (16). Carbon dioxide should be even more favorable in this respect. The value a/D, for instance (calculated to 1 atm and 0°C), for small quantities of benzene in various carrier gases gives the values: H_2: 4.9; N_2: 2.6; CO_2: 0.028.

Turbulent flow give the equation for Lewis's number:

$$\epsilon_{turb} = \frac{\alpha}{\sigma \cdot \gamma \cdot c_p} \qquad (4.5)$$

where α = heat transfer number; σ = substance transfer number; γ = specific weight; and c_p = specific heat. With fully developed turbulence, and thus very high Reynolds numbers, $\epsilon_{turb} = 1$. Whereas heat and substance transfer in the laminar range take place by molecular impulse exchange or diffusion, these processes are determined for turbulence by induced convection and therefore finish with equal velocity ($\epsilon_{turb} = 1$). Turbulent flow occurs very seldom, however, in collection vessels suitable for preparative gas–liquid chromatography (GLC) in the laboratory. Purely laminar flow is usually encountered. Eddy formation, which often occurs in the lower part of the collection vessel, should not be confused with turbulence.

The basic thermodynamic law of heat and substance transfer, together with the conditions discussed above, gives a series of suggestions regarding the proper construction of collection vessels and their most favorable use.

1. From a geometric point of view, a collection vessel should have as large a surface as possible, adjusted to the quantity of condensate, whereby intensive contact with the eluate is ensured. The area of the surface is limited by the undesirability of obtaining the substance as a thin film over a large area. One should aim at a shape whereby the substance collects at the lowest point of the trap and can be taken out by syringe. Verzele (16) investigated collection vessels of different geometry regarding their collecting properties and has found that model *E* (Fig. 4.6) gives the best results.

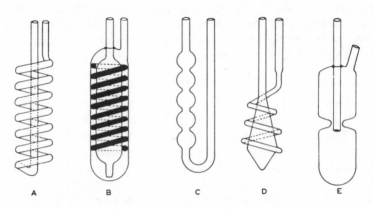

Fig. 4.6. Collection vessels with different geometries.

2. The temperature of the vessel wall (cooling bath temperature) must be significantly lower than the condensation temperature of the substance, as the difference between these temperatures is the driving force for the heat transfer. Ice water, dry ice with acetone or alcohol, liquid air or liquid nitrogen are used as cooling media, depending on the condensation temperature of the substance. The temperature difference can be chosen very large for low-molecular-weight substances with small ϵ value and thus little tendency to mist. Figure 4.7 shows how the degree of separation rises with increasing temperature difference. However, it should be generally noted that heat and mass transfer take place at possibly the same speed. The use of several cooling vessels at different temperatures is therefore recommended when separating mixtures of a wide boiling range.

3. The degree of separation improves with increased concentration of substance in the carrier gas, higher column temperature (16), and reduced pressure in the collection vessel (17). This statement, however, has little practical value as these conditions are fixed by the separating process. A high concentration of substance is unfavorable in any case from the point of view of mist formation because of quickly reaching the superheated range according to Fig. 4.8; overheating, however, is favorable. (The cooling curves shown in Fig. 4.8 and subsequent figures have only a qualitative significance. The exact curve depends, as previously noted, on flow condition and on prop-

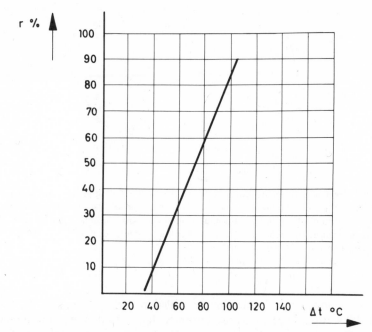

Fig. 4.7. Degree of separation depending on temperature difference Δt between eluate and cooling bath. Substance, 200 μl ether; column temperature, 70°C, carrier gas, hydrogen.

erties of the sample. Both may show local differences so that the cooling curves show only an average value.)

4. A very effective method to increase the degree of separation is to fill the collection vessel wholly or partly with an adsorbent. This shifts the condensation end point from the equilibrium curve to the unsaturated range (Fig. 4.9). The larger difference in partial pressure between the eluate and the surface of the adsorbent favors substance transfer, and heat transfer is improved at the same time through the intensive contact of both media.

Aluminum oxide (18), silica gel (19), molecular sieves (20), active carbon (21), and potassium bromide (22–24) are used as adsorbents. Potassium bromide serves at the same time for the IR analysis that often follows. Sinters, sieves, or fiber pellets from glass, metal, or plastics are used as well (25–28). The improved heat exchange is more important here because of their small adsorption capacity. Apart from adsorbents, liquids are also used. Solvents that do not interfere during subsequent use and which thus need not be separated or are easily distilled off are suitable. These solvents are either used by allowing the eluate to bubble through them (29,30) or they are frozen to form sinters in which the substance is adsorbed as in a filter (31). In a simi-

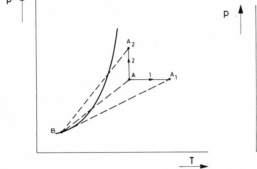

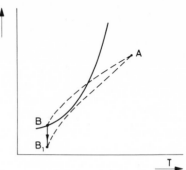

Fig. 4.8. Influence of eluate concentration and temperature on the cooling process. *A*, Initial condition; *A₁*, initial condition at higher temperature; *A₂*, initial condition at higher concentration; *B*, final condition (wall).

Fig. 4.9. Cooling process in the presence of an adsorbent. *A*, Initial condition; *B*, final condition without adsorbent; *B₁*, final condition with adsorbent.

lar way columns are suitable that are filled with uncoated (32) or coated (22,33–35) stationary phase as used in the column itself. When the temperature is sufficiently low, such a column retains the substance until the separation is finished. This is then followed by expulsion, using a small gas flow and a somewhat higher temperature into a cooling vessel. The literature cited also contains information about suitable setups of transfer instrumentation for subsequent NMR, IR, UV, MS, and gas chromatography (GC) analysis. Transfer devices have been developed, especially for IR spectroscopy, which can be used as collecting vessels and optical cells at the same time without the use of adsorbents (36,37). It must be stressed again that the problems of the methods given are connected with the subsequent separation of the substance from the adsorbent which must be accomplished by distillation, extraction, elution, or centrifuging and which is not always completely successful. Apart from those cases in which the adsorbent does not interfere, the problem posed can only be considered solved when the substance is obtained in pure form.

5. A method has proved suitable for the separation of very-high-molecular-weight materials, such as steroids and alkaloids, in which the substances are condensed together with the carrier gas. When argon is used as carrier gas (b. p. $-186°C$), the total eluate can be collected in vessels cooled with liquid nitrogen (b. p. $-196°C$) (38). This removes the diffusion resistance of the carrier gas which is otherwise present. As the decrease in volume is ca. $1:1000$, normal collection vessels can be used. Reduced pressure occurs in the collection

vessel because of total condensation. In order to prevent the surrounding air from being sucked in, the trap outlet is kept below liquid nitrogen. The evaporation of the argon must take place very slowly, as sudden evaporation pulls the substance with it. It is advisable for evaporation to take place in a vessel with liquid oxygen ($-183°C$) (39). Total condensation is also possible with carbon dioxide as the carrier gas and liquid nitrogen as the coolant (40).

A similar effect is obtained when the eluate is mixed before entering the collection vessel with vapor of a substance that condenses easily (water, acetone, etc.). The material added favors the substance transfer during condensation from the gas phase to the vessel wall. Jones and Ritchie (41), who recommend this method, have collected aromatic amines with it for later UV spectroscopy, using a ratio of 1000:1 between added vapor and material.

The collection of volatile materials can be undertaken in the following way when using carbon dioxide as carrier gas. After leaving the column, the eluate is led through potassium hydroxide which absorbs the carbon dioxide while the material collects in a space above it. This method proved suitable for C_2–C_4 hydrocarbons (42) and insecticides (43) but obviously can be used only for materials that do not react with potassium hydroxide.

6. When the separation problem essentially concerns mist formation, several strictly thermal methods can be used. Their modes of operation can be derived from Eq. 4.2, and all are concerned with keeping the substance molecules in a state of overheating until final contact with the vessel wall.

a. It has already been pointed out that the flow in the collection vessel is usually laminar, whereby a parabolic temperature partition sets in between channel axis and wall. When the heat transfer occurs faster than the substance transfer ($\epsilon > 1$), the local concentration rises above saturation concentration and mist forms. This probability is greater the further the point considered is removed from the channel wall. Mist formation starts therefore at the channel axis. With this state of affairs in mind, Colburn and Edison (44) were the first to specify a condenser with a heating element in its axis while the outer wall is cooled. This lifts the temperature level so much that, with simultaneous heat removal, a cooling curve is reached which lies in the overheating range (Fig. 4.10). For the purpose of GLC, the inner part of the collection vessel can be heated with air (45) or electrically (46) as shown in Fig. 4.11. The eluate from the column flows through a spiraling ring cleavage which is heated inside with a heating cartridge and is cooled on the outside. The inside wall has a temperature of 150–200°C. By using dry ice as a coolant, a degree of separation of 95% was obtained for ethyl caproate and limonene. The opinion often expressed that the described effect depends on turbulent flow induced by temperature difference between the middle and outside of the channel which prevents mist formation is certainly questionable.

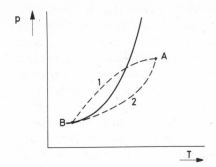

Fig. 4.10. Cooling process when heating the center of the trap. *A*, Initial condition; *B*, final condition; *1*, normal process (mist forming); *2*, process when heating.

b. As has been shown, mist is always caused when the cooling of the eluate takes place too quickly and the heat transfer thus becomes greater than the substance transfer. The cooling between outlet and collection vessel is often several hundred degrees over a distance of only a few centimeters. This must inevitably result in mist formation of mixtures for which $\epsilon > 1$. However, when cooling is slow over a distance with steadily dropping temperature, the

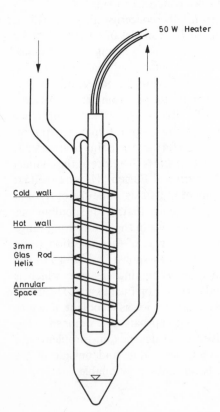

Fig. 4.11. Collection vessels with heating of trap center.

cooling curve again runs into the overheated range until condensation begins. Very small temperature gradients can cause the cooling curve to follow the saturation curve very closely (Fig. 4.12). A collection vessel based on these principles is shown in Fig. 4.13. The eluate enters a metal tube at *A* which is connected to the outlet system in a heat-conducting way so that its temperature slowly decreases from top to bottom. The condensate drips into a cooled container *C*, while the carrier gas flows away through ring cleavage *D*, which also helps to maintain the temperature gradient. With this arrangement, a degree of separation of 99% for *cis*- and *trans*-decalin was obtained (47).

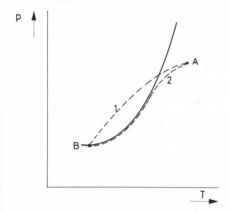

Fig. 4.12. Cooling process with temperature gradient in direction of flow. *A*, Initial condition; *B*, final condition; *1*, normal process (mist forming); *2*, process with temperature gradient.

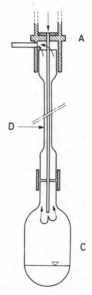

Fig. 4.13. Collection vessel with temperature gradient.

Similarly good results were obtained with the separation of long-chain fatty acid esters, however, a suitably designed oven was used to obtain the temperature gradient (48). Very simple but effective arrangements for temperature gradients are given by Teranishi et al. (49) and Brownlee and Silverstein (50).

  c. Daniels (51) described a collection vessel in which the eluate flows through a horizontal glass spiral which dips into hot oil at the lower part while the rest is cooled by an air stream. The oil is heated so that lengthwise along the spiral a temperature gradient of ca. 100°C exists. The eluate stream is thus alternatively cooled and heated so that mist that may form in the cooled part of the spiral is always returned to the overheated zone. As a temperature gradient acts at the same time, the cooling process corresponds to Fig. 4.14. This arrangement gave a separation degree of 95% for the separation of methyl oleate and β-ionone.

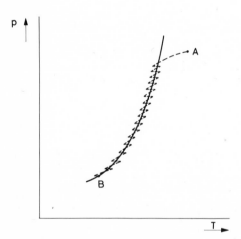

Fig. 4.14. Cooling process with intermittent cooling and heating. A, Initial condition; B, final condition.

When the methods to prevent mist formation are not successful or cannot be used for other reasons, one can try to separate the mist from the carrier gas. The difficulties connected with this process have already been pointed out.

  7. Centrifugal forces brought about when the eluate is led through a spiral are far too small to cause separation. As the mist droplets are very small and have little mass, their aerodynamic behavior is very much like that of gases. Separation in this way is successful, however, when using a centrifuge with velocities of 7000 rpm (52). Apart from the high degree of separation, this method has the advantage that condensed material is collected in a very small space. Because of the high cost involved instrumentally, centrifuges have seldom been used in practice for this purpose.

Finally, the mist may be passed through an electric field where the droplets with their electrical charge are directed toward the wall (53–57). Figure 4.15 shows a collection vessel in which the electrical field is set up between two electrodes; one lies as a straight wire in the inlet spiral, while the other one is

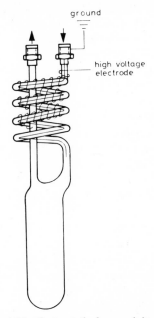

Fig. 4.15. Electrostatic fog precipitator.

wound around it. The inner electrode is grounded via the connecting nut, while the other is connected to an alternating current. The current can be changed continuously between 0 and 20 kv and is set so that the mist just disappears. When the current chosen is too high, the danger exists that the materials are decomposed. Direct current is more favorable in this respect and less dangerous for the operator, but its separating power is smaller (55).

IV. Detection

The detector of a preparative gas chromatograph serves, together with a recorder, to record continuously the chromatograms that appear one after the other and forms at the same time the basis for the automatic control. The detector must, for this purpose, satisfy the following requirements.

1. A good baseline constant in respect to changes in temperature, pressure, and flow of gas. Temperature changes may be caused by control fluctuation or by working with temperature programs. Pressure and flow changes can never be prevented completely when changing collection vessels or during injection.

2. A high sensitivity to indicate the separated substances is required, as the separation problem often consists of separating very small impurities (parts per million) from the gas mixture. A means for testing the purity of the fractions collected should also be present.

3. A wide range of linearity is as advisable as it is for analytical instruments because preparative separations are generally carried out with overloaded columns and the concentration within a single peak changes over a wide range. When the linearity range is exceeded, deformation of peaks occurs, which usually simulates a worse separation than has actually been attained.

4. Because a high percentage of liquid coating is generally used in preparative columns and part of this constantly evaporates, the detector should be insensitive to dirt and be easily cleanable in any case.

Further requirements could be added but they suffice to show that the point of view often expressed that the worst detector is good enough for a preparative GLC apparatus is untrue. All GLC detectors may in principle be used for preparative GLC. However, apart from a few exceptions, only the thermal conductivity detector (TCD) and flame ionization detector (FID) have found common acceptance.

The TCD is the most popular because it is simple and sturdy and can detect a wide variety of compounds. Unfortunately, the TCD often causes peak inversion when nitrogen, which is preferred in preparative GLC for reasons of economy, is used. Furthermore, the TCD is sensitive to changes in temperature, pressure, and gas flow and the effort necessary to stabilize these parameters often offsets the advantages.

The TCD can be placed in the main eluate stream as well as in a split stream. The arrangement in the main stream requires a correct connection of the gas lines outside the detector to reduce the sensitivity to interference. In Fig. 4.16, B is the worst method of connection although it is most often used; A is better and C is optimum, requiring, however, that the substances be completely collected in the cooling traps. Pure diffusion cells are generally used for the arrangement in the main stream. When using flow-through cells, ring-shaped reflectors (3) and glass bead filling (59) are recommended to prevent turbulence at the heating wires.

For the use of the TCD in a split stream, a small part of the flow is bled off the main stream between the column and outlet system (Fig. 4.17) and led to the measuring cell of the detector. The measuring stream, as well as the reference stream [the volumes of which depends on the geometry of the detector (5–50 ml/min)], are adjusted with needle valves. Before entry into the detector, both are brought into thermal contact to ensure temperature

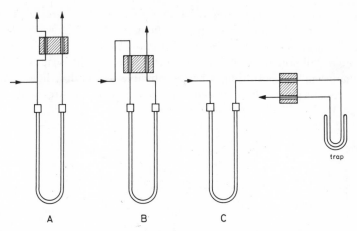

Fig. 4.16. Gas baths arrangement for thermal conductivity cell.

equilibrium. Restriction capillary *3* isolates the detector pneumatically from the outlet, which largely eliminates possible interference caused by pressure changes. This arrangement is particularly suited to the collection of unstable compounds.

In order to prevent erroneous measurements, the split stream must always have the same concentration of material as the main stream. The splitter must

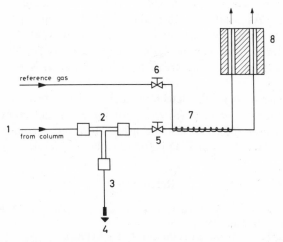

Fig. 4.17. Thermal conductivity cell in split stream. *1*, Column; *2*, splitter; *3*, restricting capillary; *4*, to outlet; *5*, needle valve for measuring stream; *6*, needle valve for reference stream; *7*, temperature exchanger; *8*, TCD.

therefore split off a representative sample from the main stream. A suitable arrangement for this purpose is shown schematically in Fig. 4.18. The end of the capillary through which the main stream enters the splitter lies closely opposite a drill hole through which the measuring stream escapes. A momentary change in concentration is hardly possible as the drill hole practically constitutes a continuation of the capillary. The main stream is reversed through 180° and led to the outlet system.

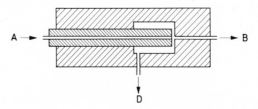

Fig. 4.18. Splitter (schematically). *A*, From the column; *B*, to the detector; *D*, to the outlet system.

Whereas the TCD was almost universally used during the early stages of preparative GC, the FID has become more popular during the last few years. The main reasons for this are extensive insensitivity to changes in temperature, pressure, and gas flow; independence of the type of carrier gas used; large linear range; high sensitivity to materials to be separated coupled with small requirements for them (not more than 1–5% of the eluate stream); short heating-up time. As disadvantages could be listed its requirement of air and hydrogen which must have accurately controlled flow volumes and the fact that only materials that can be ionized can be measured. The FID can naturally be used only in a split stream for which a splitter is used as shown in Fig. 4.18.

Other gas chromatographic detectors, such as the ionization cross-section detector (60), the β-ray detector (8), the argon ionization detector (61), and the adsorption detector (62), can also be used. However, recourse to them is made only when the TCD or FID is found to be unsuitable.

References

1. E. P. Atkinson and G. A. P. Tuey, in *Gas Chromatography 1958*, D, H. Desty, Ed., Butterworths, London, 1958, p. 270.
2. Anonymous, Nester/Faust, Sales Bulletin No. 867 (1967).
3. K. I. Sakodynskij, S. A. Wolkow, W. W. Brashnikow, and N. M. Shaworonkow, in *Gas Chromatography 1963*, H. P. Angele and H. G. Struppe, Eds., Akademie-Verlag, Berlin, 1963, p. 164.

4. J. J. Kirkland, in *Gas Chromatography*, V. J. Coates, H. J. Noebels, and I. S. Fugerson, Eds., Academic Press, New York, 1958, p. 203

5. P. A. Bushong, in *Lectures on Gas Chromatography 1962*, H. A. Szymanski, Ed., Plenum Press, New York, 1963, p. 247.

6. E. Bayer, K. -P. Hupe, and H. G. Witsch, *Angew. Chem.*, **73**, 525 (1961).

7. R. Kaiser, *Chromatographie in der Gasphase*, 1. Teil, Bibliographisches Institut, Mannheim, 1960, p. 159.

8. E. Heilbronner, E. Kovats, and W. Simon, *Helv. Chim. Acta*, **40**(2), 2410 (1957).

9. L. Blom (discussion remarks), in *Gas Chromatography 1958*, D. H. Desty, Ed., Butterworths, London, 1958, p. 284.

10. K. -P. Hupe, U. Busch, and W. Kuhn, *J. Gas Chromatogr.*, 92 (1965).

11. K. -P. Hupe, *Chromatographia*, **1**, 57 (1968).

12. J. M. Kauss, J. Peters, and C. B. Euston, *Develop. Appl. Spectry*, **2**, 383 (1962).

13. H. Hausen, *Angew. Chem. B*, **20**, 177 (1948).

14. A. Klinkenberg (discussion remarks), in *Vapor Phase Chromatography*, D. H. Desty, Ed., Butterworths, London, 1957, p. 211.

15. G. W. A. Rijnders, in *Advances in Chromatography*, Vol. 3, J. C. Giddings and R. A. Keller, Eds., Marcel Decker, New York, 1966, p. 215.

16. M. Verzele, *J. Chromatog.*, **13**, 377 (1964).

17. K. O. Kutschke, in *Vapour Phase Chromatography*, D. H. Desty, Ed., Butterworths, London, 1957, p. 96.

18. D. Ambrose and R. R. Collerson, *Nature*, **177**, 84 (1956).

19. K. Widmark and G. Widmark, *Acta Chem. Scand.*, **16**, 575 (1962).

20. M. Cartwright and A. Heywood, *Analyst*, **91**, 337 (1966).

21. C. M. Drew and J. H. Johnson, *J. Chromatog.*, **9**, 264 (1962).

22. M. D. P. Howlett and D. Welti, *Analyst*, **91**, 291 (1966).

23. K. -H. Kubeczka, *Naturwissenschaften*, **52**, 429 (1965).

24. H. W. Leggon, *Anal. Chim. Acta*, **33**, 83 (1966).

25. A. B. Carle and G. Perkins, *Anal. Chim. Acta*, **34**, 83 (1966).

26. J. Haslam, A. R. Jeffs, and H. A. Willis, *Analyst*, **86**, 44 (1961).

27. A. K. Hajra and N. S. Radin, *J. Lipid Res.*, **3**, 131 (1962).

28. P. J. Thomas and J. L. Dwyer, *J. Chromatog.*, **13**, 366 (1964).

29. R. Hardy and J. N. Keay, *J. Chromatog.*, **17**, 177 (1965).

30. D. D. Neiswender, *J. Gas Chromatog.*, **4**, 426 (1966).

31. K. Witte and O. Dissinger, *Anal. Chem.*, **236**, 119 (1968).

32. D. A. Sheares, B. C. Stone, and W. A. McGugan, *Analyst*, **88**, 147 (1969).

33. W. Kemp and D. Ronge, *Chem. Ind. (London)*, 418 (1965).

34. I. A. Fowlis and D. Welti, *Analyst*, **92**, 639 (1967).

35. B. A. Burl, M. Beroza, and J. M. Ruth, *J. Gas Chromatog.*, **6**, 286 (1968).

36. M. Beroza, *J. Gas Chromatog.*, **2**, 330 (1964).

37. J. T. Ballinger, T. T. Bartels, and J. H. Taylor, *J. Gas Chromatog.*, **6**, 295 (1968).

38. H. M. Fales, E. D. A. Haahti, T. Luukkainen, W. J. A. Vandenheuvel, and E. C. Horning, *Anal. Biochem.*, **4**, 296 (1962).

39. P.A.T. Swoboda, *Nature*, **199**, 31 (1963).

40. I. Hornstein and P. Crow, *Anal. Chem.*, **37**, 170 (1965).

41. J. H. Jones and C. D. Ritchie, *J. Assos. Offic. Agr. Chem.*, **41**, 753 (1958).

42. H. Hachenberg, *Brennstoff-Chem.*, **43**, 9 (1962).

43. C. A. Bache and D. J. Lisk, *J. Chromatog. Sci.*, **7**, 296 (1969).

44. A. P. Colburn and A. G. Edison, *Ind. Eng. Chem.*, **33**, 457 (1941).

45. R. K. Stevens and J. D. Mold, *J. Chromatog.*, **10**, 398 (1963).

46. R. Teranishi, J. W. Corse, J. C. Day, and W. G. Jennings, *J. Chromatog.*, **9**, 244 (1962).

47. G. Guiochon, M. B. Dixmier, B. Roz, and J. Taboureau, in *Gas Chromatography 1968*, H. G. Struppe, Ed., Akademie-Verlag, Berlin, 1968,p. 261.
48. H. Schlenk and D. M. Sand, *Anal. Chem.*, **34**, 1676 (1962).
49. R. Teranishi, R. A. Flath, T. R. Mon, and K. L. Stevens, *J. Gas Chromatog.*, **3**, 206 (1965).
50. R. G. Brownlee and R. Silverstein, *Anal. Chem.*, **40**, 2077 (1968).
51. H. W. R. Daniels, *Chem. Ind.* (*London*), 1078 (1963).
52. A. Wehrli and E. Kovats, *J. Chromatog.*, **3**, 313 (1960).
53. A. E. Thomson, *J. Chromatog.*, **6**, 454 (1961).
54. W. D. Ross. J. F. Moon, and R. L. Evers, *J. Gas Chromatog.*, 341 (1964).
55. L. Borka and O. S. Privett, *J. Am. Oil Chem. Soc.*, **42**, 459A (1965). Abstr.
56. D. W. Fish and D. G. Crosby, *J. Chromatog.*, **37**, 307 (1968).
57. J. L. Bloomer and W. R. Eder, *J. Gas Chromatog.*, **6**, 448 (1968).
58. Anonymous, Dr. Hupe Apparatebau, Technical Bulletin 8.68 (1968).
59. M. Modele, *Anal. Chem.*, **40**, 1444 (1968).
60. J. E. Lovelock, *Anal. Chem.*, **33**, 162 (1961).
61. D. Welti and T. Wilkins, *J. Chromatog.*, **3**, 591 (1960).
62. H. -P. Hupe Lecture, 5, Colloquium on Gas Chromatography, October 4, 1963, München.

CHAPTER 5

Programming Process and Automatic Control

U. Busch, *Hupe + Busch, Karlsruhe, Germany*

I. Reasons for Automation

Preparative gas chromatography (GC) separations are in most cases performed as repetitive operating cycles, as is explained in Chapter 1. The use of an automatic control is thus justified, as in each separating cycle the same process occurs in a fixed time sequence. Operator errors can thereby be prevented and operating costs can be considerably reduced as the instruments do not need continuous supervision. The possibility of utilizing night hours can increase the economy of a preparative gas chromatograph decisively. The advantages of automatic, compared with manual, operation can only be realized fully when the controls are so effective that no special consideration need be paid to their functioning when choosing the separating conditions. The purity and amount of trapped substances should not suffer, and it should be possible, furthermore, to run the individual separating cycles with the shortest possible time between each cycle.

II. Scope and Types of Control Devices

The following steps are included in the automatic programming of a separating cycle.

1. Injection of the sample (dosing).
2. Opening and closing of the cold traps (programmed collection).

These operations can be further supplemented by:

3. Delayed start of the collecting program.
4. Change of column temperature (temperature programming).
5. Change of carrier gas volume (flow programming).
6. Baseline correction.
7. Termination of the operating cycle after running a preselected number of separations.

For details of (*4*) and (*5*) see Chapter 6.

The control systems used for performing a programmed separation can be arranged in two groups as follows.

1. Time control only (system without feedback of the actual separating sequence).
2. Control dependent on concentration and time (system whereby program is affected by separating sequence).

III. Time-Controlled Programming Devices

Retention time is the qualitative means of identifying substances in analytical gas chromatography and this suggests the use of a time-based control both for the injection of the sample and for the opening and closing of the cold traps (1,2). This technique allows the use of a very simple type of control system (Fig. 5.1). A counting device receives counting pulses with a certain frequency from a pulse generator. Connected to the counting device are the preselection switches or plug sockets through which, after the programmed time has expired, a specific operation is started. Before a program is started, the counting device is set to zero. After the first period t_D expires, a command

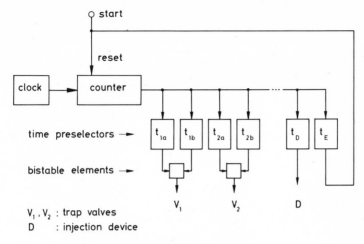

Fig. 5.1. Time control.

is issued by the control unit to the injection device which then starts dosing.

After time t_{1a} expires the first cold trap (V_1) is opened; after time t_{1b}, it is closed again. Further cold traps can be opened and closed in accordance with the components to be separated during the program, and after time t_E has expired the counting mechanism is reset to zero and a new cycle can be started. The times required for programming t_D, t_{1a}, t_{1b}, and t_E are determined by running a test chromatogram as shown in Fig. 5.2. An advantage of this simple automatic control is that no further recordings are required once the test chromatogram has been run. Set against this advantage is the disadvantage that the method is very sensitive to variations in retention times. This can be overcome by increasing the resolving power to such an extent that individual peaks are not only completely separated but also have a certain distance between them. However, one is then a long way from making optimum use of the separating column in preparative gas–liquid chromatography (GLC) work. In general, it can be stated that an automatically operated unit with pure time control makes very heavy demands on the ability to maintain constant retention times (see also Section IX.B).

IV. Programming Devices for Concentration and Time Control

The disadvantage of pure time control mentioned can be avoided when additional use is made of the detector signal for opening and closing the cold traps and only the remaining commands are purely time-controlled. A series

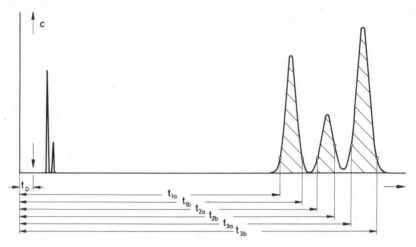

Fig. 5.2. Test chromatogram for time control. All times are calculated from the start of the program. t_D Injection; t_{1a}, t_{1b}, start and finish of fraction 1; t_{2a}, t_{2b}, start and finish of fraction 2; t_{3a}, t_{3b}, start and finish of fraction 3.

of suggestions has been made for such programming devices (3–7). Only a few of the principles involved are explained further here. Basically, not only the height of the detector signal can be used, but also its differential quotient can be included for the purpose of information.

A. Programming Device with Only One Threshold Value

For simple problems with completely separated peaks, a device is sufficient that opens a cold trap at the column outlet when concentration c_s is exceeded and closes it again when the concentration falls below this value. Should the threshold value c_s be reached a second time, the second cold trap will be opened and will be closed on the trailing edge of the peak concerned (Fig. 5.3). To permit separation independent of the number of peaks, the control device for cold traps must be supplemented by a preselecting reset circuit which brings the program back to its initial position after elution of all peaks present in the sample. The injection can function together with the reset device so that no time control is necessary for this type of equipment. Various procedures are possible to initiate the control command when the detector signal exceeds or falls below a certain threshold value. The simplest method is to use a microswitch fitted to the recorder, as it is generally desirable for checking purposes to record the entire automatic sequence of the separating program. Contacts with purely mechanical operation can be used as micro-switches, but because the corrosive atmosphere often found in the laboratory,

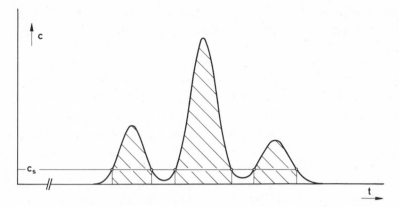

Fig. 5.3. Test chromatogram with one threshold value. c_s, Threshold value for opening and closing the cold traps.

magnetically operated reed relays or photoelectric switches are preferred. When the detector signal is not continuously recorded, a mirror galvanometer can be used in conjunction with a photocell (5). The accuracy of setting the switch and the reproducibility of the switching point should be about 2% of the range of measurement.

Figure 5.4 shows a block diagram for such a type of threshold value control. The detector signal of gas chromatograph A is fed to the recorder fitted with an adjustable microswitch B. The control device C is operated via this switch. The number of peaks to be trapped is programmed by means of the preselector switch D, thus defining the end of the separating cycle.

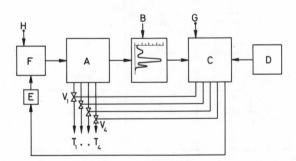

Fig. 5.4. Block diagram threshold value control. A, Gas chromatograph; B, microswitch on recorder; C, control device; D, preselected input for end of program; E, time relay for substance injection; F, injection device; G, start button; H, hand release for substance injection.

The cold traps $T_1 \ldots T_4$ are opened and closed via valves $V_1 \ldots V_4$. At the end of the separating cycle, a command is given to the injection device F via time relay E. A program is started via input G. In order to perform a test separation the injector may be operated via input H.

B. Example of Construction of Threshold Value Control

The following circuit is useful when employing solenoids for opening and closing cold traps. It is assumed here that the solenoid governing intermediate fractions is open when the current is cut off and closed when it is energized, while the solenoids for pure fractions operate inversely, i.e., are closed when the current is cut off and open when it is energized. This use of the valves has the advantage that should the feed current be shut off the cold trap for intermediate fractions will open without fail.

To operate the control device, a switch is required which is closed for a short time when the detector signal passes a predetermined level. Such a switch can, for instance, be a reed relay operated by a magnet connected with the recorder pen; alternatively, it may be a photoconductive cell which is struck by a light beam whereby a vane attached to the recorder pen passes between the light source and the cell, thus breaking the light beam. A further relay should be provided for this, with the photoconductive cell being connected in its coil circuit.

The circuit (Fig. 5.5) includes a single-pole stepping mechanism S with at least $2n + 1$ steps (where n equals the number of possible fractions) together

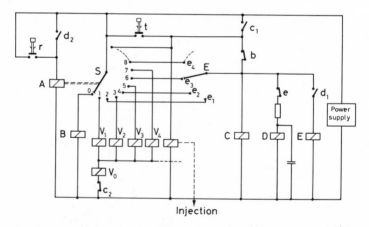

Figure 5.5. Circuit diagram for threshold value control. A, Drive coil for stepping mechanism; B, C, D, and E, relays; b, c, d, and e, accompanying contacts; E, Switch for end of cycle; S, stepping mechanism; V_0, $V_1 \ldots V_4$, solenoid valves; r, microswitch; t, start button.

with four relays, one of which, relay D, is delayed by about 0.1–0.2 sec, operating via a parallel capacitor and serial resistor.

To explain the way in which the device functions, it is assumed that the stepping mechanism S is at position 0. Relay B is then operated and all other relays, including drive coil A of the stepping mechanism, are dead. When button t is pressed, the injection device receives a control pulse and the sample is injected. When the first peak appears, the recorder signal passes the set threshold value and the stepping mechanism receives a pulse via contact r and moves into position 1. As contact c_2 is closed by relay C when in the rest position, solenoid V_1 is opened and the first peak is trapped in cold trap 1. When the sample concentration falls below the set threshold value at the end of peak 1, the stepping mechanism receives another pulse and moves into position 2. Each peak therefore initiates a forward movement of two steps. The end of the collection program is given via selector switch E. For instance, should E be brought into position e_3, then relay C will operate following the closing of the third trap. Relays D and E make and break, alternatively, at intervals of approximately 0.1 sec owing to the delay of relay D. This means that the stepping mechanism receives a sequence of pulses via contact d_2 and returns to the initial position. The remaining cold traps are not opened during the return movement, as the circuit for the solenoids is interrupted via contact c_2. While the final position of the stepping mechanism is being passed, an injection command is given via contact d and a new separating cycle starts.

When the zero position is reached, relay B is made and relay C is broken as contact b opens and the pulse sequence for the trapping mechanism is interrupted. All further separating cycles follow the same sequence.

C. Threshold Control with Various Values

For difficult separation problems, a concentration-dependent control with only one threshold value is no longer sufficient. Should a chromatogram contain a small amount of another substance apart from two incompletely resolved peaks, various threshold values must be adjustable for control purposes if all three components are to be trapped in one operation. It should be possible, furthermore, to select the switching levels for opening and closing the cold traps independently of one another in order to obtain the highest possible yield when peaks are incompletely resolved (Fig. 5.6). For the purpose of comparison, yields and purities are given for both cases.

A further requirement of multiple threshold value control is the possibility of cutting several fractions from peaks that appear to be homogeneous in the chromatogram. This is required when the separating power of the instrument cannot be raised any further and a two-peak resolution is unobtainable. As an

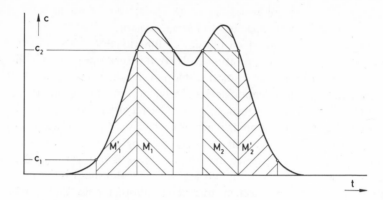

Figure 5.6. Yield of substance of various threshold values. $t_{R2}/t_{R1} = 1.05$; (t_R = retention time) $M_1 = 0.5 \cdot M_{01}$; $M_1 + M_1' = 0.78 \cdot M_{01}$; $M_2 = 0.5 \cdot M_{02}$; $M_2' + M_2 = 0.78 \cdot M_{02}$; $M_{01} = M_{02}$ = quantity injected; R_p = percentage of purity; $n = 3000$; $R_{p1} = 95\%$; $R_{p1}' = 96.8\%$; $R_{p2} = 94\%$; $R_{p2}' = 96.2\%$.

example, Fig. 5.7 shows a separation with $n = 3000$ plates, a ratio of distribution coefficients of 1.025, and a quantity ratio for the substances to be separated of 1:2. The following values are obtained for fractions 1 through 4 as shown in Table 5.1.

TABLE 5.1

Fractions of a Homogeneous Peak

Fraction	Yield, %	Purity, %
1	20.3	54.3
2	15.7	68.7
3	8.9	79.0
4	3.6	85.7
1 through 4	48.5	63.5

The yield is defined as the ratio between the collected and the injected quantity of the same substance, given complete collection in the cold trap. The trapped fractions may be subjected to further separation if the purities obtained are not yet sufficient.

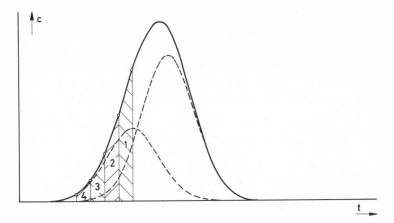

Fig. 5.7. Taking fractions of a uniform peak.

D. Peak Depression during Threshold Value Control

The majority of all presently available instruments that do not operate with a turntable valve are constructed to trap a maximum of six fractions. For the automatic operation of such an instrument, when using samples containing more than six components, two counting devices should be used, one to count the peaks eluted and a second to open each individual cold trap in sequence. The programmed collecting of fractions is then based not only on setting the threshold values but also on counting how often a threshold value must be passed before a cold trap is opened or closed (Fig. 5.8). An automatic control may be considered to be flexible enough to cope with almost every separating problem when four peaks can be discarded between any two peaks to be collected, or when up to four peaks can be collected in the same cold trap.

Such a control can be constructed in various ways. Programming can be performed with the aid of additional switches, plug-in connecting leads, punched program cards, or other supplementary parts. A detailed description is therefore not considered necessary here. The one restriction for controls that depend on threshold value only is that at points where a valve must be operated the concentration curve should not be too flat so that the position of the switching point is defined quite accurately during unavoidable variations of the quantity of sample injected. Figures 5.9a and 5.9b show, in exaggerated form, the effect of varying quantities of sample on a shallow trough and on an only faintly indicated sequence of minimum and maximum.

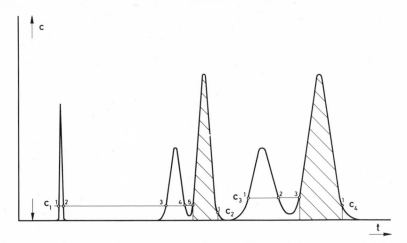

Fig. 5.8. Threshold value control when collecting individual peaks. $c_1 \ldots c_4$, Threshold values. The numbers at peak intersections with threshold values indicate how often a threshold value is passed until a cold trap is opened.

E. Combined Threshold Value–Time Control

A combined threshold value–time control (3) avoids the disadvantage of pure threshold value control of flat troughs and badly defined minima, as mentioned in Section IV.D. When the substance concentration reaches a set threshold value c_s, the cooling trap concerned is not opened immediately but a timer is set in motion first, which opens the cold trap after a set period of

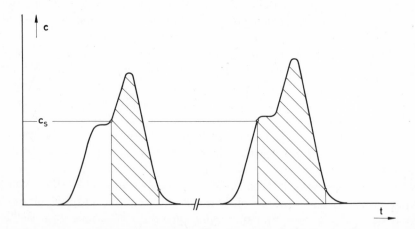

Fig. 5.9a. Shift in fraction limit with varying size of injected sample.

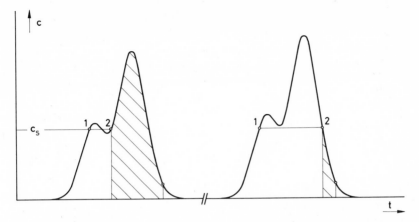

Fig. 5.9*b*. Program trouble caused by varying size of injected sample.

time t_{11}. A second timer determines the period of collecting time. Figure 5.10 shows the effect of such an arrangement of the peak shapes shown in Fig. 5.9. In this arrangement variations of retention time have only a very slight effect on the purity of the fractions collected.

V. Delayed Release of Collecting Program

Difficulties arise occasionally during programming with threshold values when solvents or other volatile parts are present in the sample because the

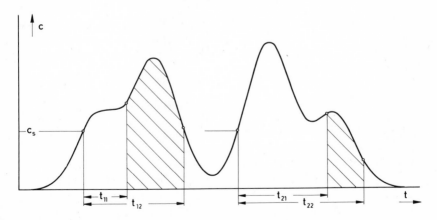

Fig. 5.10. Threshold value–time-control. c_s, Threshold value starting time relay; t_{11}, t_{12}, time delay in opening cold traps; t_{21}, t_{22}, time delay in closing cold traps.

composition of the remaining sample may alter during a large number of separating cycles. For instance, this is the case when dosing devices are employed in which a small stream of carrier gas flows continuously through the sample so that a very volatile component gradually disappears. Figure 5.11

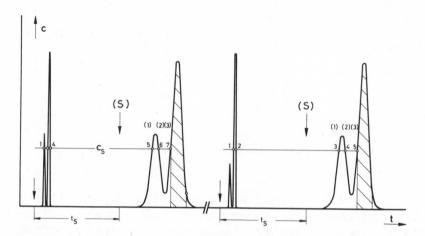

Fig. 5.11. Delayed start of collection program. ↓, Injection, t_s, time delay collecting; c_s, threshold value and indication of the number of intersections for nondelayed and for delayed start of collection; S, start of collection.

shows an example of such a change in a chromatogram. At the beginning of the separation the threshold value for fraction 1 is passed by two solvent peaks; later, however, only by one.

A remedy can be found in delayed release of collecting control. The passing of the threshold values is not signaled at the start of a program cycle for a selected time t_s, so that changes in this part of the chromatogram do not influence the rest of the program.

VI. Baseline Correction

In order to use the detector signal for cooling trap control, a stable baseline must be obtained so that the preselected switching sequence can be adhered to. This condition is fulfilled when a flame ionization detector (FID) is used at low sensitivity. When a heat-conductivity cell is used, the stability of the baseline depends on the amount of carrier gas flow-through, the detector temperature, and the ambient temperature. Different conditions may arise depending on the construction and the parameters set for operation. Frequently, a baseline drift is produced which could cause interference in the program sequence, such as during unattended running overnight. For this reason an instrument fitted with a heat-conductivity detector should therefore have a device for automatic baseline correction. It should be possible to select the start of baseline correction anywhere in the program, possibly shortly before the collection program starts at a point where no detector signal is given. Automatic baseline correction may be necessary in special cases when working with an FID, e.g., when separating and trapping small impurities from a sample. In these cases the impurity recorded on the chromatogram should be present as a sufficiently large peak which requires working at high sensitivity. At the same time, a large number of separating cycles is necessary to obtain enough substance, i.e., the total separating time may take several days. When the column temperature used is also near its highest limit, the quantity of continuously evaporating stationary phase changes over several days and so does the level of the baseline. This cannot be eliminated completely by heating the column for a long time before separation, apart from the fact that no time is needed at all when there is an automatic correcting device. The latter should be constructed so that baseline variations are corrected when they exceed 1% of the recorder span.

VII. Limitation of the Number of Separating Cycles

The use of a concentration-dependent cooling trap control requires the injection of the same quantity of sample each time. This can be guaranteed by the dosing apparatus only when enough sample is present. To prevent a reduced quantity of available sample causing an injection with only part of the set quantity (which would cause incorrect sequencing and thus spoil the fractions already collected), a means should be provided to prevent further injections after a set number of separation cycles has been reached. Simple mechanical counting devices are suitable for this purpose which, upon reaching a certain point, open a contact in the feedline of the control command leading to the injection device.

VIII. Reducing the Duration of the Separating Cycle

Technical control devices have been discussed in previous sections, and the selection of control parameters to obtain optimum separation now investigated. Basically, it can be said that the operator ought to spend as little time as possible in obtaining a certain amount of substance. This condition is fulfilled when individual separations are repeated as rapidly as possible without causing a loss in purity of the desired components. It is quite permissible for sections of the chromatogram to overlap, provided no collection takes place. Figure 5.12a shows an example of a separation in which the injection for the

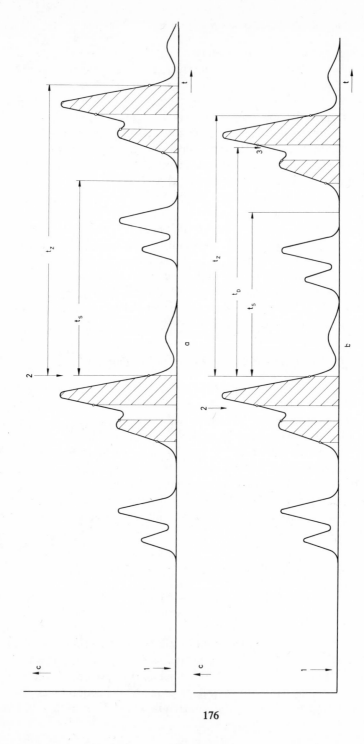

Fig. 5.12. (a) Example of a separation for various cycle lengths. ↓, Injection; t_D, time delay from end of program to injection; t_s, time delay from end of program to start of collection program; t_z, cycle time; (b) reduced cycle time.

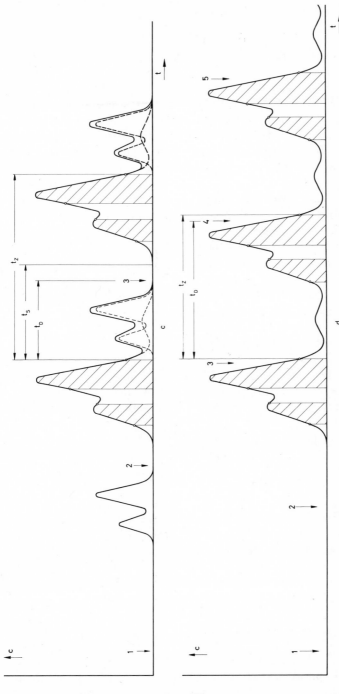

Fig. 5.12. (c) Minimum cycle time; (d) cycle time for previously cleaned sample.

177

next cycle takes place after the trapping of the last peak of the current separation. Figure 5.12*b* shows the possible reduction in time t_2 when elution of the last peak of the separation and elution of the first peak of the next separation follow each other. Figure 5.12*c* shows the maximum possible reduction, i.e., when elution of the first peak of the separation follows immediately after the trapping of the last peak of the previous separation. The time shift t_d that must be programmed for the injection is also entered. It is assumed here that a simple automatic control is being used which issues an injection command at the end of the program that reaches the injection apparatus delayed by t_d via a time relay. It can be seen that the minimum possible cycle time cannot be obtained by reducing t_d to zero. It is necessary, however, to perform a second injection at the beginning of the program by operating the injection device manually before the first collection program ends and then to choose a time t_d only slightly below t_z. The injection is therefore not given for the next separating cycle but for the one after that. The time t_s indicates how long the collection program must be interrupted so that peaks *1* and *2* are not trapped (8). This example of a separation also shows that for preparative work the separating time required depends not only on the retention time t_R of the substances to be separated but also on the elution time t_E. For all separating problems that require a large number of cycles, an attempt should be made when preparing the sample to omit low- and high-boiling impurities. Figure 5.12*d* illustrates how the time cycle can be reduced even further when peaks *1* and *2* are missing.

Occasionally, chromatograms are obtained with a relatively large interval between two groups of peaks (Fig. 5.13), e.g., for the separation of a substance that is solid at room temperature so that a solvent must be used. When the time t_p between the two groups of peaks is smaller than the sum of the eluting times:

$$t_p < t_{E_1} + t_{E_2} = t_E$$

it is possible to combine two separating cycles and thus to halve the separating time. Figure 5.13 shows the appropriate chromatogram.

If the condition $t_p < t_E$ is not quite met in such a case, it is appropriate to alter the separating parameters, i.e., have a lower column temperature, increase the retention times slightly in order to enable grouping of separations, and thereby save time.

IX. Selection of Fraction Limits

The aim of a GLC separation is to obtain from mixtures single substances with a certain purity. The required degree of purity depends on the further use of the substance and therefore embraces a very wide range. For instance,

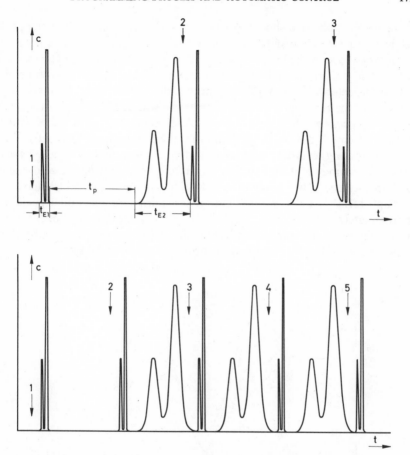

Fig. 5.13. Overlapping of separating cycles. t_{E1}, Eluting time of peaks *1* and *2*; t_{E2} eluting time of peaks *3* and *4*; t_p distance between peaks *2* and *3*.

in a mixture in which an impurity of only a few parts per million must be separated for identification, a purity of 50% may be sufficient. For radiation chemical determinations, however, a degree of purity of at least 99.995% may be required. The selection of fraction limits therefore always depends on the individual application and no general rules can be made.

A. Completely Separated Peaks

Where peaks are completely separated, the selection of fraction limits presents no difficulties. When setting the points for opening and closing the cold traps at 10% of the peak height, 97% of the peaks are collected in the traps.

As far as possible, one should not go below the value of 10% of the peak height so that baseline shift and tolerances of the control devices have no significant effect on the location of the fraction limits. From a mathematical point of view, there are no completely separated peaks, as the Gauss function only approaches the zero line asymptotically. When the chromatogram approaches the zero line between two peaks to within 3% or less of the peak height, the peaks can be considered completely separated, provided of course that the column is not overloaded and that the peak shape is symmetrical.

Compensating recorders used for recording chromatograms usually have a degree of accuracy of 0.25% so that a variation of 2.5% from the zero line is easily recognized. Two peaks of equal height, for example, which have an overlap of 1.5% of the peak height, give a purity for individual peaks of approximately 99.995% when keeping the fraction limits at 10% of the peak height (Fig. 5.14). The impurity of the smaller peak increases in proportion to the height of the larger peak when the peaks are not of equal size.

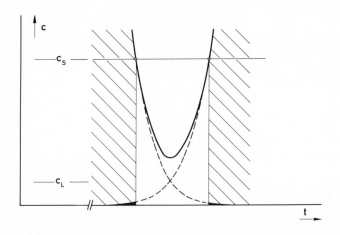

Fig. 5.14. Purity of completely separated peaks. c_0 = peak height; c_L = $0.015 \cdot c_0$; c_s = $0.1 \cdot c_0$.

B. Peaks that Are Not Completely Separated

The selection of fraction limits in the case of peaks that are not completely separated is more difficult. Three factors influence the degree of purity obtainable:

1. The ratio of the retention times.
2. Peak width.
3. The ratio of substance quantities.

Because these parameters allow for a large number of different combinations, it is difficult to indicate characteristic peak shapes that will permit a simple estimation of the degree of purity obtainable. Basically, three methods can be used for the purpose of establishing the fraction limits:

1. Estimations based on experience.

2. Calculation based on an analytical chromatogram and the plate number of the column used.

3. Analytical determination of the purity, using fractions obtained after one separating cycle.

1. Estimation of Obtainable Purity

To estimate the degree of purity that can be obtained for individual fractions of overlapping peaks, Figs. 5.15*a*, 5.15*b*, and 5.15*c* each show a group of peaks with different ratios of quantity but with the same ratio of retention times. Two fraction limits are entered in each case, namely, at 0.98 t_R and at 0.99 t_R. The graphs are calculated using the equation given by Kaiser (9):

$$c = c_0 \cdot e^{(-n/2)[((t-t_R)^2/t_R^2]}$$

where $n = 3000$. The ratio of the retention times is entered each time at the side of the graphs. The degrees of purity obtainable are given in Table 5.2, but it should be noted that the values are valid only for symmetrical peaks and not for overloaded columns. The yield is 86% for fraction limits of $0.98t_{R2}$ or $1.02t_{R1}$, and 70% for fraction limits of $0.99t_{R2}$ or $1.01t_{R1}$, all based on the quantity injected.

TABLE 5.2

		Fig. 5.15*a* $t_{R2}/t_{R1} = 1.06$		Fig. 5.15*b* $t_{R2}/t_{R1} = 1.05$		Fig. 5.15*c* $t_{R2}/t_{R1} = 1.04$	
		Purity, %					
		Fraction limit		Fraction limit		Fraction limit	
Ratio of Quantities	Fraction	0.99	0.98	0.99	0.98	0.99	0.98
1:1	1	99.6	98.3	97.9	94.2	92.8	81.4
	2	99.7	98.5	98.2	94.8	93.5	81.5
4:3	1	99.7	98.7	98.4	95.6	95.6	86.0
	2	99.6	98.0	97.6	93.0	93.4	81.7
2:1	1	99.8	99.1	98.9	97.1	96.4	90.7
	2	99.4	97.0	96.4	89.5	87.0	71.0
4:1	1	99.9	99.6	99.5	98.5	98.2	95.3
	2	98.8	93.2	92.9	79.0	74.0	42.0

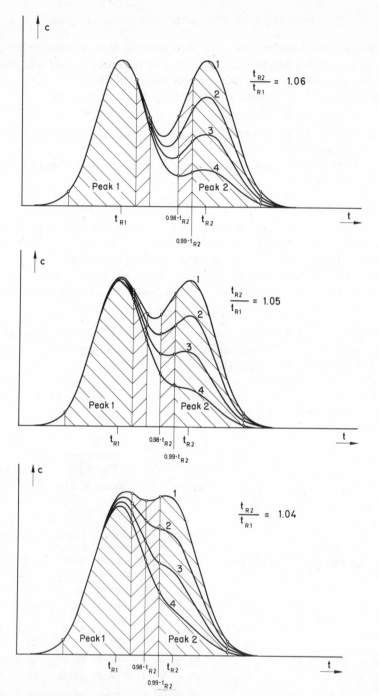

Fig. 5.15. Incompletely separated peaks. Quantity ratio $m = M_1 : M_2$; $m = 1$ for graph 1; $m = 1.33$ for graph 2; $m = 2$ for graph 3; $m = 4$ for graph 4.

These values also show the change in purity of the fractions to be expected when pure time control is used. A change in column temperature of only 0.3°C at an operating temperature of 70°C causes a change in retention time of 1%, i.e., in the examples given, the fraction limits change from 0.99 to 0.98.

2. Calculation of Purity

To calculate the purity of the fraction cuts decided upon, the equation previously given for time and concentration in a peak can be used, namely,

$$c(t) = c_0 \cdot e^{(-n/2)[((t-t_R)^2/t_R^2]}$$

The result is accurate only when the peaks actually appearing are not deformed because of column overload. Data required for this purpose are: the peak heights c_{01} and c_{02}; the plate number n; the retention times t_{R1} and t_{R2} for both overlapping peaks. t_{R1} and t_{R2} cannot easily be determined with accuracy for overlapping peaks. It may therefore be necessary to determine the exact ratio $t_{R1}:t_{R2}$ on an analytical column with a higher plate number but similar in other respects. When peak 1 is collected in the range t_{11} to t_{12}, and peak 2 in the range t_{21} to t_{22}, the following equations are valid for the percentage of purity R of the fractions.

$$R_{p1} = 100 \cdot (M_{21}/M_{11}) = 100 \left[\int_{t_{11}}^{t_{12}} c_2(t)\, dt \Big/ \int_{t_{11}}^{t_{12}} c_1(t)\, dt \right]$$

$$R_{p2} = 100 \cdot (M_{12}/M_{22}) = 100 \left[\int_{t_{21}}^{t_{22}} c_1(t)\, dt \Big/ \int_{t_{21}}^{t_{22}} c_2(t)\, dt \right]$$

The integral $\int c(t)\, dt$ is commonly tabulated as a probability integral. It can, however, also be calculated to a good approximation from the ordinate values when the ordinates are calculated at an interval of

$$\Delta t < 0.1 \cdot t_R \cdot \sqrt{(6/n)}$$

(See Fig. 5.16.)

3. Experimental Determination of Purity

Purity determinations using an analytical chromatogram after one or a few separating cycles are recommended when separations of large quantities with prescribed purity are needed. This is the only means of ensuring that the control program chosen yields the desired degree of purity for each individual fraction.

X. Increase of Throughput by Overloading the Column

When quantities of varying size are separated under otherwise similar conditions, the column efficiency usually decreases only very slightly with increased quantities until an upper limit is reached. The column efficiency

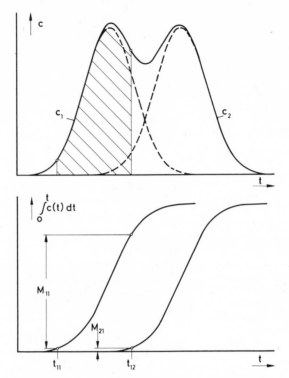

Fig. 5.16. Lapse of time for peak height and peak area.

decreases markedly, however, when the upper limit for the quantity dosed is exceeded. It can be shown (10,11) that the yield of separated material with unchanged purity still increases per unit time, although the column efficiency is actually reduced. The optimum quantity to be separated per cycle must be determined for each individual separating problem because peak deformation in this range cannot be explained easily. It should be noted that peaks not only become wider, but that retention times also become longer. Not only the quantity of substance obtained per cycle, but also the time required for each separating cycle should be taken into account when assessing the optimum quantity to be dosed.

XI. Transmitting the Detector Signal to the Control Device

When a threshold value control is used, the detector signal must be fed into the control device where it is compared with a preselected threshold value. This comparison should be reproducible with a deviation of ca. 1% or less.

The threshold values themselves should also be adjustable with a deviation of less than 1% based on the recorder span for the test chromatogram. These demands can be fulfilled only with the help of extensive electrical instrumentation when a thermal conductivity cell is used as a detector. In this case the signal voltage in the most sensitive range of the detector is only 0.5 or 1 mV for the total span of the recorder. A deviation of 1% equals a voltage of 5 μV and this condition can be met only by high-performance differential amplifiers. Furthermore, difficulties may arise when a compensating recorder and a differential amplifier are run in parallel from the same signal voltage source. For these reasons, the use of a compensating recorder for transmitting the signal is preferred. It is in any case needed for the test chromatogram. In addition, this has the advantage that one can check continuously whether or not the selected threshold values are being reached. It is most appropriate to incorporate a retransmitting potentiometer, which is possible for most types of recorders. Very few types of recorders provide a sufficient number of easily adjustable microswitches.

When a FID is used, the ionization amplifier present can be constructed in such a way that a second outlet for the control device is available which delivers a sufficiently large signal voltage (1 V or more). When using an electronic control device, line disturbance is sometimes fed into the signal lead and may cause an interruption of the program sequence. This may also be caused by "spikes" in the detector signal when small dust particles are burned in the FID. These interferences can be prevented by reducing the speed of response of the control device so that values exceeding a threshold value are valid only when the condition is maintained for more than 0.1 sec.

XII. Reliability of the Control Device

The advantages of automatic operation of a preparative GLC unit may be nullified if the control device is not sufficiently reliable. A suitable control device may contain up to 1500 construction parts, one-third of which are semiconductors. It is impossible to calculate the reliability of the complete circuit from the breakdown rate of the construction elements used because details of the special operating conditions are generally not available. As a guide to the possible reliability, a breakdown rate of 10^{-6} h may be assumed for the 500 semiconductors present. Trouble-free operation for 2000 hr, or 3 months uninterrupted operation of the instrument, can be expected. This breakdown rate is approximately valid for plastic-covered transistors under partial load, and it decreases with increasing operating time. Elevated environmental temperatures influence the reliability of the control device in two ways. Permissible limit values may be exceeded, leading to immediate faults; or the life of those elements that undergo aging is shortened. Control devices should therefore always be arranged in such a way that they are not influenced by any foreign heat source.

XIII. Consistency of Separating Parameters

Automatic control of the separating process is dependent on the consistency with which the individual separating cycles repeat themselves. When pure time control is used, the retention time must be kept constant; when using threshold value control, the sample size injected should only vary very slightly. Greater variations are permitted for the retention time provided the peak form does not change too much.

The primary requirement for perfect automatic operation is the accurate adjustment of column temperature and carrier gas pressure, and that they be independent of extraneous influences such as environmental temperature, main voltage, and carrier gas inlet pressure. Furthermore, the sensitivity of the detector used should not change. For thermal conductivity cells, this means long-term stability of the bridge current, while for the FID, stability in the stream splitter and the auxiliary gas flows is important.

References

1. D. Ambrose and R. R. Collerson, *Nature*, **177,** 84 (1956).
2. E. P. Atkinson and G. P. A. Tuey, in *Gas Chromatography*, D. H. Desty, Ed., Butterworths, London, 1958.
3. E. Heilbronner, E. Kovats, and W. Simon, *Helv. Chim. Acta*, **40,** 2410 (1957).
4. S. Sideman and J. Gilladi, in *Gas Chromatography*, U. Brenner, Ed., Academic Press, New York, 1962, p. 339.
5. H. Wiegleb and H. Prinzler, *Chem. Tech.*, **15,** 98 (1963).
6. H. Boer, *J. Sci. Instr.*, **41,** 365 (1964).
7. K. -P. Hupe, U. Busch, and W. Kuhn, *J. Gas Chromatog.*, **3,** 92 (1965).
8. K. P. Dimick and E. M. Taft, *J. Gas Chromatog.*, **1**(3), 7 (1963).
9. R. Kaiser, in *Chromatographie in der Gasphase*, Part I, Bibliographisches Institut, Mannheim, 1960, p 40.
10. P. C. Haarhoff, P. C. van Berge, and V. Pretorius, *J. South African Chem. Inst.*, **9,** 82 (1961).
11. E. Bayer and H. G. Witsch, *Z. Anal. Chem.*, **170,** 278 (1959).

CHAPTER 6

Temperature and Flow Programming

Rudolf Kaiser, *Ammoniaklaboratorium, Badische Anilin- & Soda-Fabrik AG, Ludwigshafen am Rhein, Germany*

I. Introduction

Investigations of preparative chromatography and applications thereof have been virtually exclusively concerned with operation at constant temperature and carrier gas flow rate. These matters are dealt with in detail elsewhere in this book.

The advantages accruing from temperature and/or flow programming have been well established in analytical gas chromatography (GC) and it is not unreasonable to expect that these techniques could be profitably exploited in preparative work.

A. Temperature Programming

1. Advantages

1. The separation (resolution) is optimized for any two substances. For the resolution of any two substances there is always an optimum temperature in a given separating system. When a mixture consists of more than two substances, more than one optimum temperature is required for optimum separation. The column temperature should therefore be increased during the separation from the lowest to the highest temperature required. Then, any two substances are separated in the best possible way at optimum temperature (for any given column).

2. Substances with different retention indices are eluted with the highest possible average concentration when linear temperature programming is used. This gives better yields when cutting fractions from the mobile phase and often leads to higher purity. The average peak width, and thus the time during which a peak must be led to a cold trap, is reduced. The additional flow of impurities from the mobile phase is therefore also reduced.

3. Temperature programming keeps a system at maximum temperature only for as long as is strictly necessary to give a desired separation in the shortest time. This strongly increases the lifetime of a column when compared with a method that holds the column at the maximum temperature. The thermal as well as the oxidative and hydrothermal decomposition of the stationary phase increases exponentially with temperature. The rate of evaporation of the stationary phase increases similarly.

2. Disadvantages

1. All advantages of temperature programming are quickly lost and even reversed when the heating of the column does not take place quasihomogeneously. Almost any, even the smallest, temporary change in temperature causes a temperature gradient in the column (local temperature change). When this is essentially greater, however, than the temperature gradient in ΔT. cm^{-1} caused by the passage of solute through the stationary phase, the separation of the adjacent two substances is not improved but worsened. The average concentration of the solute in the mobile phase does not rise but decreases. Only the maximum elution time is finally reduced.

The maximum permissible heating rate h in °C min^{-1} of the temperature program in practical preparative GC must be carefully balanced with the diameter, the specific heat, and the thermal conductivity of the chromato-graphic column; h is not determined solely by the length of the column, the flow of the mobile phase, and the separating problem, as is the case in tem-perature programming in analytical GC.

2. The instrumentation for temperature programming of large preparative columns is expensive. A relatively large flow of heat must be transported to the stationary phase using air for the heat transfer. After the final temperature is reached, this head must be dissipated again as quickly as possible.

The costs, on the one hand, and the possible gain in time, yield, and purity, on the other, must be carefully balanced. The range in which temperature-programmed GC pays becomes smaller the greater the dimensions of the column, especially those transverse to the direction of separation. The lower the heat conductivity of the column and the higher the specific heat of its contents, the lower the permissible maximum heating rates.

3. Unless a preparative instrument has been equipped from the start with facilities for temperature programming, the use of the technique is practically impossible or at least does not give good results. The column filling especially can be unsuitable for the purpose of temperature programming because of the kind of support and chromatographically active phase used. This also concerns the column geometry.

A critical examination of all pros and cons of temperature-programmed GC is absolutely necessary before investing in a preparative instrument. These are discussed below.

B. Locally Limited Temperature Programming

The disadvantages of temperature programming that are especially serious in preparative GLC with columns of large diameter (over 10 mm) can be com-pletely overcome (and are sometimes even of great use) when only part of the column, e.g., the first 30–60 cm can be temperature programmed or when the first part of the column can be subjected to a temperature gradient in the direction of the separation (Fig. 6.10). This method, which is important for all enrichment problems and especially for preparative GLC of sensitive substances, is discussed in detail (see also Ref. 17). Enrichment and separa-tions of heat-sensitive substances, as well as separations of substances that are very similar to each other can be performed effectively in this way. Sections of preparative columns of less than 60-cm length are mechanically easy to make in such a way that the advantages of temperature programming and the use of temperature gradients are fully exploited, while their disadvantages can be largely overcome. The use of locally limited temperature programming of the column is not confined to the column inlet. The temperature gradient technique at the column outlet may also be used to great effect for many difficult practical problems. The sharpness of separation, the purity of the eluate, and the concentration of solute collected in the cold traps are particularly improved by its use.

C. Flow Programming of the Mobile Phase

The second choice in optimizing the separation with respect to peak resolution and separation time is programming the gas flow or carrier gas pressure.

1. Advantages

Programming the gas flow does not require extensive special equipment. Changes in flow take place quickly, i.e., one can increase the gas pressure by a factor of 2 to 10 either linearly or in steps, which essentially shortens the elution time for any substance. Thus the retention time of a solute i drops from, e.g., 30 to 6 min when the gas pressure is programmed linearly from 2 to 18 atm in 6 min. The flow can also be reversed rapidly. Column backflushing is possible for cleaning the column or for keeping undesirable substances away from the collecting system. Partial column backflush, whereby only part of the column is flushed while the elution of the substance to be separated is not interrupted, plays an important practical role here. A very special advantage for preparative purposes is the column-switching technique which can replace temperature programming of the entire column length in most practical cases. This technique has hardly been used until now in preparative GLC although it is exactly here where it is most effective. The use of mechanical reversal systems which often cause problems becomes superfluous, however, when one carries out column switching according to Deans (1,2).

The equipment needed for flow programming, column backflushing, heart cutting, and multidimensional separation is limited to pressure regulators, flow resistances, and on/off valves which all operate at room temperature. All these parts can be installed outside the thermostated space. The pressure gradients that occur with changes in gas flow act practically only in the direction of separation and are harmless compared with the harmful temperature gradients that usually act transversely to the direction of separation. The former hardly influence the resolution of the adjacent peaks.

Flow changes of the mobile phase do not, or hardly, change the effective lifetime of the stationary phase, provided the packing is always correct, since the rate of evaporation depends exponentially on the temperature but only linearly on the gas flow.

An instrument that has not been designed for flow programming or column switching can readily and inexpensively be altered for this purpose.

2. Disadvantages

Apart from backflushing and column switching, radical changes in chromatographic behavior, e.g., full substitution for the effect of a temperature program, are only possible when a radical increase in gas flow is used. Pressure programming can have only the same effect as a linear temperature program when the gas flow is increased exponentially. This requires large amounts of carrier gas, reduces exponentially the concentration of the solutes to be trapped in the collecting system, and can lead to a permanent reduction in the column efficiency, especially when the packing of the stationary phase is not homogeneous. The optimum gas flow rate for maximum column efficiency is fixed and is always exceeded to a smaller or larger extent by flow programming.

The advantages of both program options, temperature and flow changes, must be used very critically, so that their disadvantages are avoided.

D. Optimum Combination of Temperature and Flow Programming

From what has been discussed before it follows that:

1. Temperature programming is suitable only for a short length of the total column, e.g., for the first 30–60 cm and for the last 30–60 cm. The technique is very valuable for optimum dosing of the sample, for effective enrichment, and for preparative GLC of chemically and thermally sensitive substances.

2. Intensive cooling of the first 30–60 cm of the column to as low as $-180°C$ is required to enrich the sample or for careful continuous dosing.

3. Fast temperature programming with heating rates greater than 7°C min^{-1} is required for optimum development of an initial chromatographic profile in the part of the column that is run isothermically.

4. The use of a temperature gradient that moves in the direction of the gas flow is required. Technically, and with the required homogeneity of the temperature area, this method is only possible over a short length of the column but more is not required.

5. Instead of temperature programming the whole remainder of the column, it is better to divide the column into two or three chemically different lengths and to carry out programmed gas flow switching. In this way separations are obtained that give the highest possible purity of the eluate in the shortest possible time.

6. With partial backflushing the preparative column is kept clean more successfully and in less time than by heating.

7. It is much more effective to separate partly overlapping peaks (e.g., low concentration of a peak on the tail of the main component) by heart cutting than by fraction cutting in the collecting system. The latter always leads to impurities, while the former leads to extensive or complete separation.

8. When using the pneumatic switching technique (Fluidik and Deans technique), flow programming and column switching do not cause pollution which is otherwise unavoidable.

The combined use of gas flow and temperature programming, which includes low temperature, column switching, and the use of a temperature gradient moving in the direction of separation, requires technical expenditure, but helps to fulfill nearly all requirements listed in the introduction, making it possible to obtain in the shortest time chemical components of the highest purity in the greatest yield with the greatest throughput. Investment in the purchase of a suitable instrument is naturally high. A suitably equipped instrument is also not yet available at the present time.

II. Temperature Programming

A. Basic Principles

1. Heating the Stationary Phase

The heating of the stationary phase can be attributed to the column wall because of the very small specific heat of the carrier gas. The heat conductivity of the solid supports customarily used in preparative GC is very small, but the specific heat may be considerable, and any change in temperature at the column wall causes a temperature gradient at right angles to the direction of separation. Giddings (3) has presented a simplified theory to calculate the temperature gradient in GC columns. The theory does not apply to short retention times or wide columns. Hupe et al. (4) have extensively investigated the temperature distribution in preparative GLC columns of larger diameter when using temperature programming. The basic equation is the Fourrier equation (5) as used in the solution of any nonstationary heat-conductivity problem, namely,

$$\delta T / \delta Z = a \nabla^2 T \qquad (6.1)$$

which when applied to the column in cylindrical coordinates has the form:

$$\frac{\delta T}{\delta Z} = a \left(\frac{\delta^2 T}{\delta r^2} + \frac{1}{r} \frac{\delta T}{\delta r} \right) \qquad (6.2)$$

where T = temperature in °C
$\quad Z$ = time in minutes
$\quad a$ = temperature conductivity in $cm^2 \ min^{-1}$
$\quad r$ = radial coordinates in centimeters

with the boundary conditions:

$\quad Z = 0$
$\quad T = T_0$ (initial temperature) for $0 \leq r < R$
$\quad r = R$ (internal radius of the column in centimeters)
$\quad T = T_W$ (wall temperature)
$\quad T_W = T_0 + hZ$ (h = heating rate in °C min^{-1})

Equation 6.2 can be solved to yield:

$$T = T_0 + h\left[Z - \frac{(R^2 - r^2)}{4a}\right] + \frac{2hR^2}{a}\sum_{n=1}^{\infty}\frac{J_0(\beta_n r/R)}{\beta_n^3 J_1(\beta_n)} \cdot e^{-a(\beta_n/R)^2 Z} \qquad (6.3)$$

The values for J_0, β_n, J_1 can be found in tables of higher functions, e.g., in Ref. 7. J_0 = Bessel function; β = zero position of J_0; $J_1 = J_0$. It was found that Eq. 6.3 could be further simplified to Eq. 6.4.

$$\Delta T = (hR^2/4a)(1 - 1.109 \cdot e^{-5.8\,aZ/R^2}) \qquad (6.4)$$

where h = heating rate in °C min^{-1}

a = temperature conductivity coefficient in cm^2 min^{-1}

R = inside radius of the column in centimeters

Z = time after the start of heating in minutes

ΔT = temperature difference between the column center ($r = 0$) and the wall ($r = R$)

The agreement between calculated values of temperature difference and experimental values was found to be less than 0.5% when Z is greater than 1 min.

The smaller the heat conductivity the greater the heating rate, and the greater the diameter of a tubular column is the greater must be the temperature difference between the center and the wall of the column. ΔT may be surprisingly large as is shown in Fig. 6.1.

A preparative column of only 38 mm i.d. packed with kieselguhr ($a = 0.2235$ cm^2 min^{-1}) shows after 5 min at a heating rate of 10°C min^{-1} a temperature difference of ca. 30°C between the center and the wall. As the migration velocity of a GC peak changes by a factor 2 with a temperature change of 30°C, this means that the peak migrates 5 min after dosing, twice as fast at the column wall as in the center. Now, columns of 38 mm i.d. are by no means especially effective preparative systems. Since the tube diameter affects the temperature distribution exponentially, it is not surprising that in another experimental test (diameter = 60 cm; heating rate 10°C min^{-1}; filled with kieselguhr, $a = 0.2235$ cm^2 min^{-1}; time $Z = 10$ min) a temperature difference of 74°C was found between the wall and the center.

2. Influence of the Packing Materials

The use of Eq. 6.4 requires a precise knowledge of the temperature conductivity a, which is defined as:

$$a = \lambda/\gamma c \qquad (6.5)$$

where λ = thermal conductivity in cal. cm^{-1} min^{-1} °C^{-1}

γ = specific weight in g cm^{-3}

c = specific heat in cal g^{-1} °C^{-1}

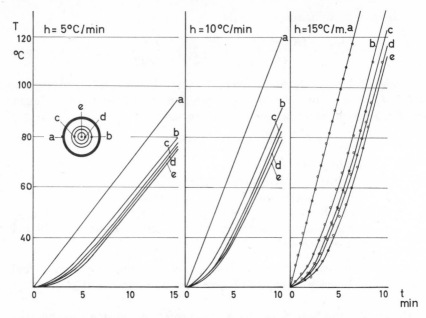

Fig. 6.1. Experimentally determined values for the temperature-time function at various points in the column packing. *a*, Wall; *e*, center; at three heating rates: $h = 5, 10, 15°C$ min^{-1}; at 38 mm in. d.; kieselguhr (Celite) as column packing. [From K. -p. Hupe, W. Bohnisch, and H. Quitt, *Chem. Ing. Tech.*, **37**, 146 (1965).]

The thermal conductivity strongly depends on the particle size of the solid support. No natural solid support is materially homogeneous; it rather consists of particles with greatly varying temperature conductivities. Furthermore, only the caloric properties of the prepared filling can be used in the calculation, so that tabulated values for density and heat conductivity coefficients of certain supports may differ appreciably from the figures pertaining to the prepared chromatographic filling. The liquid phase essentially alters the heat flow from particle to particle as well as the values of the density and the specific heat.

Hupe et al. (4) measured the relation between thermal conductivity and temperature for kieselguhr and firebrick (Fig. 6.2). They found the following values.

	λ(100°C)	d	γ	c	a
Kieselguhr	0.0113	0.1–0.2	0.253	0.20	0.223
Firebrick	0.0109	0.1–0.2	0.45	0.28	0.0865

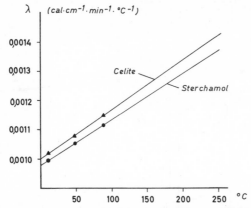

Fig. 6.2. Dependence of the thermal conductivity number of two carriers on the temperature. [From K. -P. Hupe, W. Bohnisch, and H. Quitt, *Chem. Ing. Tech.*, **37**, 146 (1965).]

where λ = heat conductivity coefficient in cal cm^{-1} min^{-1} °C^{-1} at 100°C

    d = particle size in millimeters

    γ = specific weight in g cm^{-3}

    c = specific heat in cal g °C^{-1} at 100°C

    a = temperature conductivity coefficient $\equiv \lambda \cdot \gamma^{-1} \cdot c^{-1}$ in cm^2 min^{-1} at 100°C

One recognizes the great influence of the properties of the solid support. The following figures (Fig. 6.3 showing temperature distribution at various heating rates and Fig. 6.4 showing temperature distribution at various column diameters) summarize qualitatively and quantitatively what can be said about the problem of temperature inhomogeneity in preparative columns as a function of the heating rate. The situation does not really improve when one heats linearly, e.g., in steps or ballistically.

A further effect must be incorporated when one considers the temperature influence attributable to the heat of adsorption of the migrating solute in the preparative column; this also occurs in isothermally run preparative GLC to the same extent, but during temperature programming the inhomogeneous temperature areas overlap. It is only now that the total influence of the thermal effects on the resolving power, the retention, and the concentration of peaks in the preparative column can be discussed.

Until now it has been assumed that preparative column filling is homogeneously distributed along the total cross section of the column, i.e., that the density and thus the temperature conductivity are homogeneous in cross section. This is not the case in normal practice.

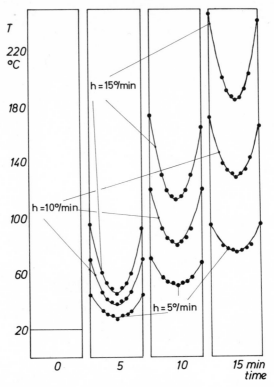

Fig. 6.3. Dependence of the temperature distribution in a preparative column. 38 mm
i. d.; Celite as column packing; at three heating rates; $h = 5, 10, 15°C \ min^{-1}$ after 0, 5, 10,
and 15 min temperature programming. [From K. -P. Hupe, W. Bohnisch, and H. Quitt,
Chem. Ing. Tech., **37,** 146 (1965).]

Neither has the mobile phase played any part in our considerations until
now, but in fact it contributes decisively to the true temperature inhomogeneity in the column; it also shows that the cross diffusion of the mobile phase
is surprisingly small.

3. Thermal and Mechanical Inhomogeneity of the Stationary Phase

Hupe et al. (8) have explored the concentration profile of migrating peaks
through preparative columns two-dimensionally and have established that
a solute peak migrates faster through the denser packing regions. This is
completely contrary to prevailing opinion. They have also established that
neither the cross diffusion of the material nor the heat transfer across the
direction of flow are large enough to prevent slanting peak fronts in preparative columns. Notwithstanding the use of filling methods considered to be

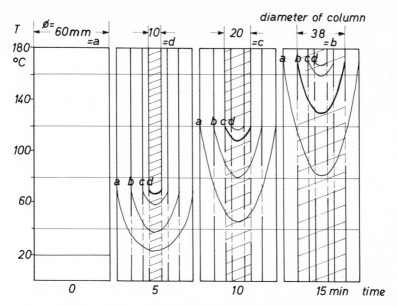

Fig. 6.4. Temperature distribution dependent on column diameter of preparative separating systems at a heating rate of 10°C min⁻¹, after 0, 5, 10, and 15 min temperature programming. The four columns have an inner diameter of 10, 20, 38, and 60 mm; Celite packing. [From K. -P. Hupe, W. Bohnisch, and H. Quitt, *Chem. Ing. Tech.*, **37**, 146 (1965).]

good according to Guillemin (9) or Spencer (10), inhomogeneities in preparative columns from 6-cm diameters upward are already so significant that, for example, after 7 min retention time relative differences of 17% of the migrating velocity can take place over the cross section of the column packing. When it is considered, for instance, that the peak migrates too fast in the center of the column and that the density of the packing there ought to be increased by mechanical pressure, the differences in velocity over the column cross section increase in reality to a relative 33%.

This means that cross-sectional packing densities vary so much in normally packed preparative columns that a 15–20% difference in peak migration velocities occurs between the center and the wall. It follows from this that the temperature conductivity too must be different over the cross section of normally packed columns, as the filling density plays its full part in the temperature conductivity coefficient. The greater the density of the column packing, the greater the temperature difference caused by temperature changes during the programming. A column that is packed more densely in the middle can therefore, to a certain extent, reduce the influence of temperature which is much lower just there. Or, expressed in another way: A prepara-

tive column packed mechanically to the highest standards (with denser packing at the walls) is quite efficient under isothermal running conditions. During
temperature-programmed conditions, depending on caloric properties of the
stationary phase, it may perform much worse than a column that has been
packed inhomogeneously over the cross section and which is denser in the
middle. Or, expressed even more simply: Preparative columns that are very
efficient isothermally may lose much of their efficiency during temperature
programming; poor columns do not lose much efficiency during temperature
programming. This may explain why there are contradictory statements in
the literature about the advantages and disadvantages of temperature programming in preparative GC.

B. Influences of Temperature Programming

1. Retention Time

The apparent retention time of the last solute to be collected is decisive in
preparative GC, as it determines the shortest possible time cycle and thus the
production capacity of preparative GLC carried out in the conventional way.
However, only the adjusted retention time t_R', or the adjusted retention
volume and the specific retention volume derived from it, are chromatographically interesting. The calculation of t_R' under temperature-programming
conditions results from statements by Said (11), which go back to Habgood
and Harris (12). The fundamental equation of Harris and Habgood (13) is
given here as it is informative:

$$h/jF_0 = \int_{T_0}^{T_R} (dT/V_R) \tag{6.6}$$

where h = heating rate in °C min⁻¹

    F_0 = gas flow in milliliters at column outlet expressed at standard
        temperature

    T = temperature in °K

    V_R = apparent retention volume

    T_R = retention temperature in °K at which the maximum concentra-
        tion of solute leaves the column

    T_0 = initial temperature

    j = pressure-gradient correction factor = $(3/2)[(p'^2 - 1)/(p'^3 - 1)]$

    p' = ratio of column inlet pressure to outlet pressure.

These investigators (13) state: Equation (6.6) "is of fundamental importance to the understanding of PTGC. It makes clear that for each compound, at any given starting temperature, the retention temperature may be
related to the program in a characteristic way as defined by $h/j \cdot F_0$. Note that

the ratio of heating rate to flow rate is the significant parameter rather than one alone; moreover, this relation may be obtained from a knowledge of the isothermal retention volumes for the temperature range in question."

It is postulated in this analysis that there are no temperature gradients within the column. This cannot be disregarded in preparative columns, however; we know that it can reach many degrees per centimeter column diameter. t_r (real) is therefore considerably greater than t_r (theoretical), i.e., more than 38% in an example under the following conditions [according to Eggert (14)]. Length of column, 2 m; column diameter, 38 mm; 20% dinonyl phthalate on firebrick; gas flow at 20°C; 760 torr measured; 1000 ml min⁻¹; $T_0 = 300°K$; $T_R = 329°K$ (real value; calculated value for T_R corresponds to the measured retention time = 339°K).

The maximum temperature difference between the wall and the center of the column packing reached 30.2°C. It follows from these data that it is expedient to spend some time on the derivation and calculation of the apparent retention time. The effect of temperature on the adjusted retention time in isothermal systems, is given by

$$t'_R = (A/F) \cdot e^{\Delta H/RT} \qquad (6.7)$$

where A = constant in ml g⁻¹
F = gas flow in ml min⁻¹
H = constant in kcal mole⁻¹
R = gas constant, 1.987 cal °C mole⁻¹

or

$$\log F \, t'_R = a + (b/T) \qquad (6.8)$$

where a, b = constants, with T in °K (Fig. 6.5).

The value of these equations in programmed systems becomes more and more doubtful the less one knows about the mechanical and thermal inhomogeneities in the preparative column, since these must be known to allow calculation of the apparent retention times as a function of gas flow and heating rate (when the temperature conductivity coefficient and the influence of the mobile phase have been measured). The overriding influence of the temperature profile in preparative columns is demonstrated by the fact that using data from Ref. 14 an error of 100% can easily occur between the calculated values and those actually found. In other words, the adjusted retention time may easily be more than twice as great as would be found from the calculation, a fact we have already established empirically above. The deviations are greater the denser the preparative column, the lower the temperature conductivity coefficient, the higher the heating rate, the more inhomogeneous the packing and the greater the retention time (Fig. 6.6).

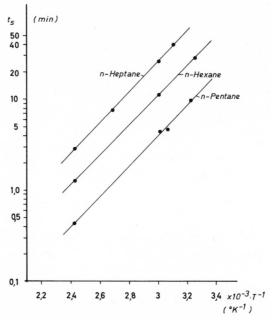

Fig. 6.5. Dependence of the net retention time t_s for three hydrocarbons on the absolute temperature, measured in an isothermally run, preparative column of 38-mm i. d. (From H. Eggert, Thesis, technical Hochschule Karlsruhe, Karlsruhe, 1964.)

2. Resolving Power

The column efficiency in chromatographic systems is generally expressed using the theoretical plate number or the HETP. For the present purpose the equation

$$TZ = (100/\Delta I) - 1 \tag{6.9}$$

is more useful. It gives the relation between the required plate number of a column and the then possible separation of two solutes that differ by retention index value 1 [Kaiser (16)]. See also Fig. 6.7.

3. Plate Number (n_{sep})

This number indicates how many peaks of equal concentration with 1.5% overlap can be separated between two homologues that differ by one CH_2 group only. Or, formulated more precisely, TZ gives the number of peaks between two adjacent n-alkanes that can be separated with a distance of 4.6 σ (overlap of symmetrical peaks 1.5%). Figure 6.7 shows the quality of a 4.6-σ separation.

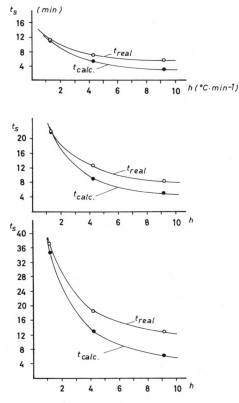

Fig. 6.6. Difference between the calculated net retention time t_{calc} and the experimentally found net retention time t_{real} in preparative columns during temperature programming ($h = 1, 2, \ldots 10°C$ min^{-1}) based on the temperature profile in the column packing. (From H. Eggert, Thesis, Technical Hochschule Karlsruhe, Karlsruhe, 1964.)

The number is illustrative, easily measured, and is valid for isothermal, isobaric, linear temperature-programmed, linearly and exponentially flow-programmed GC. To measure it one injects two adjacent n-alkanes whose C-number times 100 most closely approach the retention index of the solutes actually to be separated. The following is taken as an example.

The substances to be separated have a retention index range of 750 to 1490. Most difficult to separate are two substances with a retention index of 1220 and 1235. From Eq. 6.9:

$$TZ = \frac{100}{\Delta I} - 1 = \frac{100}{1235 - 1220} - 1 = \frac{100}{15} - 1 \cong 5.7$$

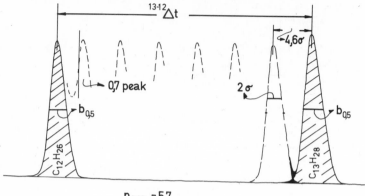

$n_{sep} = 5.7$

Fig. 6.7. To define the plate number n_{sep}, the 4.6-σ overlap, and as numerical example $n_{sep} = 5.7$ between retention index 1200 and 1300.

The problem can therefore be solved only when one uses a column under conditions that reach at least a plate number of 5.7 for the index range from 1200 to 1300. This can be tested easily. A sufficient quantity of the hydrocarbon mixture, n-heptane to n-pentadecane, is injected. The plate number for the range between n-dodecane and n-tridecane (retention index range 1200 to 1300) is measured. The equation for this is simple (16):

$$^{12-13}TZ = \frac{^{12-13}\Delta t}{^{12}b_{0.5} + {}^{13}b_{0.5}} - 1 \qquad (6.10)$$

where $^{12-13}\Delta t$ = difference in retention time between n-alkane C_{13} and n-alkane C_{12} (to be measured in millimeters from the chromatogram at peak maximum)

$^{12}b_{0.5}$ = peak width of C_{12} peak measured at half height (in millimeters)

$^{13}b_{0.5}$ = peak width of C_{13} peak measured at half height (Fig. 6.7).

Equation 6.10 bears no relation to the formula for the resolution of the n-alkanes. It rather assumes a *constant* resolution for peaks separable between both n-alkanes with 4.6-σ peak distance. A 4.6-σ separation means that the maxima of adjacent peaks are separated from each other by 2.3 times the peak width at 60.7% of the peak height. At 60.7% of the peak height a peak is 2σ wide. σ = standard deviation (see, e.g., Ref. 6).

This definition makes sense. It allows the use of a very simple and clear formula for the column efficiency (Eq. 6.10) and may be used in practice. For both preparative separation and quantitative analysis a 4.6-σ separation proves to be the best compromise between expenditure and result.

4. Column Efficiency Measured by Plate Number

Provided analytical columns are used that are heated homogeneously, the plate number decreases slightly with increasing temperature. In preparative columns the plate number and thus the resolving power decreases greatly, as shown in Fig. 6.8. The decrease, however, is not as great as one would suppose when in discussing only the reduction in theoretical plate number or the increase in HETP with the heating rate in preparative columns (13).

Although temperature programming, because of inhomogeneity of the heated area in preparative columns, greatly falsifies the time of analysis when compared with those values obtained with error-free (homogeneous) heating, the resolving power does not decrease as drastically. The use of temperature

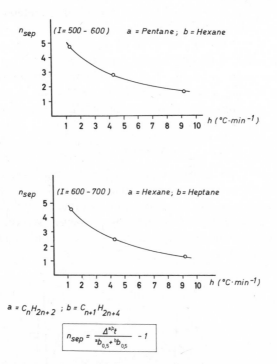

Fig. 6.8. Decrease of column efficiency with increasing heating rate h (between 1 and 12°C min⁻¹) measured as plate number n_{sep} in the index range of 500–600 (upper) and 600–700 (lower) on a preparative column of the same kind as described in Fig. 6.6. At heating rate $h = 1$°C min⁻¹, one can still completely separate 4.5 peaks in the index range 600–700 between n-hexane and n-heptene; at heating rate $h = 5.5$°C min⁻¹, only two peaks. Or, at $h = 1$°C min⁻¹, solutes with an index difference $\Delta I = 100/(n_{sep} + 1) \simeq 15$ can be separated completely; at $h = 5.5$°C min⁻¹, only with $\Delta I \simeq 33$. (Modified from H. Eggert, Thesis, Technical Hochschule Karlsruhe, Karlsruhe, 1964.)

programming in preparative GC is therefore not as useless as it appears when one works with the ambiguous values of theoretical plate number or HETP.

5. Retention Index

A useful application of Eq. 6.9 presupposes knowledge of the retention index under temperature-programming conditions. According to Harris and Habgood (13), the retention index changes during temperature programming by the increment:

$$\Delta I_T = \frac{h}{F_{(T_0)}} \left(\frac{R}{\Delta H'} \right) \left(\frac{RT_0}{\Delta H'} \right)^{-0.7} \cdot V_{M(T_0)} \tag{6.11}$$

where h = heating rate in °C min^{-1}

R = gas constant, 1987 cal °C^{-1} mol^{-1}

ΔH^1 = value of H (see Eq. 6.7) obtained for retention volumes expressed at column temperature

T_0 = initial temperature of program

$F_{(T_0)}$ = flow rate corrected for pressure gradient and expressed at standard temperature for column at temperature T_0

$V_{M(T_0)}$ = dead space volume for column at initial temperature corrected for pressure gradient and expressed at standard temperature.

6. Loading Capacity

Mikkelsen (15) points out that preparative columns are usually overloaded with solute. In these cases the correct temperature plays a decisive part. The loading capacity naturally depends on many other factors that are not discussed here. Temperature programming, however, succeeds in obtaining optimum conditions for loading capacity as is shown in Fig. 6.9.

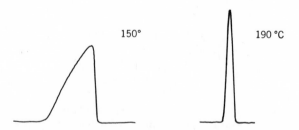

Fig. 6.9. The loading capacity of preparative columns increases with rising temperature as long as the solubility of the solute in the stationary phase increases. Peak form of methyl-octanoate at 150 and 190°C in overloaded column 20% Carbowax 20 M on Chromosorb P. (Modified from L. Mikkelson, in *Advances in Chromatography*, Vol. 2, J. C. Giddings and R. A. Keller, Eds., Marcel Dekker, New York, 1966, p. 353.)

C. Summary and Conclusions

Let us now summarize the present knowledge and add to that the facts from Ref. 13 that we have not yet discussed.

The viscosity of the carrier gas increases because of temperature programming and the gas flow therefore decreases at constant inlet pressure P_i to the 1.7 power of the absolute temperature, i.e., the reduction in pressure over the column $p_i^2 - 1$ increases at constant gas flow "according to the 1.7 power of the absolute temperature."

"The log of the net retention volume is linear with the inverse of the temperature. Gaseous diffusion coefficients increase as the 1.8 power of the temperature. Liquid diffusion coefficients also increase strongly with temperature. It is concluded that at any one temperature, flow conditions cannot be chosen that will give maximum column efficiency for all solutes. In general, column efficiency should improve, as temperature is increased."

Any heating rate, however small, causes a considerable temperature gradient across the direction of flow in preparative columns because of the poor heat conductivity coefficient of conventional stationary phases. This reduces the resolving power and considerably decreases the migrating velocity at the column center as compared with the column wall. This becomes significant for columns with diameters larger than 10 mm. Conventional cylindrical columns of large diameter reduce all advantages of temperature programming so strongly that it is better in practice to run isothermally, *unless one uses very long and therefore narrow columns with a diameter of < 10 mm.*

The round column must be abandoned to utilize the advantages of temperature programming. A rectangular column, as flat as possible, can be programmed with much less temperature inhomogeneity than a round column of equal diameter (Example: round column = 9 cm corresponds to rectangular column of 3 × 20 cm). The column needs to be only a few centimeters long, e.g., 60 cm. It serves as a precolumn which can be cooled to a low temperature for sample dosing and quickly heated to the highest temperature necessary, e.g., at 7–8°C min^{-1}. This would solve the problem of sample evaporation and the problem of enrichment of traces from gases (e.g., the unknown peaks often eluted from analytical columns). With such a steeply programmed precolumn, one could nicely solve some difficult preparative problems of analytical interest and some difficult tasks of production GC. Such a short precolumn could also be run according to the principle of the temperature gradient technique [Kaiser (16)], which gives excellent focusing effect in the enrichment and dosing of diluted materials. One could in this way even dose the sample continuously into the preparative column and evaporate it under the most sparing conditions technically possible.

The normally run part of the column nevertheless separates discontinuously
and isothermally. The above-mentioned temperature gradient acts here in
the direction of flow and not across it. The former is useful; the latter very
harmful. Figure 6.10 gives a proposition for this. The mechanical problems
encountered in such a construction of the preparative column truly are not
simple but can be solved completely.

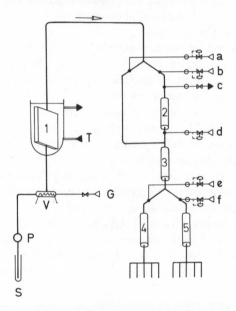

Fig. 6.10. Combination of the reversion-temperature gradient (17) with flow programming in preparative collumns [see also Deans (1,1a,2)]. The preparative column consists
of the sections *1* (flat), *2, 3, 4*, and *5* (round). Carrier gas *G* enters steadily via vaporizer *V*;
sample *S* is steadily transported via pump *P*. Column section *1* is very cold but is heated
for each analysis cycle via *T* using a silicon heating oil which flows in with a temperature
gradient, so that a temperature gradient migrates from below, upward (this is possible
using suitable technical design). Via additional carrier gas in *e* or *f*, column *4* or *5* is switched
in; *4* or *5* may be fashioned as *1*, then strong local accumulations of concentration occur
there. When one wants to backflush *2*, gas comes from *d* and *c* is opened. Little gas from *b*
closes upward. With this system, all techniques of flow and temperature programming
are possible.

III. Gas Flow Programming; Pressure Programming

A. Basic Principles

Kelley and Walker (18) have summarized the main features of pressure
programming, based on papers by Purnell (19), Scott (20), Zlatkis et al. (21),
Halasz et al. (22,23), Vergnaud (24), Mazor et al. (25,26), Dosta Neto et al.
(27), and Guiochon (28).

The basic equation is that describing the gas flow through a packed column is (29):

$$\delta c/\delta z[p\cdot(\delta p/\delta z)] = \delta p/\delta t \qquad (6.12)$$

where c depends upon gas viscosity and column permeability
  p = pressure
  z = distance along the column
  t = time

Equation 6.12 cannot be solved in general.
  In isobaric GC $\delta p/\delta t = 0$ and an approximate solution is:

$$p^2(z) = (z/L)(p_i{}^2 - p_0{}^2) + p_0{}^2 \qquad (6.13)$$

where L = column length
  p_i = inlet pressure ($z = L$)
  p_0 = outlet pressure ($z = 0$)

Kelley and Walker assume further that the equilibrium pressure along the column during a pressure program ($\delta p/\delta t = 0$) is reached so quickly with increasing p_i that no appreciable pressure rise takes place at the inlet, i.e., no additional pressure gradient is created, which depends only on the pressure increase p_i at the column inlet. This simplified assumption is certainly not valid for preparative GC, especially in those cases in which the column inlet and outlet of a number of column segments show an additional flow resistance based on unfavorable geometry. In practice the calculated reduction in retention time is less than that actually measured. The adjusted retention time becomes greater and the actual column efficiency becomes worse than that calculated when assuming an undelayed, quickly adjusted pressure equilibrium. However, the additional pressure gradient, which depends on the rate of pressure increase, interferes far less than the temperature gradient in temperature-programmed preparative columns.
  Kelley and Walker finally reduce the equations for pressure programming to

$$r = \frac{2p_o(p'_s - \alpha p'_{in})}{\alpha^2 \cdot t_s} \qquad (6.14)$$

where

$$t_s = \frac{t_R}{\alpha} = \frac{4L^2}{3ck_ip_o} \cdot \frac{[(p_s/p_o)^3 - 1]}{[(p_s/p_o)^2 - 1]^2}$$

where r = rate of pressure increase in atm min^{-1}
  p_o = initial pressure in absolute atmospheres
  $p'_s = p_s/p_o$
  p_s = constant inlet pressure that would elute solute i with retention time t_s (in isobaric mode)

$p'_{in} = p_{in}/p_o = $ initial pressure from which the pressure program starts

$\alpha = $ reduction in time when using pressure programming which shortens the retention time, according to $t_R = t_s\alpha$; $\alpha \leq 1$

$t_R = $ real retention time obtained by pressure programming

$L = $ column length

$c = $ a function of gas viscosity and column permeability

$k_i = t'_{Ri}/t_M$

B. Application

With Eq. 6.14 one can calculate any pressure program rate r necessary to obtain a definite reduction in retention time. A numerical example for a desired reduction in retention time to 20% follows. $t_s = 30$ min; $\alpha = 0.2$, i.e., desired new retention time $t_R = 6$ min; $p_i = 2$ atm; $p_0 = 1$ atm.

$$r = \frac{2.1[(2/1) - 0.2\cdot(2/1)]}{0.2\cdot0.2\cdot30} = 2.8 \text{ atm min}^{-1}$$

When using a pressure increase of 2.8 atm min^{-1}, the pressure will have risen after 6 min by $6 \times 2.8 = 17$ atm to 19 atm; this would reduce the retention time, however, by a factor of 5. The gas flow through the column has then risen naturally to its 81-fold value. The concentration in which the solute leaves the column correspondingly decreased. One must therefore pay a steep price indeed for the gain in time of a factor of 5. This includes:

1. Enormously increased gas consumption (a factor of 81).

2. Correspondingly increased cooling capacity in the cold traps.

3. Reduced yield (as higher losses in the separation of the solute from the mobile phase are unavoidable with increased gas flow).

4. A small loss in column efficiency.

This loss is naturally trivial compared with that sustained in a temperature program that gives the same time gain. Practical experiments by Kelley and Walker (18) show that pressure programming does not yet recognizably reduce the column efficiency expressed as plate number t_z at a reduction in retention time t_R by the factor $\alpha = 0.4$. This is in contrast with the increase in plate number during pressure programming indicated by the authors (of course only found qualitatively). The latter is further in accord with results obtained by Gordon et al. (31), according to which the resolving power of preparative columns even increases with higher inlet pressures. At higher inlet pressures and thus at higher flow velocity, the optimum lies as valid for analytical columns.

C. Summary and Conclusions

The gain in time is considerable in preparative GC when pressure programming is used. It can be calculated quite easily. The loss of column efficiency is less than one would suppose. However, the cost of the gain in time (gain in output in g sec^{-1}) is very high. It increases quadratically with linear time gain, expressed as gas consumption.

From this it follows that pressure programming over the total length of the preparative column is only useful in exceptional cases. It is, however, always possible to employ pressure programming effectively on short lengths of column. It is especially advantageous to obtain time gain using column switching, a technique that has not yet been introduced in preparative GC, but which combines all the advantages of temperature and pressure programming. The technique introduced by Deans (1,2) for analytical columns is of advantage here as all switching takes place pneumatically in T-pieces without mechanically moving parts and without problems of gas tightness. A combined use of locally limited temperature programming/pressure programming together with column switching, heart cutting and backflushing (see Section I), is finally most advantageous. Figure 6.10 shows some suggestions.

References

1. D. R. Deans, *J. Gas Chromatog.*, **4**, 34 (1966).
1a. D. R. Deans, *J. Chromatog.*, **18**, 477 (1965).
2. D. R. Deans, *Chromatographia*, **1**, 18 (1968).
3. J. C. Giddings, *Anal. Chem.*, **34**, 722 (1962).
4. K. -P. Hupe, W. Böhnisch, and H. Quitt, *Chem. Ing. Tech.*, **37**, 146 (1965).
5. K. Nesselmann, *Angewandte Thermodynamik*, Springer-Verlag, Berlin, 1950, p. 260.
6. B. Baule, *Partielle Differentialgleichungen*, Leipzig, 1962.
7. O. Jahnke-Emde, *Tafeln hoherer Funktionen*, Leipzig, 1960.
8. K. -P. Hupe, U. Busch, and K. Winde, *Advances in Chromatography 1969*, A. Zlatkis, Ed., Preston Technical Abstracts Co., Evanston, Illinois (1969), p. 107.
9. G. L. Guillemin, *Advances in Gas Chromatography 1965*, A. Zlatkis and L. S. Ettre, Eds., Preston Technical Abstracts Co., Evanston, Illinois, 1965.
10. S. Spencer, private communication to K. -P. Hupe, see Ref. 8.
11. A. S. Said, *Gas Chromatography 1962*, O. Brenner, O. Callen, and O. Weiss, Eds., Academic Press, New York, 1962, p. 84.
12. H. W. Habgood and W. E. Harris, *Anal. Chem.*, **32**, 450 (1960).
13. W. E. Harris and H. W. Habgood, *Programmed Temperature Gas Chromatography*, John Wiley, New York, 1966, p. 53.
14. H. Eggert, Diplomarbeit, Technische Hochschule Karlsruhe, Lehrstuhl für Thermodynamik, 1964, p. 37.
15. L. Mikkelsen, *Advances in Chromatography*, J. C. Giddings and R. A. Keller, Eds., Vol. 2, Marcel Dekker, New York, 1966, p. 352.

16. R. Kaiser, *Chromatographie in der Gasphase*, Band II, Kapillar-Chromatographie, Bibliographisches Institut, Mannheim, 1966, p. 47; R. Kaiser, *Z. Anal. Chem.*, **189**, 1 (1962).
17. R. Kaiser, *Chromatographia*, **1**, 199 (1968).
18. J. D. Kelley and J. Q. Walker, *J. Chromato. Sci.*, **7**, 117 (1969).
19. H. Purnell, *Gas Chromatography*, John Wiley, New York, 1962, p. 387.
20. R. P. W. Scott, *Gas Chromatography 1964*, A. Goldup, Ed., The Institute of Petroleum, London, 1965, p. 25.
21. A. Zlatkis, D. C. Fenimore, L. S. Ettre, and J. E. Purcell, *J. Gas Chromatog.*, **3**, 75 (1965).
22. I. Halász and F. Holdinghausen, *Advances in Gas Chromatography 1967*, A. Zlatkis, Ed., Preston Technical Abstracts Co., Evanston, Illinois, 1967, p. 23.
23. I Halász and G. Deininger, *Z. Anal. Chem.*, **228**, 321 (1967); **229**, 14 (1967).
24. J. M. Vergnaud, *J. Chromatog.*, **19**, 495 (1965).
25. L. Mázor and J. Takács, *J. Gas Chromatog.*, **4**, 322 (1966).
26. L. Mázor and J. Takács, *J. Chromatog.*, **34**, 157 (1968); **36**, 18 (1968).
27. C. Costa Neto, J. T. Koffer, and J. W. de Alencar, *J. Chromatog.*, **15**, 301 (1964).
28. G. Guiochon, *Chromatography Reviews*, Vol. 8, M. Lederer, Ed., Elsevier, New York, 1966, p. 1.
29. P. C. Carman, *Flow of Gases through Porous Media*, Academic Press, New York, 1956, p. 5.
30. A. T. James and A. J. P. Martin, *Biochem. J.*, **50**, 679 (1952).
31. S. M. Gordon, G. J. Krige, and V. Pretorius, *J. Gas Chromatog.*, **2**, 241 (1964).

CHAPTER 7

Applications in Organic Chemistry

M. Verzele, *Laboratory of Organic Chemistry, State University of Ghent, Belgium*

I. Introduction

The preparative possibilities of gas chromatography (GC) are probably more important to organic chemistry than to any other field of chemistry. In our laboratory about half of all GC separations are carried out with the object of collecting the separated fractions. These separations can be characterized by the amount of a particular compound that must be recovered. The most common type of separation concerns a relatively simple mixture from which milligram quantities of a compound must be isolated for further study by means of UV, visible, or IR spectrometry, MS, or NMR spectroscopy. Many of the problems in preparative-scale GC in the organic chemistry laboratory belong to a second type of technology in which the recovered sample size must be somewhat larger (± 50 mg), or the mixture contains a larger number of components. Occasionally, there is also a need for still larger samples (milliliter), which again necessitates slightly different techniques.

II. General Considerations

When a separation cannot be carried out with the available columns, the usual way to solve the problem is to increase the length of the most promising looking column. This often works if the separation is not too difficult. Within not too large a range, column efficiency (plate number) is indeed a linear function of column length.

In this context it is implicit that, whatever the detector, the sample size can always be small enough not to affect the column efficiency markedly. This column efficiency, usually expressed as plates per foot or plates per meter, can vary somewhat according to the stationary phase, the nature of the substances analyzed, the support material and of course according to the coating procedure and the experience of the operator; but it is usually fairly constant and only variable over a factor of 2, e.g., between 1000 and 2000 plates per meter.

The most important difference between analytical and preparative GC is that in the latter, sample sizes are large enough to affect column efficiency and that this can cause variations in the plate number by a factor of 10 to 20. This efficiency/sample size dependence in preparative GC must be taken into account together with the variables usually encountered in analytical GC. It is therefore necessary, in order to describe preparative-scale GC correctly, to know the efficiencies of a given GC column over the whole range of possible sample sizes. These efficiencies are variable and for small samples can be as high as the analytical efficiency of the column. For larger samples, because of the normal band-broadening effect with larger samples, the efficiency is, however, lower. Thus there is a preparative-scale efficiency or a "preparative-scale plate number" and it can be symbolized by n'. Examples of such n'-values are given in Table 7.1. The usefulness of such tables becomes clear from the later discussion.

TABLE 7.1

n' Values for a 6 m $\times$ 9 mm coiled Aerograph 700 column.[a]

Sample size, μl	Preparative plate number, n'
10	2340
50	1380
100	710
200	400
500	190

[a] Chromosorb W 30–60 mesh with 25% SE-30 and 200 ml H_2/min. Cumene m ples at 145°C.

III. Some Theoretical Considerations

A two-component mixture that can be separated when using small samples eventually shows peak overlap with larger and larger samples. There is thus a limit to the sample size that can be separated in one run. There is a "maximum allowable sample size" or MASS.

MASS depends on the following factors.

1. The difficulty of the separation, or the proximity of the two peaks to each other, or the α-value ($\alpha = t'R_2/t'R_1$).

2. The efficiency of the column, or the n'-values in this case.

3. In a lesser way, the time of the separation, best expressed as a function of the partition ratio k.

In analytical GC, α, n, and k are related by Eq. 7.1.

$$n = 16(\alpha/\alpha - 1)^2(1/k_2 + 1)^2 \qquad (7.1)$$

This equation is used to find the necessary column length to give a 4σ separation of a mixture with given α and k. When k is not too small, as it usually is in packed columns, e.g., $k > 3$, Eq. 7.1 can be simplified to

$$n = 16(\alpha/\alpha - 1)^2 \qquad (7.2)$$

This equation can now be used to find the sample size that can be separated in one run on a given column (MASS). This is done as follows. From analytical GC the α-value is deduced and with Eq. 7.2 the number of plates needed for the separation is calculated. An example will clarify this. *trans-* and *cis-*decalin give an α-value of about 1.25 on a SE-30 column, and this separation therefore requires 400 theoretical plates. Table 7.1 gives 400 plates for a sample of 200 μl. Therefore a decalin sample may not contain more than this amount of one of the isomers. For a 1:1 decalin mixture, MASS is thus 400 μl on this column. This is indeed found experimentally to be the case, and for each peak about 400 plates are calculated. For the general case of a complex mixture, the two components with the lowest α-value determine MASS. Of these two components the percentage (x) of the one present in the largest amount is estimated. The sample size that can then be separated in one run of this complex mixture is given by

$$MASS = 100Sn'/x \qquad (7.3)$$

where Sn' is the sample size of a single substance giving that n'-value on the column to be used which is also the n-value determined with Eq. 7.2.

IV. Columns

To separate larger amounts of material, the column volume must be enlarged. This can be done by increasing the column diameter, or its length, or both. An indication of the preferred solution may be obtained from the data in Table 7.2 listing the number of plates needed for commonly encountered α-values.

TABLE 7.2

Plate Numbers Necessary to Separate Mixtures with the Indicated α-Values

α-Value	n Required	α-Value	n Required
1.03	17,496	1.13	1,221
1.05	7,056	1.15	940
1.07	3,745	1.20	576
1.08	2,915	1.25	400
1.09	2,342	1.30	300
1.10	1,936	1.40	195
1.11	1,632	1.50	144
1.12	1,384	2.00	64

From this table it is obvious that α-values below 1.1 should be avoided in preparative GC because it simply is not possible to obtain the necessary 2000 and more plates with larger samples. This table also shows that it is worthwhile to try several stationary phases in order to find the one giving the highest α-value, since even a small increase in α-value simplifies the problem markedly.

Whatever the column type, the plate number drops with larger samples. With smaller-diameter columns this drop is sharp because overload effects are rapidly introduced, but with smaller samples the efficiency of the column is high. An example of this is found in Table 7.1. To possess higher plate numbers for larger samples the column must be made so much longer. It is thus obvious that a long, narrow column can solve a broad range of problems although the sample size must be adjusted to each situation.

With larger-diameter columns overload is less to be feared and indeed over a wide range of sample sizes the plate number falls relatively slowly. However, the plate number of wide-bore columns is not high even with the smallest samples. Indeed, with columns of 5- or 10-cm diameter, the plate number is usually only a few hundred per meter. Although such columns can handle large samples, their use is limited to relatively simple mixtures (i.e., large α-values). Although in many laboratories much effort has been put into large-diameter column technology, few details have been published. This is most probably because the results did not meet expectation.

The great versatility of small-bore, long columns is a strong recommendation for their use in an organic laboratory.

The problems of large-diameter columns are, however, interesting and intriguing. While earlier efforts in large-diameter column work were directed toward obtaining more uniform packing, recent research has been more concerned with homogenizing the gas flow (1, 2). For a column of 2 m $\times$ 10 cm, published chromatograms give plate numbers of 600 for 1-ml samples (3). This efficiency would not warrant the introduction of this type of preparative GC into the research laboratory. The more so since this figure is very difficult to reproduce. It is also a fact that almost all the commercial instruments capable of handling large-bore columns are actually used with long, narrow columns. In a recent publication describing GC on a semiindustrial scale (4) with 1-ft = diameter columns, the efficiency deduced from the published chromatograms is of the order of 50 to 150 plates per meter. This is again only useful for simple separations. The drop in plate number with increased column diameter is an interesting problem. The main reason for this efficiency drop is the fact that radial diffusion is not able to remix flow profiles in larger-diameter columns. That this is true is shown by the optimum linear gas velocity which decreases as the column diameter becomes larger and also by the following observed differences for nitrogen and hydrogen. With small column diameters (4 mm and below), nitrogen gives better results than helium or hydrogen because radial diffusion is high enough to compensate for gas flow profiles with all gases. Axial diffusion is therefore most important, and this is lowest with nitrogen, as is well known, thus explaining the superiority of nitrogen.

With columns of intermediate diameter (5–20 mm) radial diffusion in nitrogen is insufficient to ensure adequate profile remixing. This is not true with hydrogen, and results in such columns are slightly better with this gas (5). Increasing the column diameter even more leads to a situation in which radial diffusion plays no further role and nitrogen again becomes superior to hydrogen. In a 9-cm-diameter column, nitrogen gave about 10–20% more plates than hydrogen (6). Column dimensions should be chosen to give reasonable results for a large variety of problems rather than perfect results for one particular application. This is especially advisable for the research laboratory. Therefore, as stated before, long and narrow-bore columns are to be preferred. Further reasons for this choice are based on considerations for the collection of the substances. In laboratories in which samples are used for identification and for exploratory chemical purposes, the necessary sample size can be narrowed to the range of 1 μl to 1 ml. To make these amounts of separated substances easily available, they should be collected in small collecting vessels. Large volumes in the collectors lead to losses for obvious reasons. In a small collecting device, the flow rate should be rather low, since good recovery would otherwise be impossible.

While some substances can be collected adequately with flow rates up to 1 liter/min in the collectors under consideration, other substances need flow rates as low as 25 ml/min (7). With large-bore columns the flow rates are obviously much larger than these values even though the optimum flow rate is not as pronounced in preparative GC, and large gas rate variations are possible without unduly affecting efficiency.

V. Practical Aspects of Preparative GC

In the preceding chapters of this book, the technologies of preparative GC have been discussed in detail. The purpose of this paragraph is thus only to indicate the solutions that have given the best results in our laboratory. For details the reader is referred to the literature (5,7–9).

On-column injection is the preferred method of injecting a sample. Injection of large volatile samples (500 μl and up) into a heated column can cause a large increase in pressure. Backflash of part of the sample into the cold carrier gas lines should be avoided. Normally, this is achieved by providing a "backflash valve." In our experience a simple needle valve is better. This valve is closed just before injection and reopened about 1 min after injection. With larger-volume columns the effect of this can barely be seen on the baseline of the chromatograms.

Column material can be stainless steel or glass, the latter being preferred. Column shape is irrelevant. Columns should be lightly filled to minimize pressure drop, and if the column is filled in its final shape (already coiled for example), this should be done from the detector end (8).

Support material should be of the best quality obtainable, acid-washed and DMCS treated, because preparative GC is usually carried out at slightly higher temperatures than analytical GC. This higher temperature catalyzes decomposition of the substances, which is a major problem in the field. The higher temperatures are needed to pass the substances through in a reasonable time since the optimum gas rate is proportionally lower than for analytical GC columns.

Indeed, if for a 0.2-mm-diameter capillary column the optimum gas rate is about 1 ml/min, it is not 100 ml/min for a 2-mm packed column but only about 20 ml/min. For a 2-cm column it is only 200–400 ml/min, and for a 10-cm-diameter column it is about 5 liters/min. With linear extrapolation from 2 mm to 10 cm, the optimum gas rate in the larger column would be about 50 liters/min. It is quite normal when using the relatively low gas rates mentioned above to avoid longer retention times by slightly increasing the working temperature. To minimize decomposition under these conditions, the support must be of the very best quality. Support materials can indeed catalyze dehydration, dehydrohalogenation, or other types of degradation at surprisingly low temperatures. This problem can be minimized by heavily coating the support material, but then retention times increase strongly and again there is a tendency to offset this by increasing temperature. It is often better to use only thin coatings in order to obtain relatively low k-values at low temperatures and to avoid support-catalyzed decomposition by using acid-washed, DMCS-treated, high-quality material. Recommended coating percentages are then 1–5 g per 100 ml support. With Chromosorb W of density 0.22, this is equivalent to 5–25% wt %; with the even less reactive Chromosorb G of density 0.55, this is equivalent to about 2–10% wt %.

The mesh size of the support must be adapted to the column length: 60–80 mesh for columns up to 3 m in length, 30–50 mesh for columns up to 10 m; and 30–40 mesh for columns up to 20 m long.

VI. Examples of Separations

As stated in the first paragraph of this chapter, there are several distinct technologies in preparative-scale GC.

A. Simple Mixtures—Micro Amounts

When an unknown reaction is carried out, it is best to follow the disappearance of the reactants and the formation of new substances by either thin-layer chromatography (TLC) or GC. If possible, the latter is preferable. The fast analytical possibilities of GC allow the rapid pin-pointing of optimum reaction conditions. The starting substances are easily recognized from their retention times when k is not too small (the temperature should be low enough—this is an error frequently made by beginners). Newly formed substances must be identified. This can nowadays be achieved by collecting a few milligrams of the separated compounds and analyzing them spectrometrically. By repeating the same separation a few times, enough material for even NMR analysis can be collected. A very simple gas chromatograph with katharometer detection and a column 1–3 m in length by 4–6 mm in diameter does this job easily. It is even surprising how much material these columns can handle. An example of this situation is shown in Fig. 7.1. It also often happens in an organic chemistry laboratory that a series of similar compounds is synthetized for comparison purposes, as reference material in MS, or NMR or IR spectroscopy, for determination of conformational equilibria in relation to small structural changes or for still other reasons.

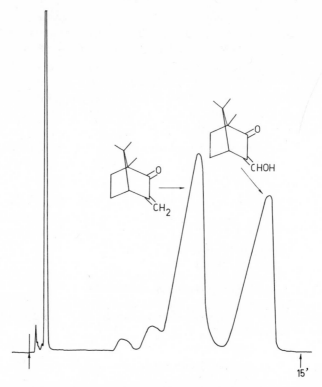

Fig. 7.1. Reaction mixture obtained from camphor. GC conditions: 3 m × 5 mm coiled column; 5 g QF₁ per 100 ml Chromosorb W 60–80 mesh in Aerograph A-350-B katharometer instrument; 60 ml H_2/min at 125°C; sample size of the mixture of substances as shown, 100 μl.

The mixtures so obtained are relatively simple, and only a very small amount of material is needed for these investigations. Before the days of preparative GC, each compound had to be prepared in milliliter amounts because of the purification difficulties, and much time was involved. Now the reaction can be carried out on a few milligrams of material and the whole reaction mixture processed by preparative GC. As yield is irrelevant, as long as enough substance is obtained, the difficulties of such work have decreased considerably. An example of an investigation followed by MS (10) is shown in Fig. 7.2. Since the necessary sample size is low, analytical column length is sufficient. Another example is the separation shown in Fig. 7.3. The substances were needed for studies on conformational equilibria (11), and their structure had to be ascertained through NMR spectroscopy. Therefore somewhat larger amounts were needed. The mixture was obtained by the Prins reaction on cyclooctadiene.

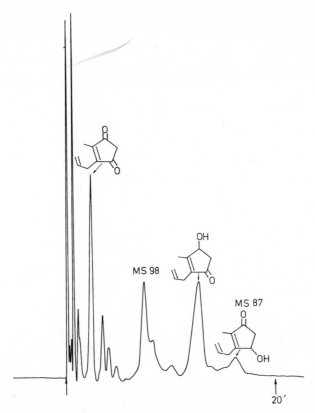

Fig. 7.2. Mixture obtained on reduction of the first labeled peak. GC conditions: 2 m × 5 mm coiled column; 5 g polytetramethylene oxide (PTMO) per 100 ml of Chromosorb W 60–80 mesh in Aerograph 90 P katharometer instrument; 30 ml H_2/min at 195°C; sample size of the mixture obtained on reduction of the allyl methyl cyclopentenedione, 3 μl. The traces of the substances collected in small tubes are sufficient for MS investigation.

Collection in the cases of Figs. 7.1, 7.2, and 7.3 and similar separations can be reduced to its simplest form. Often it is sufficient to insert a glass tube (5–10 cm long, 2-mm diameter) in the gas chromatograph outlet at the moment the peak appears. Condensation in the tube can easily be seen. As soon as some deposit appears, this is sufficient for MS analysis and the collection tube can be transferred as such to the heated inlet of the mass spectrometer. For NMR analysis a real drop of liquid must be collected. If this is not possible in one run, the same separation is repeated using the same collector until the amount is sufficient.

It is of course important that the gas chromatograph exit up to the collector is at least at column temperature, otherwise premature condensation pro-

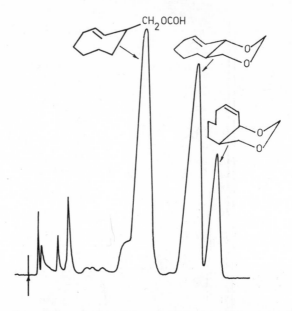

Fig. 7.3. Reaction mixture from a Prins reaction on cycloheptene. GC condition: 6 m + 5 mm coiled column; 5 g SE-30 per 100 ml Chromosorb W 60–80 mesh in Aerograph 90P katharometer instrument; 35 ml H_2/min at 185°C; sample size of the mixture as shown, 45 μl.

duces remixing of the peaks. It is therefore good practice to check the purity of the collected substances by analytical GC.

When cooling is necessary (the collector should be about ±150°C below column temperature) simple methods are still possible. Devices such as that shown in Fig. 7.4 cooled in liquid nitrogen are effective. In Fig. 7.4a, a small test tube (7 cm × 7 mm) is the collector. Carrier gas escapes through a furrow in the stopper. The glass device shown in Fig. 7.4b is used when slightly more material must be collected.

Occasionally, aerosol formation renders collection difficult. This usually occurs with very high-boiling or with crystalizing substances. In this case adsorption on coarse drying silica gel is the best solution. This was, for example, true in the case of the substances in Fig. 7.5 having structures 1, 2, and 3. Collection of five injections of 150 μl of solution at 50% was obtained by passing the effluent gas through traps of 2 g of acid and solvent-washed silica gel (10–20 mesh) in glass tubes 12 × 40 mm inserted in the chromatograph outlet. The three tubes used in this case were washed with ether, and solvent removal gave 133 mg of 1, 78 mg of 2, and 173 mg of 3. Substance 2 is isomeric with 3 but of unknown structure. Collection is quantitative. The same silica gel tubes can be used repeatedly.

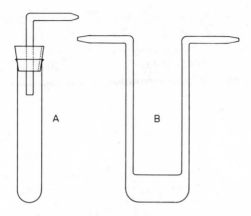

Fig. 7.4. Collecting devices for preparative GC with cooling.

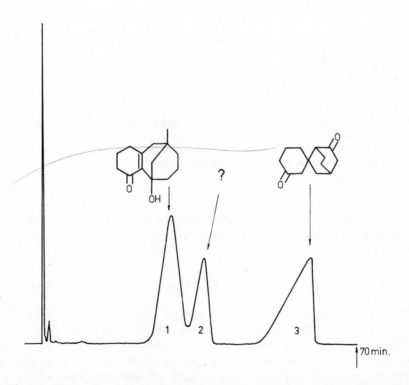

Fig. 7.5. Dimers of cyclohexenone derivatives. GC conditions: 3 m × 9 mm coiled glass column; 5 g Versamid per 100 ml Chromosorb W 30–60 mesh in Aerograph Autoprep 700 katharometer instrument; 150 ml H_2/min at 230°C collection on silica gel of 150-μl sample of the mixture as indicated.

For really high-boiling substances it is also only possible to use narrow-column preparative-scale GC. In this case the carrier gas rate must be high in order to get the substances through the column, and this is impossible in larger volume columns without losing too much efficiency. An example of this is given in Fig. 7.6. The separation of the two Diels-Alder myrcene

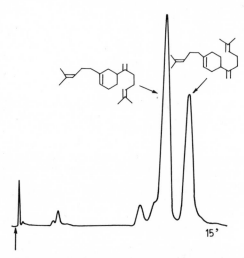

Fig. 7.6. Diterpene mixture obtained by Diels-Alder dimerization of myrcene. GC conditions: 3 m × 5 mm coiled glass column; 5 g Apiezon per 100 ml Chromosorb W 60–80 mesh in Aerograph A-350-B katharometer instrument; 300 ml H_2/min at 250°C; sample size, 20 μl, α-value, 1.15. On SE-30 as stationary phase the α-value is only 1.11.

dimers, which are obtained on heating myrcene (13), is shown. Chromatography is carried out at 250° and with 300 ml/min hydrogen in a 5-mm diameter column.

B. Difficult Separations—Complex Mixture—Semimicro Amounts

When the α-value is too low (α < 1.10) preparative GC becomes impossible on analytical columns. The efficiency of the column must be increased, and as indicated before this is best achieved by increasing the column length. A good compromise is a column 10–20 m long with a diameter of 1–2 cm. Plate numbers as a function of sample size for such a column filled with Chromosorb W 30–60 mesh are shown in Table 7.3.

TABLE 7.3
n'-Values for a 20 m $\times$ 9 mm Coiled Column

Sample size, μl	n'-Value
10	7270
50	3240
100	1700
200	1040
500	480
1000	240

ᵃ 20 m $\times$ 9 mm column; 5 g SE-30 per 100 ml; Chromosorb W 30–60 mesh. Trans-decalin samples at 180°C.

Such columns can indeed separate mixtures with low α-values. An example is shown in Fig. 7.7. The α-value between the second and third peaks is only 1.08. On this chromatogram obtained with 100 μl of the mixture, plate numbers are about 3000–4000.

Longer columns are also indicated for complex mixtures, as low α-values are unavoidable in this case. Although complex mixtures are mostly of natural origin, they can be the result of synthesis. It is obvious that the longer columns under consideration can separate larger amounts of simpler mixtures.

A series of further examples of simple mixtures is now presented in Figs. 7.8 through 7.13, showing larger and larger samples as the α-values increase. Figures 7.14 through 7.18 show the separation of complex mixtures of synthetic or natural origin.

C. Large Samples

When still larger amounts of mixtures have to be separated the same technology can be followed in larger volume columns. The most practical columns seem to be of 20-m length and about 2-cm diameter. When the column is coiled, ovalization of the column section is advisable (8). Examples of such separations are shown in Figs. 7.18, 7.19, and 7.21. It is obvious that this type of column can separate very large amounts of simple mixtures. These very large volume columns can also be used for separation or enrichment of low-concentration impurities. In one case it was possible, using 40-ml samples, to isolate 1,1-dichloroethane present as a 0.1% impurity in $CDCl_3$. In another case, volatile impurities at even lower percentages in methylene

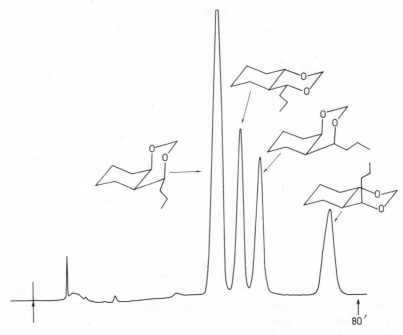

Fig. 7.7. Disastereoisomeric ethyldioxadecalans as shown. GC conditions: 20 m × 9 mm coiled glass column; 5 g QF₁ per 100 ml Chromosorb W 30–60 mesh in modified Aerograph Autoprep 700 katharometer instrument; 150 ml H_2/min at 165°C; sample size of the four stereoisomers as shown, 100 μl.

chloride were concentrated first, using 20-ml samples and collecting the first ml of the eluted substance. After this, normal preparative separation of the enriched mixture was possible.

VII. Miscellaneous Points of Interest

A. The Use of Glass Beads

The tail fraction of essential oils usually contains high-boiling substances, e.g., diterpenes. High-boiling compounds are very difficult to separate by preparative-scale GC. While steroids can be successfully chromatographed analytically, this becomes more difficult as the sample size is increased. This problem was raised earlier in the chapter. The reason for this is not only insufficient thermal stability, but with high k-values adsorption effects become more and more important, causing excessive band broadening. By reducing adsorption, preparative-scale GC of high-boiling compounds may be im-

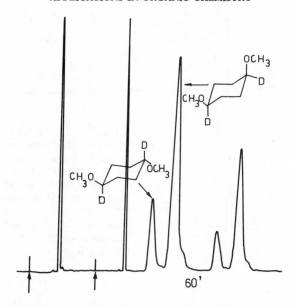

Fig. 7.8. Deuterated dimethoxycyclohexane stereoisomers. GC conditions: the same as in Fig. 7.7 but with 5 g QF$_1$ per 100 ml Chromosorb G 30–40 mesh; 300 ml H$_2$/min at 140°C; sample size, 50 μl.

proved. This is achieved by using glass beads as column support material. An example of this is shown in the chromatography of jasmine essential oil. Figure 7.22 is such a separation on a Chromosorb W–filled column. The tail fraction shows insufficient resolution. The same separations on a glass bead column is shown in Fig. 7.22. The high-boiling substances show much better separation and more substances managed to pass through the column.

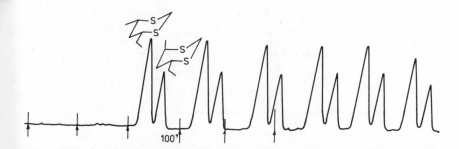

Fig. 7.9. Dithiane stereoisomers. GC conditions: the same as in Fig. 7.7 but with 450 ml H$_2$/min at 130°C; repeated (multiple) injection procedure with 100-μl samples.

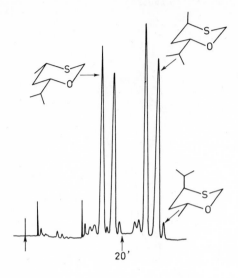

Fig. 7.10. Diastereoisomeric thioxans as shown. GC conditions: the same as in Fig. 7.7 but with 5 g PTMO per 100 ml Chromosorb W 30–60 mesh; 200 ml H₂/min at 200°C; sample size of the methyl isopropyl thiooxane as shown, 300 μl; the α-value for the two major peaks is 1.14. A second sample was injected before the first had been eluted (multiple injection).

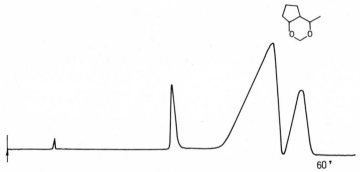

Fig. 7.11. GC conditions: the same as Fig. 7.7 but with 5 g SE-30 per 100 ml Chromosorb G 30–40 mesh; 300 ml H₂/min at 130°C; sample size, 350 μl; α-value of last peaks, 1.10; plate number for the last peak, 2000.

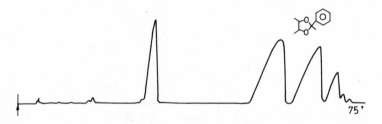

Fig. 7.12. The four diastereoisomeric trimethylphenyldioxolans. GC conditions: the same as Fig. 7.7 but with 5 g SE-30 per 100 ml Chromosorb W 30–60 mesh; 200 ml H₂/min at 210°C; sample size of the dioxolan diastereoisomers was 500 μl with α-values of 1.16 and 1.10 for the last peaks.

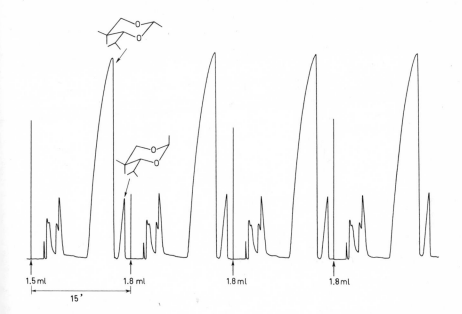

Fig. 7.13. GC conditions: the same as Fig. 7.7 but with 5 g PTMO per 100 ml Chromosorb W 30–60 mesh; 200 ml H_2/min at 180°C; sample sizes of the consecutive separations as shown on the graph, 1.8 ml; α-value, 1.4.

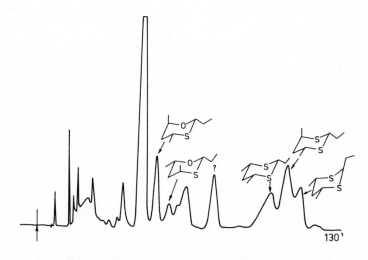

Fig. 7.14. Oxathian and dithian mixture obtained by synthesis. GC conditions: 20 m × 9 mm coiled glass column; 5 g QF_1 per 100 ml Chromosorb G 30–40 mesh; 450 ml H_2/min at 145°C; sample size, 100 μl.

227

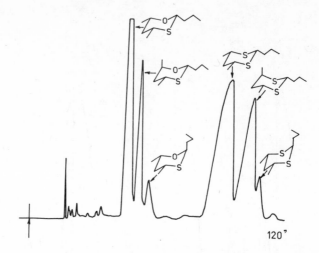

Fig. 7.15. GC conditions: the same as Fig. 7.14 but with 350 ml H$_2$/min and with 150 μl of a mixture, obtained by synthesis, of isomeric thiosans and dithians.

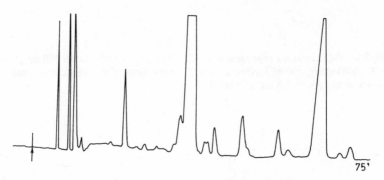

Fig. 7.16. Natural mixture of hop oil alcohols. GC conditions: the same at Fig. 7.14 but with 5 g PTMO on Gas-Chrom P 30–60 mesh; 250 ml H$_2$/min at 170°C; sample size, 200 μl. Identification of the numbered peaks was possible by spectrometry and is discussed in Ref. 13.

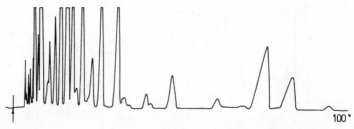

Fig. 7.17. Methyl esters of the acids isolated from hop oil. GC conditions: the same as Fig. 7.16 but at 180°C with a 300-μl sample. Identification of the collected peaks is discussed in Ref. 13.

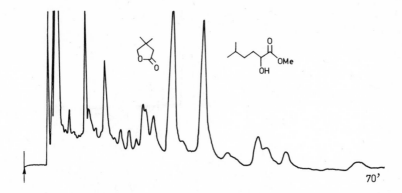

Fig. 7.18. Ozonization reaction mixture of tetrahydrohumulone. GC conditions: the same as Fig. 7.16. but at 190°C with a 350-μl sample. The interesting point here is the peak of the hydroxy acid containing an unchanged chiral center and which has enabled determination of the absolute configuration of (−)-tetrahydrohumulone.

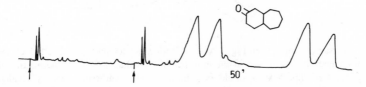

Fig. 7.19. Mixture of *trans* and *cis* bicyclic ketones. GC conditions: 21 m × 20 mm coiled aluminum column; 5 g Carbowax 20M per 100 ml Chromosorb 20–30 mesh; 600 ml H₂/min at 200°C; sample size, 2 ml. Spectrometric analysis revealed the structures as shown in the figure.

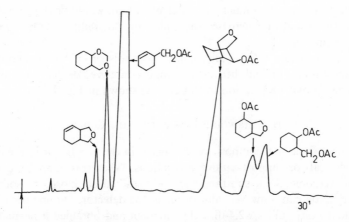

Fig. 7.20. Mixture obtained by a Prins reaction on cyclohexene. GC conditions: 20 m × 20 mm coiled glass column; 5 g SE-30 per 100 ml Chromosorb G 30–40 mesh; 300 ml H₂/min at 180°C; sample size, 2 ml. Spectrometric analysis revealed the structures as shown in the figure.

229

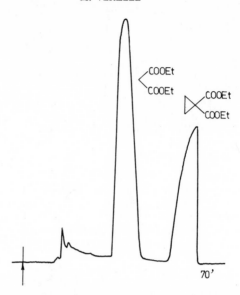

Fig. 7.21. Same column as in Fig. 7.20. GC conditions: 400 ml H_2/min at 195°C; sample size of the substances as shown in the graph, 7 ml; α-value, 1.6. It is obvious that much larger samples of this mixture could be separated. The synthesis had only yielded the 7 ml mentioned.

With such complex mixtures temperature programming is indicated. This not only allows chromatography of a mixture with a wide boiling range, but each peak is eluted in a much narrower band than would be the case isothermally. This decidedly increases the collection possibilities. Other examples are given in Ref. 12.

Preparative-scale GC of high-boiling alcohols or of halogen-containing substances is also difficult because of support-catalyzed decomposition. Glass beads can provide a solution in this case as shown in Fig. 7.23.

B. Catalytic Effects outside the Column

Changes in sample composition can occur in the injector or in the column through heat- or support-catalyzed reactions. When this happens, it is difficult to overcome. Some substances are very easily decomposed or isomerized. Isomerization can, however, also occur in the detector. On one occasion we separated *trans*- and *cis*-2,4-dimethyldioxolan and obtained a normal result with $\alpha = 1.08$ on Chromosorb G coated with Carbowax 20M. The collected substances, however, showed the same composition as the starting material. The reason for this was found to be an acid-fouled hot wire detector causing

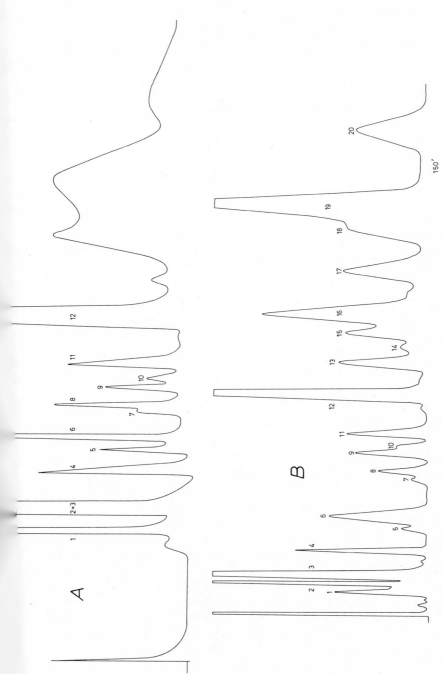

Fig. 7.22. Jasmine essential oil. GC conditions: 11 m × 13 mm coiled glass column in Varian Aerograph 713 preparative-scale gas chromatograph with FID; 750 ml H₂/min starting at 130°C and programmed at 4°C/min to 245°C, then isothermal; sample size, 150 μl. *a*: On Chromosorb G 30–40 mesh coated with SE-30 (2 g per 100 ml support). *b*: On DMCS-treated glass beads 60–70 mesh coated with SE-30 (1 g per 100 ml support). The peaks in both Figs. 7.9*a* and *b* are numbered for easier comparison of the chromatograms. Peak *12*, for example, is benzoyl benzoate.

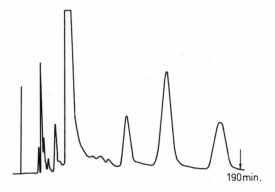

Fig. 7.23. Mixture obtained from alkaline treatment of dichloromethyl isobutyl ketone. The three last peaks contain chlorine. GC conditions: 20 m × 20 mm coiled (ovalized) glass column with 2 g Carbowax 20M per 100 ml glass beads of 0.8-mm diameter; laboratory-made instrument; 600 ml H_2/min at 150°C; sample size, 2 ml.

the substances to isomerize again after the separation. Washing the detector with tetralin, bicarbonate, water, and acetone removed the acid contamination and subsequent separation of the same mixture proceeded in a normal way.

C. Automated Collection

Much effort has gone into automating the collection of preparative-scale GC separations since it is very often necessary to repeat the same separations several times in order to obtain enough material. Unfortunately, automation is only possible for relatively simple problems carried out on a routine basis. In the research laboratory where mixtures are very often priceless, the loss of the material needed to trigger the automatic collector is often unacceptable. Successful automation is dependent on a perfectly steady baseline. Such baselines are usually only to be found in chromatograms reserved for publication (sic). With temperature programming the difficulties are compounded. Consider, for example, Fig. 7.22 showing the essential oil separation. If the trigger for collection is set at 5% of total recorder deflection (which is really not high) several small peaks would be completely lost. Different trigger levels can be installed for different peaks, but it is obvious this can only be achieved after a thorough study of the chromatography of the particular mixture. By that time the problem, as far as a research laboratory is concerned, is solved. So indeed automation is only for those problems for which the same separation has to be carried out a very large number of times.

For these reasons all separations discussed in this paper were carried out manually, or with a semi-automatic collector.

References

1. F. Debbrecht, F and M Company Tech. Paper No. 28 (1966).
2. *Chem. Eng. News*, *43*, 46 (1965); *Chem. Week*, March 20 (1965).
3. *Facts Methods*, **6**, No. 5, F and M Company (1965).
4. A. Carel, R. Clement, and G. Perkings, *J. Chromatog. Science*, **7**, 218 (1969).
5. M. Verzele, *J. Chromatog.*, **15**, 482 (1964).
6. J. Albrecht and M. Verzele, *J. Chromatog. Science*, **8**, 586 (1970).
7. M. Verzele, *J. Chromatog.*, **13**, 377 (1963).
8. M. Verzele, *J. Gas Chromatog.*, **4**, 180 (1966).
9. M. Verzele, *J. Chromatog.*, **19**, 504 (1965).
10. M. Vandewalle and co-workers, unpublished results, this laboratory (1969).
11. M. Anteunis and co-workers, unpublished results, this laboratory (1969).
12. M. Verzele and M. Verstappe, *J. Chromatog.*, **26**, 485 (1967).
13. H. Lammens and M. Verzele, *Bull. Soc. Chim. Belges*, **77**, 497 (1968).

CHAPTER 8

Applications in Flavor Research

C. Merritt, Jr. *Pioneering Research Laboratory. U. S. Army Natick Laboratories, Natick, Massachusetts*

I. Introduction

Gas chromatography (GC) is widely used to separate the very complex mixtures of volatile components encountered in studies of food flavor and aroma. The choice of the method is obvious. The separation in some cases requires the resolution of mixtures containing several hundreds of components, and the components of such mixtures may consist of members of a wide variety of functional group types having in addition a wide range of volatility. The concentration range of the components in the mixtures also may vary several thousandfold.

The difficulties of separation are compounded by attendant difficulties in isolation of the flavor components from the food, and in their subsequent detection and identification. All the problems constitute a formidable task in trace analysis.

Although in preparative-scale analysis one is inclined to think in terms of larger-bore columns and relatively large amounts of sample, the analyst is seldom afforded such luxury in food flavor analysis. Rather, because the components of interest may be present in the food in less than part per billion amounts, only very small amounts are available for analysis, even after isolation from or concentration in the large amount of diluent usually present. Nevertheless, since most analyses require some ancillary method of identification following GC separation, the eluates must provide clean-cut fractions of very high purity. It is this requirement that categorizes most flavor analyses in the domain of preparative-scale GC.

235

Since the isolation of the flavor components, their concentration, and preparation for GC separation can influence the analysis as much as the separation procedures themselves, and since the results are meaningless without consideration of the ultimate identification of the component or its sensory contribution to flavor, these aspects as well as pertinent GC behavior are discussed in this chapter.

Nearly all workers agree that food flavor is somehow related to the nature of the volatile organic components present in great abundance and variety in most foods and that a knowledge of the composition of those components may lead to a means of understanding the nature of the flavor, or of controlling or improving it, or of preventing its deterioration on treatment or storage of the food.

However, widely different methods of analysis have been employed by investigators in their studies of flavor in the various types of food products. In general, two main approaches have emerged. In some cases a knowledge of the composition of all the volatile compounds that can be isolated from the food, i.e., so-called "total volatile analysis," is needed, whereas in other instances the components found in the vapor in equilibrium with the food, i.e., direct vapor or "headspace" analysis, is considered to be related more significantly to the flavor characteristics of the food. Each of the two approaches has its advantages and limitations and in general tends to serve a different purpose.

Since it is assumed in flavor research that only the volatile compounds are significant in assessing the flavor quality of a product, only those components are of interest that have sufficient volatility to provide vapor concentrations above thresholds of sensory response. At present, these quantities are very uncertain, if known at all, and for the most part are relative and dependent upon the various circumstances under which the measurements are made. Total volatile analysis, therefore, is intended to provide a background of the overall composition of the volatile components from which the investigator may subsequently select those that relate to the special problem under study. The description of composition provided by total volatile analysis, however, depends on the methods of isolation and detection employed as well as the physical state and gross composition of the food to be analyzed. The relative amounts of protein, fat, and carbohydrates; the ratio of solids to liquid; the amount of water; and the physical structure of the food all may influence both the nature and amount of the volatile components ultimately identified. It is to overcome these problems that direct vapor analysis is also employed on the assumption that these conditions more closely simulate the circumstances under which flavor and odor are observed. Various drawbacks may be cited in this approach as well, since it is equally difficult to relate headspace composition to sensory response. Frequently, an analysis of both the total volatile constituents and the headspace vapor is performed to provide the investigator with information about the relative quantitative relationships of the flavor components in the food and their possible contribution to sensory response.

In spite of its great significance in flavor research, a discussion of the relationship of the chemical composition of foods to organoleptic response is beyond the scope of this chapter. Rather, the emphasis is upon the nature and quality of the analytical results obtained. The reader who may be interested in the sensory aspects can find the subject well treated in the current literature (1–6).

II. Isolation and Concentration Methods

There are two criteria that are vital to the success of any analysis for flavor compounds. Since the presence of the most minute quantity of a substance may be significant, the procedures employed must neither contribute impurities nor permit the loss of constituents. Even prior to the isolation procedure, the food may require some special preparation such as cutting, grinding, mincing, pressing, or centrifuging in order to convert the product to a suitable physical state. These operations are best performed in closed systems and at cold temperatures, preferably subambient with dry ice or liquid nitrogen, to insure that no loss of volatile components occurs and to repress any chemical or enzymic side reactions that could result in the introduction of artifacts.

Since the product to be analyzed may vary so widely in its initial constitution, the isolation procedure may also involve a great variety of methods. Various forms of extraction, adsorption, etc., may be used as a first step, but since volatility is the ultimate criterion, and a separation of volatile from nonvolatile substances must eventually be achieved, it is customary when possible to start with a distillation.

An interesting study of the choice of isolation techniques has been made by Weurman (3). By reviewing the procedures employed in a random selection of 234 flavor and odor studies conducted during the period from 1960 to 1967, he found that in 78% of them distillation was used as the primary means of isolation. Although a greater use of distillation was anticipated, closer examination of the investigations revealed that a number of studies were concerned with products that had been previously distilled, such as alcoholic beverages, fruit odor concentrates, essential oils, etc., thus making a prior distillation unnecessary. Extraction was used in 60% of the studies, but in many investigations this was a preliminary step which was followed by a distillation to separate the nonvolatiles and remove the solvent.

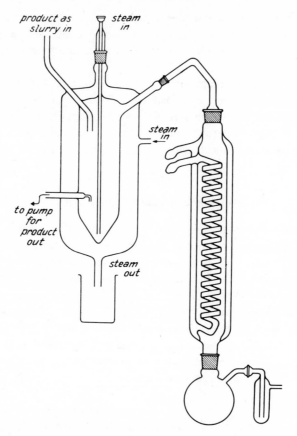

Fig. 8.1. Continuous steam distillation apparatus. [Reprinted with permission from *J. Agr. Food Chem.*, **17**, 375 (1969).]

Distillation procedures can assume many forms (7–9). Most of them can be placed in one of the categories outlined in Table 8.1. The great variety of specific distillation procedures that have been developed cannot be described here. Schematic diagrams of some typical types of apparatus used are shown in Figs. 8.1 through 8.4. Many are cited in current reviews of techniques for flavor and odor research (1–4). There are several important factors, however, that affect the analytical results. Among these are the size of the sample, the temperature, the pressure, whether or not a closed or open system is used, and whether or not steam is used.

The size of the sample and, accordingly, the size of the apparatus employed depends largely on the nature of the product, and probably, in the final analysis, on the amount of flavor compounds needed for detection and

TABLE 8.1

Forms of Distillation Used in Food Flavor Research[a]

Open system, single plate, total takeoff
  At atmospheric pressure
  At reduced pressure ("vacuum distillation") ("stripping")
  High-vacuum degassing and stripping (of fats or high boiling solvents)

Open system, multiplate, fractional distillation
  At atmospheric pressure
  At reduced pressure

Open system steam distillation
  At atmospheric pressure
  At reduced pressure ("vacuum steam distillation")

Open or closed system air or inert gas entrainment

Closed system transfer
  At atmospheric pressure
  At high vacuum ("low temperature sublimation")

[a]Adapted from Weurman (3).

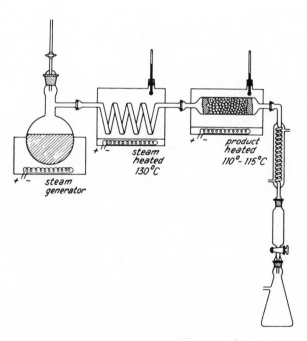

Fig. 8.2. Apparatus or distillation with dry steam. [Reprinted with permission from *J. Agr. Food Chem.*, **17**, 373 (1969).]

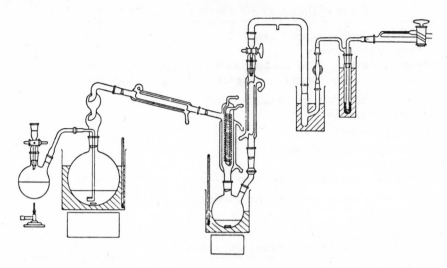

Fig. 8.3. Apparatus or combined reduce pressure steam distillation and simultaneous concentration of the volatiles in the distillate. [Reprinted with permission from *J. Am. Oil Chem. Soc.*, **44**, 572 (1967).]

determination. In practice, flavor investigations have been conducted on material ranging in amount from several tons to a few grams. However, as techniques for isolation, concentration, and separation become more efficient, and detection and identification methods more sensitive, smaller-sized samples can be employed.

The choice between an open or closed system seems to be rather obvious. A closed system precludes any possibility of loss of volatile components. Closed transfer systems, however, although demonstrably very effective, have been manifestly neglected as a means of isolation of flavor compounds. Applications to date have been primarily for relatively small food samples, and although this in itself is a testimonial to its efficacy the main reason closed systems are not widely used is probably because of difficulties in constructing large-scale systems. The advantages of closed system distillation, particularly at low pressure, are described in detail below.

The problems concerned with the choice of the temperature and pressure to employ in a distillation frequently place the investigator in a dilemma. Although the criterion of volatility suggests that the temperature should be high and the pressure low, consideration must be given to the dangers of artifact formation. If the temperature is too high, thermal degradation may occur. In other circumstances, when enzymic reactions may occur, it is perhaps better to use a higher temperature to repress them. In any case, at atmospheric pressures, normal distillations frequently yield a mixture of

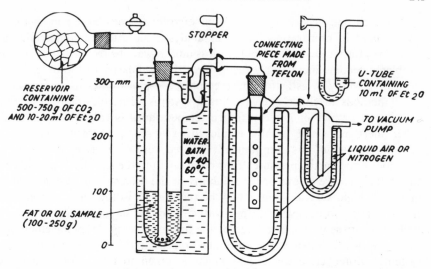

Fig. 8.4. Schematic diagram of carbon dioxide distillation apparatus for removal and collection of flavor compounds from fats and oils. [Reprinted with permission from *Acta Chem. Scand.*, **20**, 2626 (1966).]

volatile components more characteristically associated with cooked food than with the fresh or raw product. It seems evident that reduced pressure is nearly always desirable, and this in turn allows lower temperatures to be used.

A final consideration in the choice of a distillation procedure is whether or not to employ steam. The procedure is usually employed with aqueous products such as beverages, fruit juice, and similar items. With this type of product, steam distillation permits the volume of product being distilled to remain constant and thus prevents any changes in salt concentration, pot temperature, etc., that might be encountered with loss of water. The transport of vapors is enhanced and the distillation may be continued until all the odorous volatile components have been distilled. The main disadvantage appears to be the relatively high temperature employed. Reduced-pressure steam distillations have sometimes been employed when thermal changes may be a problem, but the difficulties of collecting the condensate without a loss of the low-boiling components in the vacuum system make this approach less desirable. In any steam distillation, or even in a simple distillation of an aqueous system, large amounts of water are collected with the flavor compounds in the distillate. Some type of subsequent concentration step must then be used to remove the excess water. Steam is also used frequently to increase the volatility of high-boiling components found in fats and oils, but the introduction of water into the system may cause additional problems in subsequent concentration steps.

Although open system distillation in some form, both large and small, is widely used in preparation for flavor analysis, the attendant uncertainties of artifact formation and component loss must always be reconciled. As these problems become increasingly recognized, the popularity of the procedures decreases. Many investigators have employed various types of gas entrainment methods in either open or closed systems with cold-trapping of the volatiles at dry ice or liquid nitrogen temperatures. Nitrogen, helium, and other inert gases are usually employed. Two basic problems, however, arise in the use of gas entrainment. The recovery of volatiles is mainly a function of the partial pressure of the component in the gas phase, and where the vapor pressure is low, or the solubility in the food solvent system (water or liquid) high, very long times may be required to isolate the components. If not carried to completion, the relative amounts of high-boiling to low-boiling constituents will be altered by the time of gassing. Trapping efficiency is also very critical in a gas entrainment procedure, particularly in open systems. Moreover, extreme care must be taken to avoid contamination from impurities in the flushing gas. A diagram of a closed gas entrainment system that has been used successfully to collect the flavor compounds from a variety of products such as corn oil, apple juice, wine, shrimp, or potato chips is shown in Fig. 8.5. In general, however, gas entrainment is probably

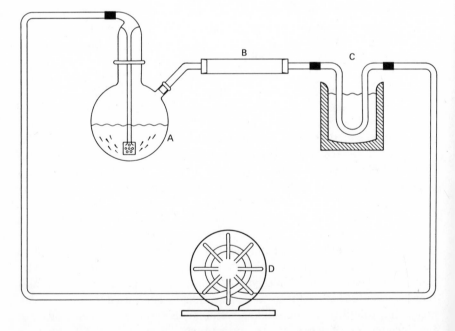

Fig. 8.5. Closed system gas entrainment apparatus for collection of volatile flavor compounds. [Reprinted with permission from *Anal. Chem.*, **32,** 1535 (1960).]

a better procedure to use for collecting samples for direct vapor analysis than for total volatile analysis (10).

Closed system, high-vacuum transfer methods (11–13) have few of the disadvantages associated with conventional extraction, distillation, or gas entrainment methods. No component loss is encountered in the condensate, and recovery is mainly a function of time and temperature. Since very high vacuum is used, temperatures are very mild. For pressures of about 10^{-3} torr, it is seldom necessary to exceed room temperature to recover components having boiling points up to 250°C. For higher boiling compounds (b.p. $\sim$300°C), warming to about 50°C is usually sufficient. Applications of the method to date seem to have been confined entirely to laboratory-scale operations where the amount of food sampled has not exceeded several kilograms. As the advantages are more widely recognized, it may be expected that suitable engineering will provide apparatus to accommodate large amounts of product if needed. In addition to providing a means of isolating volatile organic compounds from the food, high-vacuum transfer at sub-ambient temperatures (sometimes called "vacuum sublimation") can provide a means of separating flavor components from diluents such as water or carbon dioxide.

A schematic diagram of a typical, simple, closed, high-vacuum transfer system is shown in Fig. 8.6. The product is contained in one flask, frozen at liquid nitrogen temperature, and evacuated to remove air and other noncondensable gases. (If the noncondensable gases are needed for analysis, they may be collected in an alternative procedure as described below for the analysis of headspace gas.) The transfer system is then isolated from the vacuum system, the product thawed, and the volatiles condensed in the receiver flask at −196°C. Collection of the volatiles proceeds until the pressure in the system drops to its initial value or until a steady-state pressure is attained. In general, it is usually unnecessary to continue the distillation until the sample is exhausted of its volatile content since the bulk of the volatile material is transferred in a relatively short time. The distillation proceeds rapidly at first and then approaches total transfer asymptotically. If the last residual traces of volatile components are not needed (as in qualitative rather than quantitative analysis), the collection may be terminated when a suitable amount has been collected. An example of the time-volume relationships encountered in a high-vacuum transfer is shown in Fig. 8.7.

The separation of the diluents in the total condensate may be accomplished as shown in Fig. 8.8. The total condensate is cooled to −80°C with dry ice, and the vapor in equilibrium with the solid phase is condensed in a receiver at −196°C. The total condensate is repeatedly thawed and refrozen and the vapor collected when no residual pressure ($\sim$$10^{-3}$ torr) is observed over the solid phase. At −80°C the separation achieved is mainly between water and components that are more volatile than water at −80°C. A distribution

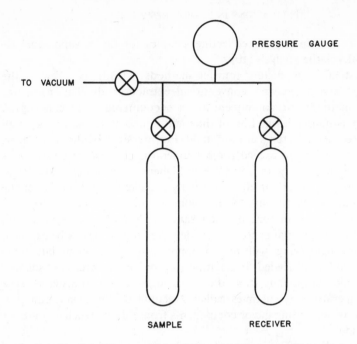

Fig. 8.6. Diagram of closed system vacuum transfer apparatus.

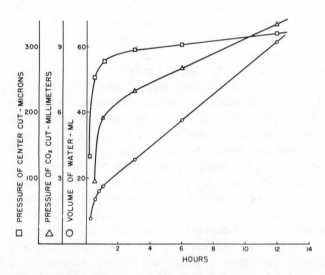

Fig. 8.7. Graph showing effect of collection time on total condensate. Ordinates refer to fractions obtained by procedure outlined in Fig. 8.8. [Reprinted with permission from *J. Agr. Food Chem.*, **7**, 786 (1959).]

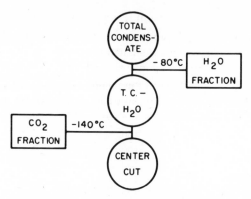

Fig. 8.8. Diagram showing fractionation of total condensate. [Reprinted with permission from *J. Agr. Food Chem.*, **7**, 786 (1959).]

of some typical compounds is shown in Table 8.2. A complete separation cannot be attained for substances having a vapor pressure near that of water, or if the solubility in water reduces the partial pressure of the component in the gas phase. Substances such as toluene or ethanol, therefore, are distributed between the two fractions. This behavior poses no problem, however, in qualitative analysis, and in quantitative work it simply requires that the component be determined in each fraction.

If carbon dioxide is a diluent, a separation from the less volatile compounds can be achieved by a similar low-temperature vacuum sublimation at −140°C. The chromatograms shown in Fig. 8.9 illustrate the separation achieved. It may be noted that a peak for carbon dioxide is not seen in the flame ionization detector (FID) chromatogram of the liquid nitrogen fraction, and for detection and identification of components by GC, or by a coupled mass spectrometer, separation of the carbon dioxide diluent would not be necessary. For other ancillary methods of detection or identification, however, such a separation may be required. The vacuum sublimation method has been applied as a concentration method to separate flavor volatiles from other diluents. For example, Whitfield and Shipton (15) succeeded in removing 95% of the ethanol from the total condensate obtained from frozen peas by vacuum sublimation at −90°C and 5×10^{-4} torr.

Vacuum transfer procedures may be carried out in a variety of equipment, although glassware is mainly employed. The size of the apparatus is determined largely by the volume of the liquid on solid phases to be contained. Condensation of the gases is enhanced by keeping the volume of the traps small. (Extreme caution must be taken, however, that the containers can withstand the pressures that may sometimes be generated upon thawing of a condensate.) A photograph of a typical vacuum transfer manifold showing a setup for both a collection and a separation is shown in Fig. 8.10.

TABLE 8.2

Distribution of Components in an Aqueous Total Condensate
after Vacuum Sublimation at −80°C

Component	Percent in liquid nitrogen trap	Percent in dry ice trap
Hydrogen sulfide	100	0
Methanethiol	100	0
Ethanal	100	0
Propanal	100	0
Diethyl ether	100	0
2-Propanone	100	0
Pentane	100	0
Hexane	100	0
Benzene	100	0
Heptane	98.4	1.6
Ethanol	59.1	40.9
Toluene	51.8	48.2
Octane	17.9	82.1
Decane	0	100
Hexanal	0	100
2-Heptanone	0	100
2-Nonanone	0	100
Benzaldehyde	0	100
Water	0	100

It is frequently desired to analyze the headspace vapor without altering its composition. The most direct way to do this is by sampling the headspace with a syringe (16–20). The concentration of flavor components, however, invariably is very dilute because of the large volume of air present. The closed vacuum transfer system allows the headspace to be sampled, the diluent to be removed, and if desired a total condensate to be collected after headspace analysis. A typical sampling apparatus is shown in Fig. 8.11. The product is contained in a sealed flask during treatment and/or storage. When the volatile components are to be analyzed, the flask is attached to a vacuum manifold with the appropriate collection flasks or tubes. The entire system is evacuated to about 10^{-4} torr and closed off from the vacuum system. The coiled tubes (1A, 1B, and 1C) in which headspace vapor is to be collected are open to the manifold; the U-tube is sealed off. The break seal is now ruptured by means of a small iron bar controlled with a magnet, and the headspace gas expands into the coiled tubes to provide three aliquot samples.

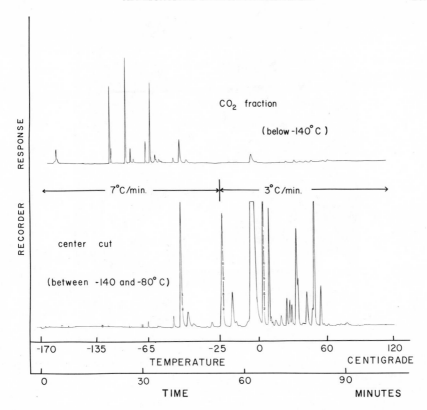

Fig. 8.9. Programmed cryogenic temperature gas chromatograms of carbon dioxide and center cut fractions from irradiated beef volatiles. [Reprinted with permission from *Anal. Chem.*, **36**, 1506 (1964).]

The headspace sample may be swept onto a gas chromatography column with carrier gas, or it may be fractionated prior to separation as described above to remove diluents such as air, carbon dioxide, or water by an appropriate choice of sublimation temperature. When the headspace sample is cooled to −196°C with liquid nitrogen, an analysis of the noncondensable gases can be obtained. Following analysis of the headspace vapor, the total condensate may be collected in the U-tube (no. *1*) and fractionated and analyzed in the usual manner. If aliquots of the total condensate are desired, they also may be prepared by splitting the sample in the gas phase. Details of these procedures are described in the original papers.

In some instances flavor volatiles have been adsorbed on charcoal (21–23) or some other adsorbent following distillation or gas entrainment rather than condensed in cold traps. Probably fewer components are lost through

Fig. 8.10. Photograph of vacuum transfer apparatus (left) and vacuum sublimation apparatus (right). [Reprinted with permission from *J. Agr. Food Chem.*, 7, 785 (1959).]

inefficient trapping when adsorption is used, but it is difficult to obtain adsorbents that do not carry trace impurities. Recovery from the adsorbent by desorption of the vapors is uncertain and solvent extraction again introduces the probability of contamination.

A comparison of various methods for isolation of flavor compounds was conducted by Heinz et al. (21). The chromatograms they obtained of their various essences are shown in Fig. 8.12. It is evident that different methods give different results. The problem for the investigator is to decide what is significant. The fact that charcoal adsorption apparently yields more components hardly means that it is a superior procedure. Many of the compounds may result from artifacts.

In nearly all the isolation methods used, various diluents are collected along with the desired flavor compounds. Some means of concentrating the flavor compounds is usually needed before separation can be undertaken. Distillation, freeze concentration, and absorption have been the methods most commonly employed. Concentration by distillation has taken the form mainly of flash evaporators or "strippers" in order to avoid excessive heat as

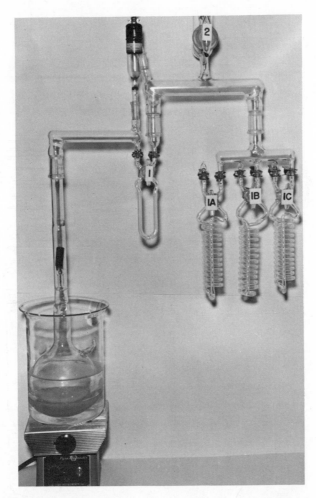

Fig. 8.11. Collection apparatus showing vacuum distillation of butter fat volatiles. [Reprinted with permission from *J. Am. Oil Chem. Soc.*, **44**, 26 (1967).]

the solvent or diluent is removed. Examples of two such devices are shown in Figs. 8.13 and 8.14.

Freeze concentration (24–27) and zone melting methods (28) have been developed which some workers believe offer a most effective way to concentrate the volatiles without introducing problems of component loss, contamination, or artifact formation, but manipulation is tedious and painstaking and the methods have been used infrequently.

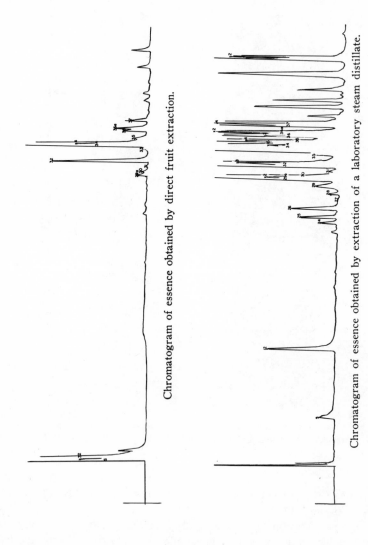

Chromatogram of essence obtained by direct fruit extraction.

Chromatogram of essence obtained by extraction of a laboratory steam distillate.

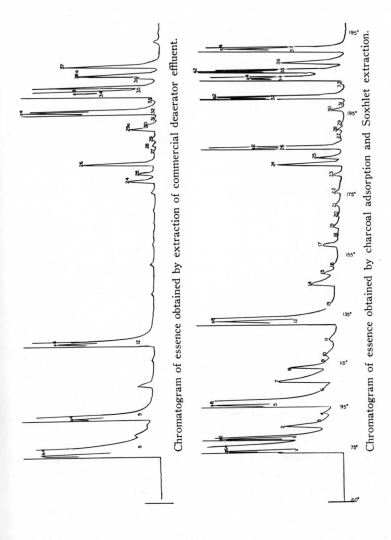

Chromatogram of essence obtained by extraction of commercial deaerator effluent.

Chromatogram of essence obtained by charcoal adsorption and Soxhlet extraction.

Fig. 8.12. Chromatograms of pear essences obtained by various isolation procedures. [Reprinted with permission from *J. Food Sci.*, **31**, 63 (1966).]

251

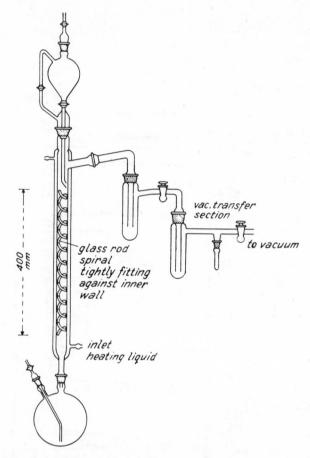

Fig. 8.13. Apparatus for stripping of volatiles from high-boiling solvents, fats, and oils.
[Reprinted with permission from *J. Agr. Food Chem.*, **17**, 377 (1969).]

Adsorption has also been used as a means of concentrating the volatiles. The procedure seems to be most effective when the adsorbent is the stationary phase of a chromatographic column contained in a small cold trap (29–31). The trap can then be used subsequently as the inlet for the gas chromatographic column. When used in this way the trap, in effect, becomes a part of the GC column when elution is begun. The trap can be warmed to enhance elution time and to minimize band spreading on the column. An improved manner of conducting this type of concentration-elution process using programmed subambient temperature chromatography is described below.

The amount of sample preparation required to prepare a sample for GC separation can vary from a single tapping of the headspace vapor with a syringe to the complex isolation and separation scheme illustrated in Fig. 8.15. The magnitude of the variety of procedures that may be employed can be appreciated by consideration of Weurman's (3) survey. In the 234 studies reviewed, 425 procedures, most of them different, were found to have been applied. How much preparation and the type of procedures to use in isolating flavor compounds are among the most challenging decisions the flavor chemist

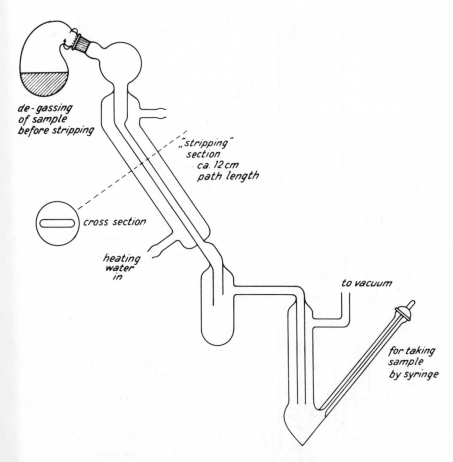

de-gassing of sample before stripping

"stripping" section
ca. 12 cm path length

cross section

heating water in

to vacuum

for taking sample by syringe

Fig. 8.14. Apparatus for stripping of volatiles from high-boiling solvents, fats, and oils. [Reprinted with permission from *J. Agr. Food Chem.*, **17**, 378 (1969).]

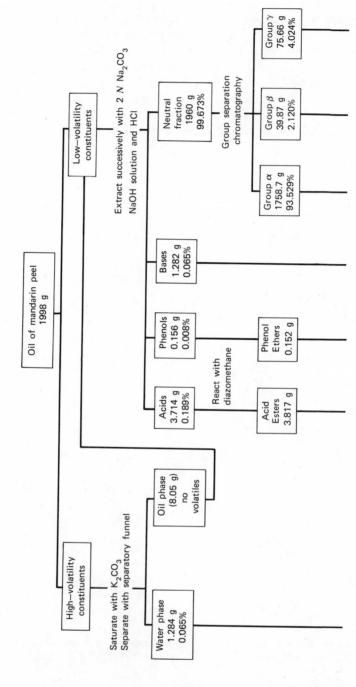

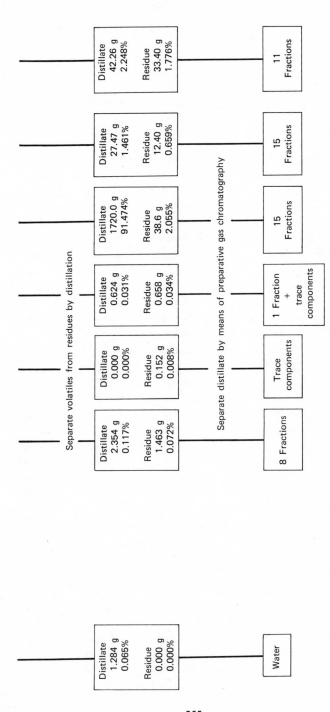

Fig. 8.15. Diagram of separation steps employed in the isolation of flavor volatiles from a citrus fruit. Adapted from Kugler and Kovats (32).

must make. His ingenuity and initiative in this regard can many times far outweigh the choices he may make in completing the separation and identi-fication of the components.

III. Gas Chromatographic Separation

The second step in the analysis of a food flavor or aroma is to separate the components of the mixtures isolated in the various ways that have been described in Section I. In modern flavor research and technology, these separations are accomplished almost exclusively by GC. Most of the fundamental aspects of GC separation that apply to the separation of food flavor compounds are discussed elsewhere in this book or in the many other excellent treatises on the subject. A particularly relevant treatment of the subject has been given by Teranishi et al. (1). This section deals only with certain specific or special considerations pertinent to flavor studies.

The choice of chromatography column type and the various operating parameters with which it is used are the main concerns of the flavor chemist. Since in nearly all instances the mixtures encountered are too complex to permit identification of the components without using ancillary methods, high-resolution separation must be achieved. The wide range of component concentration encountered requires that the detectors be sensitive to minute traces of component, while the columns accommodate relatively large sample size. Moreover, a wide range of compound types ranging from very polar to nonpolar and encompassing all the functional group classes are found in the mixtures to be separated. Efforts to relate selectivity characteristics of stationary phases to separability are therefore usually unrewarding. Un-fortunately, it appears that the choice of a suitable column can be made more successfully on the basis of experience than on the basis of principle.

Some practical precepts have been developed, however. The choice of column type, for example, is overwhelmingly in favor of the open tubular column, because of its superior resolving power. The relative advantage is illustrated in Fig. 8.16, which shows a comparison of packed and open tubular column chromatograms of identical samples of headspace vapor. The disadvantages of the small sample size accommodated by the 0.01-in.-wall coated column have been overcome in current column technology by the use of wide-bore columns (34,35) 0.03 and 0.04 in. (i.d.) in diameter and several hundred to a thousand feet in length. Such columns provide a separability completely comparable to the performance of smaller-bore columns and accommodate sample sizes of the same magnitude as $\frac{1}{8}$-in. diameter (0.04 in. i.d.) packed columns; see Table 8.3. A disadvantage of the wide-bore open tubular column, perhaps, is the very long time required for component elution. From 30 to 45 min, depending on flow rate, are frequently needed to elute a nonretentive gas from a 1000-ft wide-bore column and the chromatograms may require several hours for completion. Nevertheless, this is frequently a small sacrifice for the excellent separation achieved.

TABLE 8.3

Sample Load per Peak[a]

	Column i.d., in.	Load mg
Open tubular (wall coated)	0.01	0.005
	0.02	0.025
	0.03	0.25
Open tubular (support coated)	0.02	0.25
Packed	0.04 (1/8 in. o.d.)	0.25
	0.1	0.5
	0.2	1.0
	0.5	50.0

[a]Adapted from Teranishi et al. (1).

Other column types, such as the support-coated open tubular (SCOT) column (36) and the multichannel open tubular column (37), likewise provide high resolution with large sample capacity. The SCOT column, in particular, is proving to be very efficacious because its compact size, usually 0.02 in. i.d. X 50 ft, imposes very few restrictions on the control of operating parameters such as temperature or carrier gas flow rate. A typical separation achieved on a 0.02 in. X 50 ft SCOT column is shown in Fig. 8.17.

Another column type that provides good separability with large sample capacity is the high-efficiency packed column (38). This column, long in length, narrow in diameter, packed with uniformly graded supports and lightly loaded with stationary phase, exhibits performance characteristics similar to the wide-bore open tubular type. A comparison of a 0.1 in. X 60 ft high-efficiency packed column with a 0.03 in. X 1000 ft open tubular column is shown in Fig. 8.18. The elution time of the components is much less on the packed column than on the open tubular column. Very high inlet pressures (~200 psi) are required by the packed column, thus creating a problem with the introduction of samples by any method except syringe injection. The higher carrier gas flow rates used also mitigate against the use of the column in combined GC–MS arrangements.

In comparing packed with capillary columns, the significance of the number of theoretical plates is often misunderstood, since it is frequently inferred

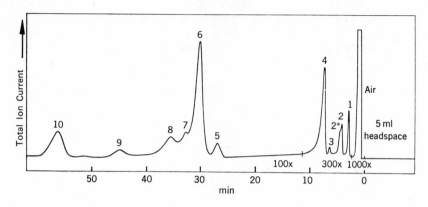

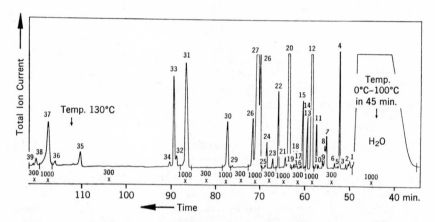

Fig. 8.16. Comparison of separations obtained with packed and open tubular columns. Sample: 5 ml of headspace gas over peppermint tea. *a*, Packed column, 3.5 m × 4 mm, 25% LAC 1-R-296 on 60–80 mesh Chromosorb W. 70°C ml/min He; *b*, capillary column, 50 m × 0.25 mm, polypropylene glycol; temperature programmed, 0.8 ml/min He. Adapted from deBrauw and Brunnee (33).

that the capillary is the better column because it has the greater number of theoretical plates. In practice, the flavor chemist is more concerned with resolution than efficiency, however. The resolution on a packed column may be as good as the resolution on a capillary with many more times the number of theoretical plates. The most difficult problem is in the preparation of the column, and it is generally easier to construct highly efficient capillary columns with good separability than it is to make equally good packed columns. One major advantage of high-efficiency columns is the sharpness of the peaks obtained. Small peaks, attributable to trace quantities in the

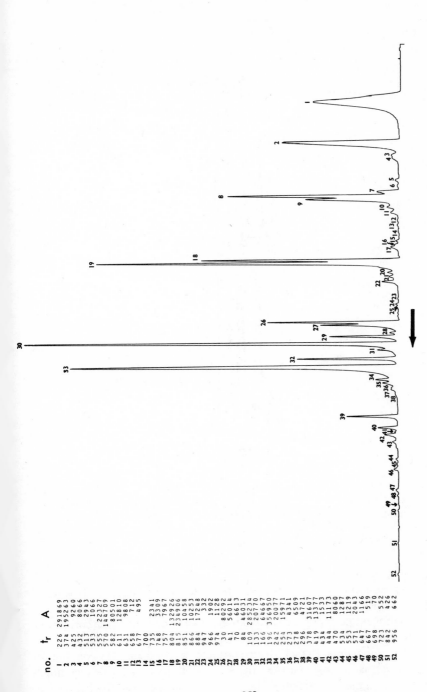

no.	t_r	A
1	226	291869
2	374	195263
3	425	9260
4	432	80066
5	513	2943
6	533	10066
7	555	22327
8	570	147709
9	582	80501
10	611	128010
11	658	9808
12	677	9712
13	677	495
14	700	
15	735	2341
16	748	3309
17	757	7967
18	804	132926
19	815	349060
20	851	10458
21	866	10253
22	884	17248
23	947	3332
24	966	1902
25	970	1128
26	41	80292
27	70	56014
28	70	66613
29	84	460031
30	109	285234
31	133	207770
32	166	646667
33	196	356950
34	242	200077
35	254	15971
36	273	43341
37	278	6509
38	296	4721
39	378	316607
40	419	133733
41	434	11533
42	440	11173
43	470	80060
44	534	1287
45	553	1219
46	571	2043
47	667	1166
48	698	519
49	698	70
50	723	552
51	842	426
52	956	662

259

Fig. 8.17. Chromatogram of volatile components isolated from chicken flesh. 0.02 in. × 50 ft SCOT column coated with tris. Temperature programmed from −80 to 125°C at 5°C/min. Printout of digital peak integrator shown in inset at left.

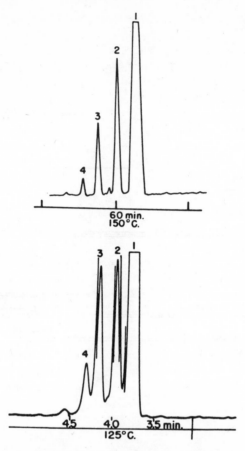

Fig. 8.18. Separation of isopulegol stereoisomers on: *a*, 0.03 in. × 1000 ft open tubular column coated with SF 96(50); *b*, 0.1 in. × 60 ft packed with 1% SF 96(50) on HMDS-treated Chromosorb G, 100–120-mesh, air particle-selected. [Reprinted with permission from *Advances in Gas Chromatography*, A. Zlatkis, Ed., Preston Technical Abstracts Co., Evanston, Illinois, 1967, p. 30].

mixture, are not only more readily detected, but are also more easily trapped for subsequent identification procedures.

The component peaks of the chromatogram shown in Fig. 8.17 correspond mainly to pure components. Many represent mixtures of two, three, or perhaps more components. In fact, the literature of flavor chemistry consistently demonstrates that a single chromatogram can hardly ever achieve complete separation of all the components. The chromatograms of volatiles from meats (39,40), fruits (41,42), beverages (43,44), coffee (45) and a host of other products frequently display 100 or more peaks and many times show as many as 300 or 400. It is probably safe to say that the composition of none of the mixtures has been completely elucidated. Moreover, it cannot reasonably be expected that such mixtures can be resolved on a single chromatography column. Some preliminary separation may be necessary and in most cases is desirable.

Although chemical procedures (46,47) and liquid column methods (48–50) have been used for preliminary separation, the packed gas chromatography column can be used to advantage (51). Packed columns can range in diameter from 0.1 in. to 1 in. or more, but in flavor studies the column diameters employed are usually at the small end of the range. Columns of $\frac{1}{8}$-in. diameter are the most commonly used. They are easy to prepare, provide moderately good resolution in convenient column lengths, and generally have sufficient capacity for the usual sample sizes encountered in flavor analyses.

The choice of column size is governed mainly by practical considerations. For preliminary separation of relatively large fractions, the packed column is ideal, and since capacity is the prime consideration size is determined mainly by the size of the sample to be processed. As the requirement for resolution becomes greater, the more likely will be the selection of an open tubular or high-efficiency packed column (0.1-in. i.d.). There is no particular advantage to the so called "packed capillaries," i.e., packed columns of 0.04-in. i.d. or less (52), since the capacities are actually less than the widebore open tubular types and the pressure drop is so high that it is difficult to prepare columns with a large number of theoretical plates.

The choice of an open tubular column is likewise a matter of sample size. The capacity of an 0.01-in. column is so low that a sample splitter is usually needed to prevent overloading of the column. The columns are frequently useful, however, for screening samples, selecting a stationary phase, and estimating the composition of the mixture. For final preparative separation of pure components, the task is most likely to be delegated to a wide-bore, open tubular column of 0.03 or 0.04 in. diameter or a 0.02-in. SCOT column.

Although the dimensions of the column may greatly influence capacity and efficiency, separability is attributable mainly to the nature of the stationary liquid phase. Few if any recommendations can be given in the choice of the liquid phase. The great number of stationary phases commercially available tends primarily to confound the analyst rather than to simplify his problem. Since any given mixture of volatiles from a food may likely consist of components ranging from very polar to nonpolar, the question arises whether to employ a polar or nonpolar liquid phase, or perhaps one of intermediate polarity. The proposed systematic polarity classification such as that of Rohrscheider (53) contribute little to the evaluation of liquid phases for mixtures of components with a range of polarity. Mon et al. (54,55) have devised an empirical approach which is simple enough to apply in practice yet provides some insight into the choice of a column. A chromatogram of a mixture of about six compounds each having a different functionality and different polarity is obtained for each column to be tested. Theoretical plates are calculated to estimate column efficiency, and selectivity is evaluated from the retention times relative to decane, which is one of the components. Typical results for certain liquid phases are shown in Table 8.4. Although no abolute predictions can be made from such measurements, indications of a preferred selection can be given. A column showing good to moderate efficiency for all types and a wide spread of relative retention times is thus probably superior to one that may demonstrate, for example, poor efficiency for a very polar type or a narrow range of relative retention times.

TABLE 8.4

Relative Retention Times of Test Mixture Components
on Various Stationary Phases[a]

	ApL	96(50)	20 M	T-20	X-305
n-Decane	1.0	1.0	1.0	1.0	1.0
Limonene	1.1	1.2	3.0	2.9	4.0
n-Amyl acetate	0.38	0.53	2.2	2.0	3.0
n-Octanal	0.58	0.94	4.2	4.3	5.6
n-Hexanol	0.22	0.44	4.5	4.9	14.0

[a]Adapted from Teranishi et al. (1).

Studies with the polarity test mixture have revealed considerable difference in the behavior of certain liquid phases in packed columns and in open tubular columns. These differences seem to be more pronounced wth nonpolar- than polar-type liquids. As shown in Fig. 8.19. another effect frequently observed is severe tailing of certain components. The behavior is more commonly observed with polar components and is believed to be attributable to adsorption phenomena. The effect can be reduced by adding a small amount of a surface-active agent when coating the column. Columns coated in this way are found to provide sharper peaks with less tailing.

Regardless of the results of column evaluation by test mixtures, the ultimate criterion of column performance is with the mixture to be separated. Some investigators prefer to use a pilot column for evaluation, i.e., a small column such as 0.01 in. × 50 ft coated with the same stationary phase as the

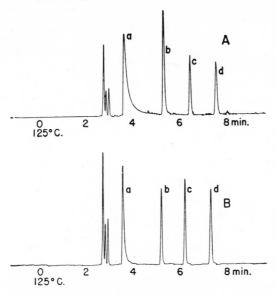

Fig. 8.19. Comparison of chromatograms on column coated with the addition of a surface-active agent. *A*, 1% Ipegal, 99% Apiezon C; *B*, 5% Ipegal, 95% Apiezon C. Components: *a*, *n*-hexanol, *b*, *n*-octanal, *c*, limonene. [Reprinted with permission from *J. Gas Chromatog.*, **5**, 498 (1965).]

larger column to be used for final separation. Others rely heavily on experience. For many mixtures of flavor components having a wide range of polarity, polar liquids are found to provide greater separability than nonpolar liquids. In Fig. 8.20 the separation of a mixture of volatiles consisting mainly of components from a homologous series of aliphatic hydrocarbons, aldehydes, ketones, esters, alcohols, mercaptans, and sulfides is shown to be more effective on β,β'-oxydipropionitrile (OPN) than on squalane. This is not an isolated case. In our laboratory many of the polar phases such as OPN, *TRIS*, and Carbowax 20M, have proved superior to nonpolar stationary phases such as Apiezon and the various silicone oils for separation of volatiles such as meats, eggs, coffee, cheese, fish, and vegetables.

It is important to remember, however, that few if any mixtures can be completely separated by a single column. Since the selectivity of a polar liquid usually differs greatly from that of a nonpolar liquid, a separation performed on a second column of opposite polarity from the first often yields an overall improvement in the separation of the mixture.

The physical properties of the liquid phase may also be important. On open tubular columns in particular the efficiency of the column depends on the diffusivity of the liquid phase which in turn is a function of its viscosity.

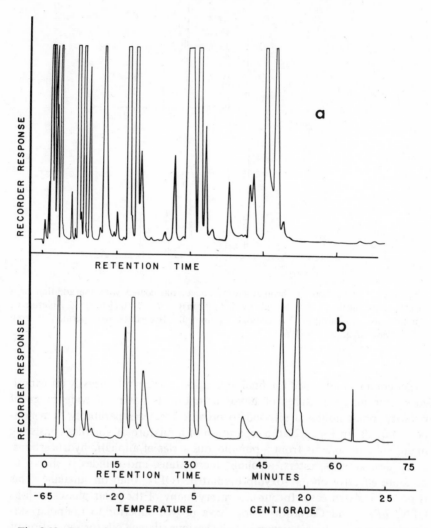

Fig. 8.20. Comparison of separation of mixture of flavor volatiles on: a) $\beta\beta'$-oxydipro-pionitrile, b) squalane. 5% liquid phase on 80–100-mesh Chromosorb W—$\frac{1}{8}$ in. $\times$ 10 ft packed columns. Temperature programmed from $-65°C$ to room temperature at 2°C/min.

The viscosity of liquids is low at elevated temperatures and no problems are encountered with columns that are heated. The viscosity increases near the melting point, however, decreasing the diffusivity and correspondingly the efficiency of the column. This behavior may be troublesome, particularly in temperature-programmed operations of the column. Liquids of low viscosity should be used whenever possible.

Other significant properties of the liquid are vapor pressure and thermal stability. Since the separated components may ultimately have to be collected for identification, contamination by the liquid phase, if volatile, or by decomposition products, if labile, must be avoided.

Because of the wide range of volatility found in mixtures of flavor volatiles, temperature programming of the column is almost always used. In flavor studies the boiling range of components can vary from below room temperature to 250°C or above. To accommodate such samples, the temperature must be programmed from below ambient (14,56) to the upper temperature limits of the column. The effect of such wide-range programming is seen in Fig. 8.21. If programming starts at ambient, low-boiling components

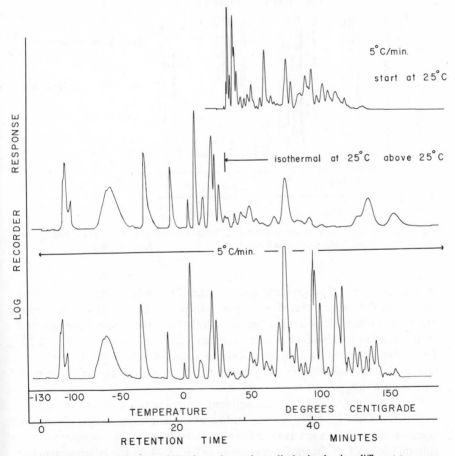

Fig. 8.21. Chromatograms of "hydrocarbon mixture" obtained using different temperature programs. Column: ⅛ in. × 10 ft, 5% SE 30 on 80–100-mesh Chromosorb W. Log recorder response used to provide attenuation for more abundant components. [Reprinted with permission from *Anal. Chem.*, **36**, 1502 (1964).]

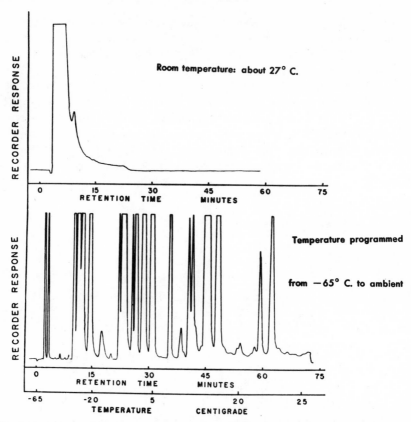

Fig. 8.22. GC separation of the volatile components isolated from ground coffee beans. [Reprinted with permission from *Anal. Chem.*, **35**, 112 (1963).]

are not separated; if programming stops at ambient, high-boiling components are not eluted or are eluted with poor efficiency. Separation is optimum when the starting temperature is below $-100°C$ and the temperature elevated during elution to 150°C or above. Another example is shown in Fig. 8.22. The chromatograms for a mixture of coffee volatiles eluted at room temperature and with subambient programming show clearly the improvement achieved by subambient programming.

The apparatus required for subambient programming can be very simple as illustrated in Fig. 8.23, or may utilize automatic programmers with the injection of coolants such as liquid carbon dioxide or liquid nitrogen into the column chamber as shown schematically in Fig. 8.24. In the latter system the column may be heated as well as cooled, thus allowing programming to take

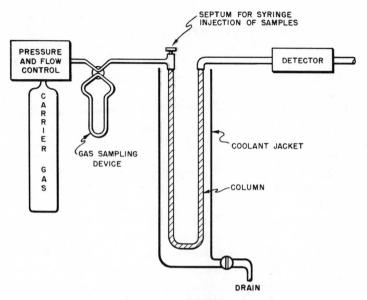

Fig. 8.23. Apparatus for subambient programmed temperature GC. [Reprinted with permission from *Anal. Chem.*, **35,** 111 (1963).]

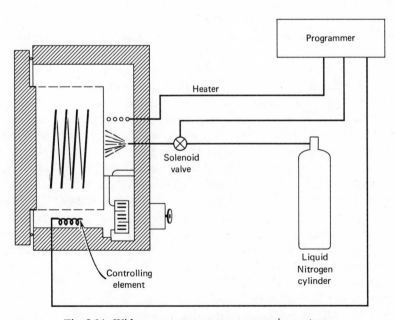

Fig. 8.24. Wide-range temperature-programming system.

place over a wide range of temperature. With the automatic system, programming rates may be selected over a wide range also, but in general the best separations are obtained when the programming rate is very slow. The rate of temperature rise is optimum in the range 1–3°C/min, and in most cases should not exceed 5°C/min.

Subambient programmed temperature GC is equally effective with both polar and nonpolar columns. Altenau et al. (57) have shown that separation is attributable to gas-solid adsorption phenomena below the melting point of the liquid phase. There is a decrease in efficiency in the region of the melting point of the liquid phase, but this is minimal if liquids are used that have sharp melting points or whose solid/liquid transition occurs over a narrow range of temperature.

The choice of starting temperature for a subambient program depends on the volatility of the components in the mixture. In principle, there is a temperature for every compound below which the compound remains adsorbed on the column and does not elute. If the starting temperature is above that temperature, separation will not be optimum. This is shown in Fig. 8.25. Conversely, if the temperature is maintained below the elution temperature of the component, it will be held at the head of the column. This behavior can be utilized to great advantage in headspace analysis.

If a sample of headspace is injected onto the column at a temperature below the elution temperature of the components but above that of the diluent, the components are condensed on the head of the column while the diluent is eluted. Air and other noncondensable gases may thus be separated from flavor volatiles. Moreover, flavor components in a dilute air sample can be concentrated at the column head from repeated injections as the diluent in each portion is eluted. When a sufficiently larger amount of flavor components have been accumulated at the column head, the program may be initiated and elution of the components proceeds as the temperature rises. The method has been used very effectively in the analysis of cigarette smoke (58), tea and coffee headspace (59), and banana volatiles (60).

When gaseous samples are flushed onto a cold column with carrier gas, the condensation of components results in a very effective method of so-called "narrow band–on column" injection. Chromatograms obtained from the subsequent programmed temperature elution of the components show a minimum of band spreading and characteristically have very sharp peaks.

A modification of the procedure for sample introduction onto cold columns has been used very successfully for separation of solvent extracts of dilute aqueous flavor concentrates (12). Because of its high boiling point and strong polarity, water is frequently troublesome as a diluent in mixtures to be separated by GC. It is seldom eluted as a narrow band (except from special

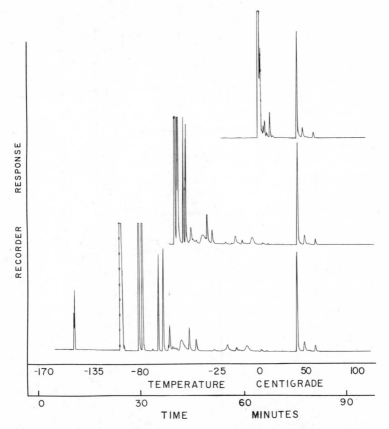

Fig. 8.25. Programmed temperature chromatograms of laboratory gas starting at different temperatures. Column: 5% tris on 60–80-mesh firebrick. Temperature program. ~5°C/min. Aerograph, F & M apparatus. [Reprinted with permission from *Anal. Chem.*, **36,** 1505 (1964).]

columns) and usually is found to elute in the middle of the range of components being separated. An example is shown in Fig. 8.26. If the volatiles are extracted from the water with an organic solvent, and the organic solvent subsequently removed, separation of the volatiles is greatly simplified.

If an aqueous solution of flavor volatiles is extracted directly with a low-boiling organic solvent, followed by evaporation of the organic solvent, the more highly volatile components of the flavor mixture may be lost. The problem is avoided by employing a prior distillation by means of closed system high-vacuum sublimation. Thus the initial aqueous solution is distilled at −80°C, producing a volatile fraction and a water fraction (Fig. 8.8).

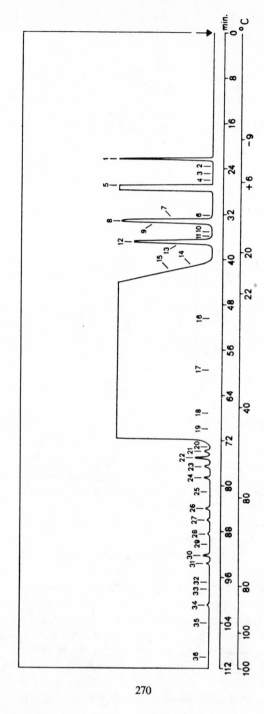

Fig. 8.26. Programmed temperature chromatogram of tea headspace vapors on an open tubular column. [Reprinted with permission from *J. Gas Chromatog.*, **4**, 397 (1966).]

270

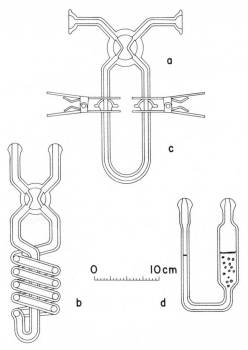

Fig. 8.27. Various gas chromatographic inlet devices. *a*, Section through inlet by-pass valve; *b*, spiral trap for introduction of large volumes of gases; *c*, capillary U-tube for introduction of known weights or volumes of liquids or solids; *d*, bulb U-trap for introduction of concentrated solvent extracts of flavors. [Reprinted with permission from *J. Gas Chromatog.*, **2**, 134 (1964).]

After collection the volatile fraction is vaporized and introduced onto the GC column by flushing from the trap in which it is collected (61). Examples are shown in Figs. 8.27*b* and 8.27*c*. The volatile fraction is separated in the usual way by subambient temperature GC.

The water fraction is extracted with a low-boiling organic solvent such as ether or isopentane. The extract is then reduced in volume by evaporation at −80°C, which is chosen as the temperature of evaporation because all the components of the mixture having sufficient vapor pressure to evaporate at −80°C have been previously distilled and collected in the volatile fraction. Only solvent is removed and no loss of components occurs. When the volume of solvent has been reduced to a small amount, e.g., a milliliter or less, the organic solution is transferred to an ebullition tube such as that shown in Fig. 8.27*d*, and the tube is attached to the chromatograph inlet. The solution is volatized by passing the carrier gas through the solution, allowing the

vapors to pass to the column which is maintained at an appropriate subambient temperature as in headspace sampling. The components are condensed on the column head, and the solvent diluent is eluted from the column.
When the solvent has been removed, chromatographic separation of the
components is initiated by starting the temperature program.

The results of the separation are a mixture of an homologous series of
ketones extracted with isopentane from water as shown in Fig. 8.28. In this
case the temperature program was begun when evaporation of the solution
was started. Solvent was eluted until the temperature reached 25°C. Subsequently, at 50°C elution of the ketones began and continued until the last
member eluted at ∼160°C.

Another example is shown in Fig. 8.29. The sample is the water fraction
from a mixture of volatiles from irradiated ground beef. The volatiles were
extracted with ether and the ether solution was concentrated at −80°C.
The residual solution was vaporized with carrier gas onto the column at
−10°C until all the ether was eluted. In this case the column was attached
to the inlet of a fast scanning mass spectrometer so that identification could
be made from the mass spectrum of the eluting component. When all the
ether had eluted, the temperature program was started and the components

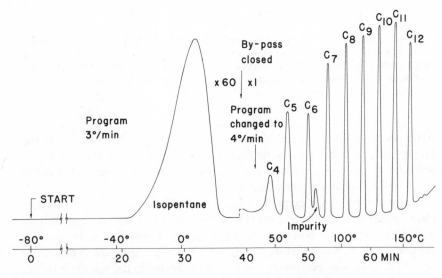

Fig. 8.28. Analysis of a dilute solvent solution of C_{3-12} methyl ketones. 6 ft, 0.25-in. o.d.
packed column containing 10 wt % stationary phase on Chromosorb W, 60–80 mesh.
Injection inlet temperature 200°C; column at −80°C during entrainment of solvent by
carrier gas and distillation of residual ketones onto column, then temperature programmed.
Sample: 2 μl ketone mixture in 1 ml solvent. [Reprinted with permission from *J. Gas
Chromatog.*, **2**, 135 (1964).]

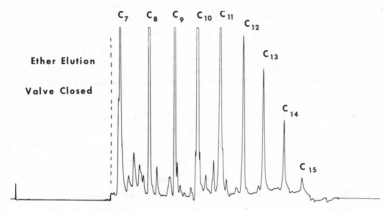

Fig. 8.29. Programmed temperature gas chromatogram of separated components from ether extract of irradiated beef "water fraction." [Reprinted with permission from *Flavor Chemistry*, Advances in Chemistry, Series No. 56, American Chemical Society, Washington, D.C., 1965, p. 232.]

of the mixtures were subsequently eluted. Analysis of the mixture showed the sample to consist primarily of a series of aliphatic hydrocarbons and smaller amounts of miscellaneous other types. Hydrocarbons of seven carbon atoms or more were found in the water fraction. Hydrocarbons with six or fewer carbon atoms were found in the volatile fraction.

IV. Identification of Components

The final step in the study of flavors is the identification and determination of the components. In flavor analysis this nearly always must be accomplished through the use of ancillary methods. GC retention data is usually inadequate for identification even when two or more columns are used. Peak areas may provide quantitative data about the purity of peaks needed to interpret the results.

Ideally, the analysis may be performed in a single unified system in which separation is achieved on a gas chromatograph column and analysis is accomplished by means of a directly coupled mass spectrometer. The efficacy of combined GC–MS analysis is well known and has been described by many authors (60,62,63). For many reasons, however, other ancillary methods may also have to be used. IR, NMR, optical rotary dispersion, and a host of other methods as well as sensory evaluations may be needed before all the conclusions can be derived.

One of the most important requirements for success in using these methods is the isolation and collection of the separated components. Techniques for these purposes are as wide and varied as the methods for which the samples are to be collected. Some of these are described in Chapter 4. Others, particularly those developed for flavor analysis using the column technology described in this chapter, have been given in an excellent treatment of the subject by Teranishi, et al. (1). Excellent treatises (64,65) are also available covering the application of the various methods used in the ultimate analysis. The tremendously broad scope of these methods precludes any adequate treatment of their application here. The flavor chemist is indeed fortunate, however, that his efforts to understand the behavior that nature has hidden in great complexity is made easier by the vast array of methods and techniques now at his disposal. His success will be commensurate with the initiative, ingenuity and intelligence used in their application.

References

1. R. Teranishi, I. Hornstein, P. Issenberg, and E. L. Wick, *Flavor Research: Principles and Techniques*, Marcel Dekker, New York, in press, 1970.
2. P. Issenberg and I. Hornstein, in *Advances in Chromatography*, Vol. 9, R. A. Keller, Ed., Marcel Dekker, New York, 1970, p. 295.
3. C. Weurman, *J. Agr. Food Chem.*, **17**, 370 (1969).
4. D. A. Forss, *Food Prod. Develop.*, **3**, 78 (1969).
5. M. A. Amerine, R. M. Pangborn, and E. B. Roessler, *Principles of Sensory Evaluation of Food*, Academic Press, New York, 1965.
6. D. C. Guadagni, R. G. Buttery, and S. Okano, *J. Sci. Food Agr.*, **14**, 761 (1963).
7. J. A. J. M. Lemli, Dissertation, University of Groningen, Netherlands, 1955.
8. D. A. Forss and G. L. Holloway, *J. Am. Oil Chem.*, *Soc.* **44**, 572 (1967).
9. E. Honkanen and P. Karvonen, *Acta Chem. Scand.*, **20**, 2626 (1966).
10. W. W. Nawar and I. S. Fagerson, *Anal. Chem.*, **32**, 1534 (1960).
11. C. Merritt, Jr., S. R. Bresnick, M. L. Bazinet, J. T. Walsh, and P. Angelini, *J. Agr. Food Chem.*, **7**, 784 (1959).
12. C. Merritt, Jr., P. Angelini, M. L. Bazinet, and D. J. McAdoo, in *Flavor Chemistry*, Advances in Chemistry Series No. 56, R. F. Gould, Ed., American Chemical Society, Washington, D.C., 1966, p. 225.
13. P. Angelini, D. A. Forss, M. L. Bazinet, and C. Merritt, Jr., *J. Am. Oil Chem. Soc.*, **44**, 26 (1967).
14. C. Merritt, Jr., J. T. Walsh, D. A. Forss, P. Angelini, and S. M. Swift, *Anal. Chem.*, **36**, 1502 (1964).
15. F. B. Whitfield and J. Shipton, *J. Chem. Ind.* (*London*), **83**, 1038 (1966).
16. R. G. Buttery and R. Teranishi, *Anal. Chem.*, **33**, 1441 (1961).
17. C. Weurman, *Food Technol.*, **15**, 531 (1961).
18. W. W. Nawar and I. S. Fagerson, *Food Technol.*, **16**, 107 (1962).
19. R. Teranishi, R. G. Buttery, and R. F. Lundin, *Anal. Chem.*, **34**, 1033 (1962).
20. R. E. Kepner, H. Maarse, and J. Strating, *Anal. Chem.*, **36**, 77 (1964).
21. D. E. Heinz, M. R. Sevenants, and W. G. Jennings, *J. Food Sci.*, **31**, 63 (1966).
22. C. S. Tang and W. G. Jennings, *J. Agr. Food Chem.*, **15**, 24 (1967).
23. H. E. Nursten and W. G. Jennings, *Anal. Chem.*, **39**, 52 (1967).
24. J. Shapiro, *Science*, **133**, 2063 (1961).

25. H. Schildknecht and F. Schegelmilch, *Chem. Ing. Tech.*, **35**, 637 (1963).
26. D. S. Bidmead, in *Recent Advances in Food Sciences*, Vol. 3, Butterworths, London, 1963, p. 158.
27. R. E. Kepner, S. van Straten, and C. Weurman, *J. Agr. Food Chem.*, **17**, 1123 (1969).
28. P. A. T. Swoboda, in *Gas Chromatography 1966*, A. B. Littlewood, Ed., Elsevier, London, 1967, p. 344.
29. I. Hornstein and P. F. Crowe, *Anal. Chem.*, **34**, 1354 (1962).
30. M. E. Morgan and E. A. Day, *J. Dairy Sci.*, **48**, 1382 (1965).
31. R. G. Arnold and R. C. Lindsay, *J. Dairy Sci.*, **51**, 224 (1968).
32. E. Kugler and E. sz Kovats, *Helv. Chim. Acta*, **46**, 1480 (1963).
33. M. C. tN. deBrauw and C. Brunnée, *Z. Anal. Chem.*, **229**, 321 (1967).
34. R. Teranishi and T. R. Mon, *Anal. Chem.*, **36**, 1490 (1964).
35. R. Teranishi, R. A. Flath, and T. R. Mon, *J. Gas Chromatog.*, **4**, 77 (1966).
36. I. Halasz and C. Horwath, *Anal. Chem.*, **35**, 499 (1963).
37. C. Merritt, Jr., and J. T. Walsh, *J. Gas Chromatog.*, **5**, 420 (1967).
38. T. R. Mon, R. A. Flath, R. R. Forrey, and R. Teranishi, in *Advances in Gas Chromatography*, A. Zlatkis, Ed., Preston Technical Abstracts Co., Evanston, Illinois, 1967, p. 30.
39. C. Merritt, Jr., P. Angelini, and D. J. McAdoo, in *Radiation Preservation of Foods*, Advances in Chemistry Series No. 65, R. F. Gould, Ed., American Chemical Society, Washington, D.C., 1967, p. 26.
40. H. W. Ockerman, T. N. Blumer, and H. B. Craig, *J. Food Sci.*, **29**, 123 (1964).
41. W. H. McFadden, R. Teranishi, J. Corse, D. R. Black, and T. R. Mon, *J. Chromatog.*, **18**, 10 (1965).
42. W. G. Jennings and M. R. Sevenants, *J. Food Sci.*, **29**, 158 (1964).
43. D. Reymond, F. Mueggler-Chavan, R. Viani, L. Vuataz, and R. H. Egli, *J. Gas Chromatog.*, **4**, 28 (1966).
44. H. Maarse and M. C. tN. deBrauw, *J. Food Sci.*, **31**, 951 (1966).
45. R. E. Biggers, J. J. Hilton, and M. A. Gianturco, *J. Chromatog. Sci*, **7**, 453 (1969).
46. E. L. Wick, T. Yamanishi, A. Kobayashi, S. Valenzuela, and P. Issenberg, *J. Agr. Food Chem.*, **17**, 751 (1968).
47. A. O. Lustre and P. Issenberg, *J. Agr. Food Chem.*, **17**, 1387 (1969).
48. R. G. Buttery and L. C. Ling, *Brewers Dig.*, No. 8, 71 (1966).
49. J. G. Kirchner and J. M. Miller, *Ind. Eng. Chem.*, **44**, 318 (1952).
50. K. E. Murray and G. Stanley, *J. Chromatog.*, **34**, 174 (1968).
51. J. Stoffelsma, G. Sipma, D. K. Kettenes, and J. Pypker, *J. Agr. Food Chem.*, **16**, 1000 (1968).
52. I. Halasz and E. Heine, in *Advances in Chromatography*, Vol. 4, J. C. Giddings and R. A. Keller, Eds., Marcel Dekker, New York, 1967, p. 207.
53. L. Rohrschneider, *Z. Anal. Chem.*, **170**, 256 (1959).
54. T. R. Mon, R. R. Forrey, and R. Teranishi, *J. Gas Chromatog.*, **4**, 176 (1966).
55. T. R. Mon, R. E. Forrey, and R. Teranishi, *J. Gas Chromatog.*, **5**, 497 (1965).
56. C. Merritt, Jr., and J. T. Walsh, *Anal. Chem.*, **35**, 110 (1963).
57. A. G. Altenau, R. E. Kramer, D. J. McAdoo, and C. Merritt, Jr., *J. Gas Chromatog.*, **4**, 96 (1966).
58. D. R. Rushneck, *J. Gas Chromatog.*, **3**, 318 (1965).
59. J. Th. Heins, H. Maarse, M. C. tN. deBrauw, and C. Weurman, *J. Gas Chromatog.*, **4**, 395 (1966).
60. P. Issenberg, A. Kobayashi, and T. J. Mysliwy, *J. Agr. Food Chem.*, **17**, 1377 (1969).

61. D. A. Forss, M. L. Bazinet, and S. M. Swift, *J. Gas Chromatog.*, **2,** 134 (1964).
62. C. Merritt, Jr., *Appl. Spectry. Rev.*, **3,** 263 (1970).
63. W. H. McFadden, in *Advances in Chromatography*, Vol. 4, J. C. Giddings and R. A. Keller, Eds., Marcel Dekker, New York, 1967, p. 265.
64. L. S. Ettre and A. Zlatkis, *The Practice of Gas Chromatography*, Interscience, New York, 1967.
65. L. S. Ettre and W. H. McFadden, *Ancillary Techniques of Gas Chromatography*, Interscience, New York, 1969.

CHAPTER 9

Biochemical and Biomedical Applications of Preparative Gas Chromatography

W. J. A. VandenHeuvel and G. W. Kuron, *Merck Sharp & Dohme Research Laboratories, Rahway, New Jersey*

I. Introduction

A great advantage in writing a review chapter on any aspect of gas-liquid chromatography (GLC) is the relatively short history of the subject matter. An article appearing in 1950 would have been very brief indeed, consisting only of a reference to the statement by Martin and Synge in 1941 that in partition chromatography, "The mobile phase need not be a liquid but may be a vapour. . . . Very refined separations of volatile substances should therefore be possible in a column in which permanent gas is made to flow over gel impregnated with a non-volatile solvent. . . ." (1). Development of theory, technique, and applications advanced considerably in the 1950s but, aside from rapid acceptance of GLC methods for fatty acid analyses (and to a much lesser extent for the determination of methylated sugars) there could be relatively little to excite any but the more perceptive and imaginative biochemist or pharmaceutical chemist in a 1960 review article.* Most endeavor appeared to be in the hydrocarbon field. That was the year, however, when the successful GLC of biologically active compounds of considerable molecular weight and structural complexity was first demonstrated. Methods and conditions developed for steroid analysis (3) were also found to be applicable to many alkaloids (4), and within several years thin-film column packings containing highly thermostable stationary phases coated on inactivated supports, combined with highly sensitive ionization detectors, were employed to separate directly—or as derivatives—a great variety of natural products and synthetic compounds of biological interest including amino acids, aromatic acids, fat-soluble vitamins, sugars, biogenic amines, various drugs, and other compounds (5). Recently, thanks to reagents capable of transforming nonvolatile compounds into derivatives amenable to GLC, the area of application of these methods at the microgram and nanogram levels has expanded further to include water-soluble vitamins (6), Krebs cycle acids (7), some nucleotides and nucleosides (8,9), gibberellins (10), prostaglandins (11), antibiotics (12), cardiac glycosides (13), phosphatidylserines (14), and others (15). Although a separation method, GLC is usually employed for "analytical" purposes, e.g., identification and quantification; excellent detection devices are available for the determination of microgram-picogram quantities of organic matter in the gas phase. The potential for using GLC as a method of isolation or purification (often referred to as "preparative" GLC) surely exists and has been demonstrated, but in practice this usage has lagged far behind analytical applications. Enthusiasm for preparative GLC must be tempered with the realization that microgram and submicrogram separations for analytical purposes may be more readily carried out than microgram-milligram separations for isolation purposes. Our objective in writing this chapter has been to choose recent, appropriate, representative examples of preparative GLC of interest to the research scientist in biochemical and related areas rather than to carry out an exhaustive literature survey. If it appears that some of the examples deal with compounds far removed from traditional medicinal chemicals, this should be taken as a reflection of the diversification that has become typical of biochemical research during this decade. The term "preparative," as used in this chapter, is employed when material is collected from a GLC

*See, e.g., the excellent review article by Lipsky and Landowne (2).

column in order to permit its further use for other purposes such as characterization by various spectroscopic methods, additional purification or analysis by other separation techniques, determination of radioactivity, biological testing, and chemical modification. Sample size can be anywhere from microgram to milligram. Perhaps the term "collection" or "isolation" is preferred to "preparative," but the latter has tradition on its side.

It is important to emphasize that GLC is most useful when it is employed in a complementary fashion with other methods of isolation, separation, and identification. The combination of thin-layer chromatography (TLC) and GLC (in this order) is particularly attractive (16). TLC separations are based principally on the polarity of the solute molecules, whereas GLC separations, especially with nonpolar stationary phases such as SE-30 and OV-1, depend mainly upon solute molecular weight and shape. Thus when the two methods are used in combination, TLC segregates the sample components into groups of compounds that possess similar functional group content; the various groups can then be subjected to GLC, in which the members of the group are differentiated on the basis of molecular weight. The normal sequence is TLC followed by GLC; however, several recent papers report the reverse— the GLC effluent is directed onto a TLC plate and the plate is then developed (17,18).

In preparative GLC the sample is normally collected in a length of glass capillary, Teflon tubing, or a special device, although it can be led directly into a counting vial for assay of radioactivity or, as indicated above, directly onto a thin-layer plate. The step following collection from the GLC effluent, whether it be the determination of a spectrum or radioactivity, can generally be carried out at any time once collection has been accomplished. The same is true for substances to be tested for biological activity or to be subjected to the transformations wrought by organic chemists; the preparative GLC and the subsequent procedure are consecutive events. This is distinct from two techniques that perhaps can be considered preparative since, in a sense, the GLC column prepares the sample for delivery to another, completely different method of analysis. We are referring to two methods of increasing popularity. The first is combined GLC–mass spectrometry (MS) in which the components eluted from the column are led directly into the mass spectrometer* while still in the vapor state (normally via a "molecular separator" to remove carrier gas) (19).

The second is simultaneous or combined gas–liquid radiochromatography, in which the column effluent is continuously monitored in a flow-through system for the detection of radioactivity (20–22).

*Unless specified otherwise, combined GLC–MS refers to low-resolution or single-focusing MS.

II. General Comments on GLC Behavior

A. Volatility

Three aspects of the GLC behavior of a compound at any sample level are its volatility, thermostability, and separability from interfering or related substances. A compound must possess a suitable vapor pressure (1–2 mm Hg) under the GLC column operating conditions in order to undergo successful chromatography. Exceedingly low volatility can result from either very high molecular weight (tristearin), high polarity because of the presence of strong intermolecular forces (sugars and amino acids), or a combination of the two (cardiac glycosides, neomycins). The use of high column temperatures and short columns packed with thin-film packing is useful in the first case, and triglycerides are routinely analyzed in several laboratories (23). When faced with a very polar compound, one must resort to chemical modification of the substance by derivative formation, both to increase volatility and to reduce irreversible adsorption ("loss on the column") during the analysis. Methylation of long-chain fatty acids is probably the best known example of derivatization (24). Methyl ethers (25), acetyl esters (26), and especially trimethylsilyl (TMSi) ethers (27), are excellent derivatives for carbohydrates. A variety of approaches are used in work with amino acids (28), including N,O-TMSi derivatives (29–31), n-alkyl-N-trifluoroacetyl esters (32,33), and phenylthiohydantoin (PTH) derivatives (34–36). The carboxyl group and hydroxyl group are derivatized in currently suggested procedures for work with bile (37,38) and phenolic acids (39,40). The most troublesome groups, then, are those that contain "active hydrogen atoms," and these can usually be converted to much less polar functions, with a concomitant increase in GLC volatility and reduction in adsorptive properties. Sugar phosphates (41) and water-soluble vitamins (6) are transformed to the TMSi derivatives, and even steroidal glucuronides (42), cardiac glycosides (13), and several of the neomycins (12) can be eluted successfully from GLC columns as derivatives. The GLC of many biogenic amines is not practicable because these amino and hydroxyl group–containing compounds undergo irreversible adsorption with most column packings; for this reason, considerable effort has been expended in seeking suitable derivatization procedures. A recent approach employs the heptafluorobutyryl derivatives (43); TMSi derivatives have also been suggested (44). Compounds such as the Krebs cycle acids and prostaglandins contain carboxyl, hydroxyl, and keto groups; for the latter function, the O-methyloxime has proven to be a valuable protecting group for GLC (45,46). Nucleosides and nucleotides are formidable challenges to the gas chromatographer, but trimethylsilylation has been found to be successful in some instances, allowing the GLC of these compounds (8,9). The success generally achieved in forming suitable derivatives of this wide variety of compounds can be ascribed to the availability in the late 1960s of reagents such as bis-trimethylsilylacetamide (BSA) (47,48), bistrimethylsilyltrifluoroacetamide (BSTFA) (30,48), TMSi imidazole (48,49), heptafluorobutyrylimidazole (43,48), and methoxylamine hydrochloride (45,46), which are employed to produce nonpolar derivatives possessing significant vapor pressures under GLC conditions. Classic physical properties (e.g., melting point, color, ease of crystallization) are of little significance and value in GLC and related techniques.

B. Thermal Stability

Certain functional groups or combinations of functional groups are known to be thermally unstable and thus may not survive GLC column conditions intact. One of the early reasons for collection of a GLC effluent was to ascertain whether or not the compound of interest survived the analysis unchanged.* Figure 9.1 illustrates the separation of a mixture of steroids at the milligram level. The retention time relationships with the nonselective stationary phase SE-30 indicate that no structural transformations have occurred. Melting points and IR spectra for the collected compounds agreed well with those of the reference compounds; indeed, several of the collected fractions exhibited sharper melting points than the starting compounds.

Instability is an ever-present danger, however, and it is often difficult if not impossible to detect by peak shape alone whether an elimination or other transformation has occurred. Further, even retention times can be misleading. As an example, early work with trifluoroacetyl derivatives can be cited (51). Cholestanyl trifluoroacetate was observed to yield a symmetrical peak with no indication of structural alteration. The retention time was considerably shorter than expected, but fluorine-containing compounds often do possess surprising volatility. It was decided to collect the material represented by the peak and subject it to IR spectroscopy in order to prove that the eluted compound was indeed the ester. This was a very informative experiment, for the IR spectrum disclosed the absence of the bands characteristic of the trifluoroacetoxy group; the collected compound was, in fact, 2-cholestene.†

Although homoallylic alcohols such as cholesterol are stable to GLC conditions, allylic alcohols are not; e.g., 4-methyl-4-cholesten-3β-ol undergoes

---

*This instability problem has proven to be far less troublesome than first thought (50).

†The source of trouble was the presence of column packing within the flash heater zone (290°C) and a column temperature of 260°C. Under more moderate and proper column conditions (flash heater 260°C, column 220°C), trifluoroacetyl derivatives are usually stable (38,52). Certain bile acid trifluoroacetates, however, are exceptionally susceptible to thermal decomposition (38).

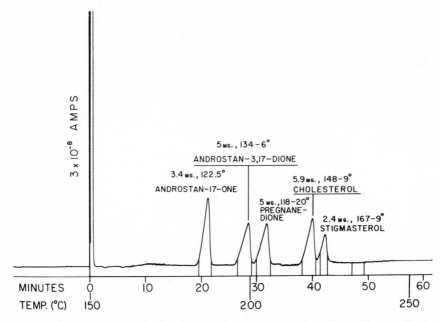

Fig. 9.1. Separation of a mixture of steroids with a preparative column. Times and temperatures for the temperature-programmed (2.1°C/min, 30 psi) procedure are indicated. Vertical lines indicate the fraction-cutting operation. Amount and melting point for each major fraction are shown. Identification of each component was completed by IR spectra comparisons. Recovery of individual components was nearly quantitative, except for the last compound (stigmasterol, recovery about 50%). Column conditions: 12 ft × 12.4 mm glass U-tube; 0.75% SE-30 on 100–140-mesh acid-washed and silanized Gas-Chrom P. Illustration taken from Ref. 111.

dehydration (53), as does allocholesterol (4-cholesten-3β-ol) (53,54); trimethylsilylation has been reported to stabilize the allocholesteryl system (55). Similarly, an allylic terpenoid diol (a component of the sex pheromone of the queen butterfly) and its diacetate could not be purified by GLC without decomposition. The ditrimethylsilyl ether was amenable to preparative GLC, and the collected ether yielded a molecular ion and M-15 fragment in its mass spectrum which confirmed the structure of the parent diol (55a). The instability of the PTH derivatives of the hydroxy-substituted amino acids theonine and serine has also been demonstrated by spectroscopic analysis of collected material (34); this too can be suppressed by derivative formation (35,36). Trimethylsilylation does not guarantee success, however, for although TMSi derivatives of aldose 6-phosphates undergo *preparative* GLC successfully, it has been reported that aldose-1-phosphate derivatives are transformed to the TMSi ether of the sugar anhydride (9). Collection of eluted compound and subsequent spectroanalysis was also of critical importance in demonstrating that the vitamins D undergo an "on-column" cyclization to a mixture of isomers [e.g., calciferol (vitamin D_2) yields a mixture of pyro and isopyro-calciferol (56)]. The unexpectedly short retention times observed for cortisone and several related compounds suggested that molecular integrity was not retained during GLC. IR spectrometry of the collected effluent of cortisone and comparison with known reference standards disclosed that the dihydroxyacetone side chain is not stable, and the compound actually eluted is the corresponding 17-ketosteroid (56a). Formaldehyde has also been trapped and identified as a product of this on-column reaction (57). Numerous approaches have been taken to stabilize the side chain and prevent the nonquantitative conversion. Stable systems are achieved by formation of (1) the 17α,21-diacetyl derivative (58); (2) the 17,20,21-bismethylenedioxy compound (59); (3) the 20-methoxime-21-trimethylsilyl ether (60); (4) the 20-methoxime-17α,21-ditrimethylsilyl ether (61); (5) the 20-methoxime-17α,21-cyclic boronate esters (62); and (6) the 17α,21-dimethylsiliconide (63), a structure analogous to an acetonide. The varied approaches illustrate the kind of chemistry that must sometimes be undertaken to prevent undesired molecular transformations. Successful preparative GLC of a steroid drug with a side chain of this type would require quantitative derivatization and conversion back to the parent compound, in addition to quantitative elution and collection.

C. Improvement in Separation

1. Stationary Phases

The ways of nature are such that we must often deal with compounds very closely related in structure and difficult to separate. A judicious choice of stationary phase may lead to separation. Stationary phases are usually divided into two types—nonpolar or nonselective, and polar or selective (64). Differences in retention behavior of solutes observed with the former are generally based on differences in molecular weight and shape rather than specific interactions between functional groups (Fig. 9.1). The increase in retention time observed when an oxygen atom is introduced into a molecule (e.g., transformation to an alcohol) is rather closely related to the resulting increase in molecular weight (the retention time may be approximately doubled) and epimeric hydroxyl group–containing compounds are often not well separated. Even with this type of stationary phase, however, what may appear to be a small structural difference can lead to separation. The nonselective stationary phases most frequently employed in "biochemical-biomedical" research are dimethylpolysiloxanes of the SE-30, JXR, or OV-1 type, or close relatives [e.g., F-60 (or DC-560), a dimethylpolysiloxane containing a low percentage of *p*-chlorophenyl groups]. These stationary phases not only exhibit useful partitioning properties, but are also thermally stable and exhibit low volatility at their normal operating temperatures, qualities very desirable in preparative separations.

Selective or polar stationary phases effect separation of solutes on the basis of their molecular weight and shape, and also functional group content. Introduction of an oxygen atom as discussed above may result in a four- to six-fold increase in retention time. Selective phases may exhibit selective retention properties for specific functional groups. The fluoroalkylpolysiloxane FS-1265 (or QF-1) shows selective retention for ketones and is effective in separating epimeric alcohols (65). This latter property is shared with the polyesters [e.g., neopentylglycol succinate (NGS)] and cyanoethylpolysiloxanes (CNSi, XE-60), and these phases also show selective retention for unsaturation (64,65). With QF-1 an olefin may be eluted faster than the corresponding saturated compound (64,65). CNSi is capable of distinguishing between isomeric ketones, and a group of very polar copolymers (66) is valuable for work with unsaturated compounds (66,67). A wide range of selective phases is available, so that it is a rare pair of compounds that cannot be separated by use of one of them. It must be emphasized, however, that volatility differences suitable for an analytical determination may not prove sufficient if the separation is to be carried out on the preparative scale, where column efficiencies are usually reduced and column overload conditions may pertain. Higher column temperatures may be required in preparative separations. As selective stationary phases are generally less thermally stable and more volatile than the frequently used nonselective phases, their use in preparative GLC is somewhat restricted if it is not possible to decontaminate the collected compound from "column bleed" by one means or another. OV-17, a phenylmethylpolysiloxane, is a very useful high-temperature, selective stationary phase (48,49).

The value of selective stationary phases is illustrated in a study by Dutky et al. (68) on the conversion of cholestanone to cholestanol by the larvae of the common housefly (*Musca domestica*). The isolated cholestanol fraction was found by GLC with the selective phases QF-1 and NGS to contain two components, the retention times for which corresponded to cholestanol and cholesterol (these sterols do not separate with nonselective stationary phases such as SE-30). The presence of the latter would be expected, for nonlabeled cholesterol was contained in the larval diet. It was necessary, however, to ascertain whether or not some of the unsaturated sterol arose from the cholestanone-^{14}C by metabolism. Use of the QF-1 column permitted separate trapping of the two peaks. The cholesterol peak (shorter retention time) contained less than 0.6% of the recovered radioactivity, whereas the cholestanol peak contained greater than 96%.

2. Derivatization

The experienced gas chromatographer often turns to or includes derivative formation in his approach to separation; this avenue, by accentuating functional group or stereochemical differences, has been widely employed for effecting otherwise difficult separations (49,64,65). A great advantage of this functional group alteration approach to separation is that derivatives of closely related compounds can often be easily separated with nonselective stationary phases (49,64,65). For example, testosterone and epitestosterone are not separated with SE-30, but their trimethylsilyl ethers separate readily under the same GLC conditions. Derivatization results in an increase in molecular weight and usually a decrease in solute polarity.

Derivatives (e.g., TMSi ethers) usually possess somewhat longer retention times than the parent compounds with nonselective stationary phases* (somewhat of a disadvantage in preparative work as the column temperature may have to be raised, leading to an increase in column bleed). With selective stationary phases, however, the increase in molecular weight is more than offset by the decrease in polarity, and the retention times of the often hydrocarbonlike derivatives may be one-third or less than those of the parent compounds. The separation of derivatives may also be greater with selective than nonselective stationary phases.

A particularly interesting example of the use of derivative formation to improve the separation of closely related compounds has been reported by Haahti and Fales (70). During an investigation of the unsaponifiable fraction of a diester wax from chicken preen glands it became evident that this fraction contained a series of homologous compounds—long-chain 2,3-*n*-alkane diols. Although GLC of the diol fraction showed three components, this was not a guarantee of the presence of only three diols, for the chromatographic system (SE-30) might not be able to distinguish between *threo* and *erythro* isomers. In order to accentuate the stereochemical differences, the authors turned to acetonide formation.† A *threo* diol is converted to a *trans* system, whereas a *cis*-substituted ring results from an *erythro* isomer. Since these two ring systems are rigid structures, one could hope for an improvement in separation compared to the parent diols. The three uropygiol peaks yielded six new peaks upon subjection to acetonide-forming conditions. Figure 9.2 shows the chromatogram of the total acetonide sample. Preparative GLC was employed to collect pure samples (see Fig. 9.2) of the six acetonides for further characterization by mass spectrometry. The compounds exhibited major fragments at M-15 (loss of a methyl group) in their mass spectra, facilitating

---

*An exception to this are trifluoroacetyl derivatives of alcohols, which even with nonselective columns may possess shorter retention times (52, 69).

†Acetonides have been suggested as useful derivatives for the GLC characterization of unsaturated fatty acid methyl esters (71).

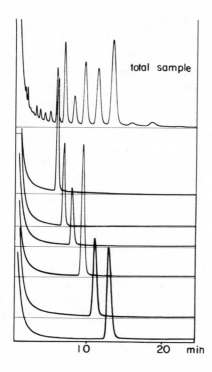

total sample

1 0   2 0   min

Fig. 9.2. GLC of uropygiol acetonides, both total sample and isolated fractions (preparative GLC). Analytical column conditions: 2 m × 3 mm i.d. glass U-tube; 3% SE-30 on 100–140-mesh Gas-Chrom P; 130°C. Illustration taken from Ref. 70.

the determination of their molecular weights. The *cis* or *trans* assignment of the uropygiols was based on the GLC behavior on nonpolar columns of *cis*- and *trans*-olefins in which the *cis* compound precedes the *trans*-olefins. The obviously high degree of purity was obtained by collecting only the center portion of each peak.

The separation of some closely related compounds can require the employment of both a selective stationary phase and derivatization to accentuate structural differences (72,73).

The discussion of means for increasing volatility, preventing molecular alterations, and improving separations should make clear the key role of derivative formation in the GLC of biologically important compounds. Fortunately, with both analytical and preparative GLC, the derivative often need not even be isolated in the classic sense prior to chromatography, for the reaction mixture can be applied to the column and the excess reagents eluted in the solvent front. With preparative GLC, however, while a derivatization step may be necessary to achieve the desired chromatography, the collected derivative may have totally different properties from the original compound. This may be a source of concern, in contrast to the case of analytical GLC, in which the effluent is led directly into a detector and then vented and forgotten. For certain purposes the chemical alteration may cause no problems, e.g., the counting of radioactivity should not be affected, and mass spectral examination of the derivatives may be equally as useful as that of the parent compound. Functional group changes may interfere with IR and NMR examination. Unfortunately, some of the derivatives widely used in GLC, while perfectly stable in the inert gas phase, are somewhat labile in the laboratory atmosphere. The partial reversion of TMSi derivatives of alcohols, and especially carboxyl and amino group–containing compounds, to the parent substances during or following collection must be considered. The collected material must of course be returned to its original state for biological testing, or if physicochemical tests or further separations are to be carried out on the parent compound.

Sjövall and Sjövall employ a method using aqueous methanolic acetic acid to hydrolyze TMSi ethers of steroids (74). In a most interesting approach to the combination of GLC and TLC (in that order), Curtius (17) has developed a method (1% hydrogen chloride in methanol) for hydrolyzing TMSi derivatives on a thin-layer plate. TMSi ethers themselves can be separated by TLC (75–77) and column chromatographic (78) techniques. Hydrolysis of derivatives is very unlikely when the GLC effluent is not handled in the laboratory but, as in the direct combination of GLC and MS, is analyzed immediately upon its elution from the column.

III. Specialized Techniques: GLC in Combination with Mass Spectrometry and Determination of Radioactivity

A. Mass Spectrometry

MS is the method par excellence for characterization of organic compounds (nonpolymeric) at the microgram level. The more commonly available low-resolution ("single-focusing") mass spectrometer often provides the molecular weight (to the closest integral atomic mass unit) of the compound under analysis, and the so-called fragmentation pattern can serve as a reliable fingerprint for identification by comparison. Further, the nature of the functional groups of a compound can sometimes be deduced from the pathways by which the compound fragments. MS functions best with relatively pure samples, and these are what a GLC column is expected to yield from mixtures. The direct combination of the two techniques circumvents the need for actual collection of the component of interest but imposes the necessity for some kind of often expensive and/or possibly temperamental device to serve as a link between the two.* If a high-resolution mass spectrum is desired, the

*A number of devices are in use (79–83); the Ryhage type molecular separator appears to have won widest acceptance for work with natural products (19).

present state of the art is such that the component is usually collected and then subjected to MS by direct probe or heated inlet system. As this type of analysis can supply the empirical formulas for the molecular ion and various fragments, it is often of great value. Collection of a pure component from a GLC column is then of vital importance. Whether combined directly or indirectly, GLC and MS are highly complementary methods, and the opportunity to correlate mass spectrometric properties with retention behavior is very useful. The following discussion is not designed to indicate a preference for directly combined GLC–MS over the serial collection/analysis approach, but to demonstrate that each has its virtues. It is the combination of techniques, not how they are combined, that is important.

Structural features that lead to differences in chromatographic behavior may not lead to dissimilar mass spectra, and thus the ability to introduce the compounds into the mass spectrometer individually and correlate retention time with mass spectrum is of great value. Conversely, the mass spectrometer may distinquish between two compounds that exhibit similar retention behavior, and thus be able to indicate nonhomogeneity of a peak and also possibly provide identification of the unresolved components (19,31,84).

Direct combination of GLC and high-resolution MS is less developed than combined GLC–low-resolution MS. When a high-resolution mass spectrum is required, the most common approach at the present time is to collect the component(s) of interest from the GLC column and then carry out the MS examination.* The report of Tham et al. (86) illustrates this point. These workers isolated a substance from the urine of patients with severe burns, and were able by GLC–MS to ascertain that the substance (as its methyl ester) possessed a molecular weight of 143. More definitive information was required, and a sample was purified by GLC and subjected to high-resolution MS; an empirical formula of $C_6H_9NO_3$ was observed (for the methyl ester), and this plus other data (TLC, GLC) made possible the identification of the unknown as pyroglutamic acid.

If a low-resolution mass spectrum of a component is desired, either approach—direct combination or collection followed by MS—can be employed. Two examples of the latter are the report by Bryce et al. (87) on the isolation and identification of triterpene methyl ethers from New Zealand plants, and the study of the structure of the methyl ester–TMSi derivative of the β-D-glucosiduronic acid conjugate of 11-ketoetiocholanolone by Jaakonmaki and associates (87a). When there is a very limited amount of sample (e.g., a few micrograms) in a mixture, one must rely upon GLC to separate the components of interest from the extraneous mass, and the direct combination technique is very attractive. An excellent example of such a situation is reported

---

*See however, Watson and Biemann (80; J. T. Watson, in *Ancillary Techniques of Gas Chromatography*, L. S. Ettre and W. H. McFadden, Eds., Wiley-Interscience, New York, 1969) and the recent paper by Krueger and McCloskey (85).

in the paper by Greer et al. (88). The urinary aromatic acid fraction from a patient with Wilms' tumor was found by GLC (following acetylation and methylation) to contain a compound which compared favorably in retention behavior to the corresponding derivative of *p*-hydroxyphenyllactic acid. The patient expired before more urine could be obtained, leaving only the aromatic acids from 3 ml of urine for further investigation. The authors considered it unlikely that they could collect the component of interest by preparative GLC and turned to combined GLC–MS which demonstrated that the unknown urinary aromatic acid was indeed *p*-hydroxyphenyllactic acid. Combined GLC–MS should prove to be extremely useful in the area of drug metabolism studies, as has been exemplified in a number of recent publications (89).

Curtius (90) studied the ring structures of ketoses (as the TMSi ethers) utilizing GLC and MS. Preparative GLC was employed to provide samples for MS analysis; however, when the samples available were very small, or the compounds exceptionally labile, direct combination GLC–MS was employed.

Jaakonmaki et al. (91) employed preparative GLC and subsequent MS to isolate and characterize ethyl β-D-glucosiduronic acid (derivatized to form the methyl ester trimethylsilyl ether) as a metabolite of ethanol in rats and an adult male human alcoholic. These authors were also successful in carrying out direct combined GLC–MS on derivatized urinary extracts. Occasions do arise, however, when even though only a few micrograms of material are available, it must be purified by preparative GLC to allow characterization. A case in point is the recent paper by Sjövall and Sjövall (74). An unknown pregnanetriol from human pregnancy plasma (actually present as a monosulfate; 3–20 μg per 100 ml) could not be completely characterized by GLC and GLC–MS data alone. A small amount of pure material was required to allow chemical transformation designed to assign the stereochemistry at C-3, C-5, and C-20. Although the quantity of material available was small, it proved possible to collect a minute amount of the steroid, as its TMSi ether derivative, in a glass capillary attached to the jet of a flame ionization detector. The parent compound (2 μg) was regenerated from the derivative, and the sample divided into two equal parts for a number of chemical manipulations. Acetonide formation was utilized to determine the orientation of the hydroxyl group at C-20. Assignment of an alpha orientation was based upon retention time ratios of the isomers. Preparation of the 17β-carboxylic acid derivative and comparison of its retention behavior and mass spectrum with that of authentic C_{20} acid indicated a 5α (A/B *trans*) configuration. The orientation of the hydroxyl group at C-3 was ascertained by GLC of the TMSi ethers of the intermediate aldehydes. These data proved sufficient to permit the identification of the compound as 5α-pregnane-3α,20α,21-triol.

A variation of combined GLC–MS, "mass fragmentography," makes possible very sensitive (nanogram-picogram) and selective detection of drug metabolites and other compounds (92). Briefly, the mass spectrometer is employed as a selective GLC detector. It is focused alternately on several m/e values for which only the compound of interest produces fragments (for high sensitivity these signals should be of high intensity). A detector response is produced only when a compound is eluted from the GLC column that yields fragments of the appropriate m/e.

B. Radioactivity Determinations

The use of radioactively labeled compounds is virtually mandatory in biosynthesis and metabolism studies. GLC of an appropriate isolate with a normal mass detector generally indicates the presence of numerous compounds, some of which can be tentatively identified on the basis of the known retention times of the expected products. The crucial question is usually concerned with the presence or absence of radioactivity in certain of the components—if radioactivity is present, the compound must be directly related in some way to the labeled starting material. The simplest and most commonly used approach to answering this question is to combine a trapping procedure for the collection of the GLC components (at those retention times when compounds thought to possibly contain the label appear) and a separate counting procedure. The collection approach allows for the possibility of long-term counting, enabling one to work with low levels of radioactivity, and is advantageous only from this important point of view. The use of a mass detector response to initiate collection to permit counting of a component is hazardous, however, since the specific activity of a radioactive compound may be such that insufficient mass is eluted to allow mass detection, and thus unless fractions are collected during the entire GLC analysis it is possible for a radioactive component to escape unnoticed. Fractions should be collected at frequent intervals of equal duration throughout the chromatographic analysis (or during that period of time when all the possible components of interest are known to elute), greatly reducing the possibility of missing a labeled component. As this may be tedious if carried out manually, the use of automated fraction collection devices is highly attractive (93). Indeed, Thomas and Dutton (94) have reported a highly automated system for dual-label (carbon-14, tritium) gas radiochromatography involving not only automated serial collection of fractions but also the use of digital computer techniques to analyze the counting (liquid scintillation) data. Computer techniques also improved the sensitivity and resolution of the analysis and were employed to plot the radioactivity profile in the GLC effluent vs. retention time.

The often remarkable resolving power of GLC columns is one of the reasons why this method of separation is so effective, but it should be kept in mind that when a technique involving collection from a column is employed for radioassay (or other purposes), resolution of the system as a whole is determined by the frequency with which the fractions are collected. It is of little value to employ a highly efficient GLC column with a multicomponent sample if the collection vials are changed only every 10 min. Some contamination may be accepted when the component is collected to allow the determination of, e.g., its mass spectrum, since this method often permits the recognition of nonhomogeneity. In the case of radioassay, however, it is not possible to distinguish between counts from one carbon-14-containing compound and counts from another. Indeed, such contamination may result in attributing the presence of a radioactive label to the wrong component.

In addition to collection resolution, collection efficiency is obviously of vital importance in the use of preparative GLC for radioassay and is a very real technological problem (see Section IV, p. 296). It should be quantitative, or at least reproducible, and within a family of related compounds, independent of structural differences.

Submicrogram amounts of polar compounds are likely to encounter adsorption effects during GLC. Such "loss on the column" may not be of great significance with large samples; however, the loss is on an absolute basis—e.g., 0.05 μg per run—and thus chromatography of 0.1 μg leads to the disappearance of 50% of the sample. The collection and counting of compounds of high specific activity is thus affected adversely, for such an effect reduces the amount of radioactivity that could be collected from a column. Addition of cold compound as a carrier has been suggested as a method for reducing the magnitude of this problem (95). Conversion to nonpolar derivatives is an effective means for reducing irreversible absorption (36,66,73).

Trapping of the sample with separate counting of radioactivity has been employed by Goad and Goodwin (96) in phytosterol biosynthesis studies. These authors recently presented data supporting the proposal that cycloartenol may be the precursor of fucosterol, the major sterol of the marine brown alga *Fucus spiralis*. In these experiments, incubation of the alga with doubly labeled (carbon-14 and tritium) mevalonic acid yielded a radioactive 4,4-dimethyl sterol fraction to which was added cold cycloartenol and 24-methylenecycloartenol. The mixture was acetylated, and cycloartenyl acetate was obtained by preparative TLC. Lanosteryl acetate was added to a portion of the cycloartenyl acetate, and this mixture subjected to preparative GLC experiments (an effluent splitter was employed; collection was in capillary tubes at ambient temperature). In one experiment, fractions were collected at 1-min intervals. Comparison of the plot of carbon-14 radioactivity (measured by liquid scintillation counting) vs. time of collection and the mass

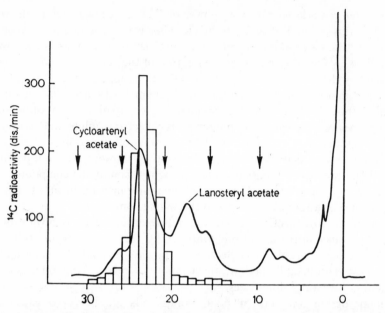

Fig. 9.3. Gas–liquid radiochromatography of the 4,4-dimethyl steryl acetates isolated from *F. spiralis* following incubation with (4R)-mevalonic 2-^{14}C-4-^{3}H$_1$ acid. Samples were trapped at 1-min intervals and assayed for radioactivity. In a second preparative run, samples were trapped between the points indicated by the arrows and assayed for radioactivity. Column conditions: 6 ft × ⅛ in. i.d.; 3 % XE-60 on 80–100-mesh HMDS Chromosorb W; 237°C; 40 ml/min. Illustration taken from Ref. 96.

record (Fig. 9.3) clearly indicates the presence of only one radioactive peak, coinciding exactly with the retention time of cycloartenyl acetate. In the other experiment involving preparative GLC, the component peaks (lanosteryl acetate and cycloartenyl acetate) observed on the mass detection record were collected during the time periods indicated by the arrows. Most of the radioactivity was associated with the cycloartenyl acetate peak and possessed the same ^{3}H/^{14}C ratio as the originally isolated radioactive triterpenes. Very little radioactivity was found in the collected lanosteryl acetate.

Preparative GLC techniques involving trapping and subsequent counting of the collected fractions for radioactivity have been used extensively by the Beltsville group (Entomology Research Division, U. S. Department of Agriculture) in their studies on insect sterol metabolism. These workers have demonstrated, e.g., that the nymphal German cockroach (*Blattella germanica* L.) and the larva of the Virginia pine sawfly (*Neodiprion pratti* D.) can convert β-sitosterol to cholesterol (97,98), whereas neither the larva nor the adult housefly (*M. domestica*) is capable of this side-chain dealkylation (99,100).

Demosterol has been shown to be an intermediate in this C_{29}–C_{27} transformation in the tobacco hornworm *Manduca sexta* J. (101).

Actual collection of the eluted compounds can be combined with concurrent counting of radioactivity, as in the methods developed by Popjak and associates (102) and Karmen et al. (103). The former permits simultaneous mass and radioactivity determination (by passage of the column effluent directly from the mass detector into a chilled chamber of circulating scintillator solution; an integral record is obtained). In the latter approach, a tube containing anthracene is arranged for direct measurement of carbon-14 radioactivity by scintillation counting. Karmen reports essentially quantitative collection because of minimized transfer and handling problems (103).

The continuous monitoring approach is in effect equivalent to collecting an infinitely large number of fractions. Although this dynamic approach requires the use of higher levels of radioactivity than collection/count techniques, a labeled component is less likely to escape detection.*

Nes and co-workers (21) have reported the successful simultaneous measurement of mass and radioactivity in a GLC effluent; in this approach the effluent molecules pass directly through a gas flow proportional counter. However, problems of condensation and adsorption of samples possessing low volatility within the flow counters or heated ionization chambers may be considerable.

One can also completely eliminate the collection aspect of these methods by combustion (hot copper oxide) of the eluted components to carbon dioxide and water (which is then reduced to hydrogen gas). An approach suggested by Karmen and co-workers (104) involves employment of a cartridge filled with anthracene crystals as a continuous "flow-through" detector of radiation. Karmen (105) has also suggested use of a proportional type counter for assaying carbon dioxide and hydrogen gas for radioactivity. Swell (20) has published a well-documented paper on simultaneous determination of mass and radioactivity using this approach. The column effluent is split 9:1, with the smaller proportion going to a flame detector; most of the effluent stream is passed through a reactor tube (copper oxide and steel wool at 630°C) and the organic compounds converted to carbon dioxide and hydrogen. The resulting gas is monitored by a proportional type detector. Efficiency for the system was reported to be at least 90% for carbon-14 and 60% for tritium. In another approach the total GLC effluent is oxidized, and mass detection is accomplished by measurement of the carbon dioxide with a microthermistor detector; radioactivity is then measured by a proportional counter (22).

*For a discussion of the relative merits of these two approaches, see *The Gas-Liquid Chromatography of Steroids*, J. K. Grant, Ed., Cambridge University Press, 1967, p. 259, and Refs. 95 and 106.

An example of the simultaneous determination of mass and radioactivity of labeled compounds is illustrated in Fig. 9.4 (20). The leaf sterol fraction from tobacco plants grown in an atmosphere of $^{14}CO_2$ contains four major components which, as can be noted, are of approximately equal specific activity. As has been indicated, such a dynamic system cannot function satisfactorily with samples of low specific activity; in this example the specific activity for the sterol fraction was 200 dpm/μg, and a relatively large quantity of mass (50 μg total) was applied to the column.

As collection and monitoring of radioactivity are not the principal subject matter of this chapter, the interested reader is advised to read the two excellent review articles by Karmen (103,107).

High-resolution columns are a desirable feature of GLC, but it should be recognized that a column that separates closely related compounds such as stereoisomers may also distinguish between a compound and its isotopically labeled analogue. Isotope (stable or unstable) fractionation effects have been observed and could prove to be confusing, if not misleading, in experiments involving collection techniques or monitoring of radioactivity. Kirchner and Lipsett (108) observed that when fractions were collected every 20 sec the tritium/carbon-14 ratio of collected steroids containing these isotopes varied across the GLC peak. Kliman and Briefer (95) also investigated this phenomenon.

The isotope fractionation effect is especially great with deuterium-containing compounds; McCloskey and associates (109) have reported good analytical GLC separation of perdeuterated fatty acid methyl esters from the corresponding normal esters with standard-length packed columns (the deuterium-containing compounds are eluted faster). Bentley et al. (110) found baseline separation of glucose and glucose-d_7 as the TMSi ethers with a 15-m column (with the latter possessing the shorter retention time). These authors used preparative GLC to isolate the β-anomer of the TMSi derivative of glucose-d_7 from a mixture of the α- and β-anomers.

IV. Operating Parameters

The chromatogram in Fig. 9.1 not only illustrates a preparative separation and demonstrates the relationship between solute molecular weight and retention behavior with a nonselective stationary phase, but also is an example of the use of temperature programming in such a separation. The method of application of the sample to a GLC column is very important, both in analytical and preparative work. So-called plug flow is highly desirable, but this requires rapid sample injection and vaporization of sample components. Whereas this condition is relatively easy to achieve for analytical GLC, the larger sample size (both the volume of the solution injected and the actual

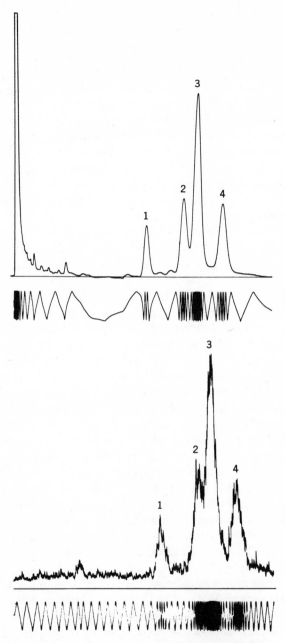

Fig. 9.4. Separation of ^{14}C tobacco leaf sterols. Upper Chromatogram is mass determination; lower chromatogram is simultaneous determination of radioactivity (as $^{14}CO_2$). Peaks 1 and 2 were not identified; peaks 3 and 4 were identified as stigmasterol and β-sitosterol, respectively. Sample applied contained 50 μg and 10,000 dpm. Column conditions: 6 ft $\times$ 4 mm i.d. glass U-tube; 3% SE-30 on 100–120-mesh Gas-Chrom Q; 250°C; 30 psi. Illustration taken from Ref. 20.

mass of sample components) in preparative GLC makes it more difficult to achieve the desired rapid introduction and vaporization of sample, especially with compounds of relatively low volatility. When temperature programming is employed, however, this problem is circumvented. The solution of the sample in an appropriate solvent can be applied to the GLC system relatively slowly, by either continuous or repetitive injection, for at the relatively low column temperature prevailing at the beginning of the analysis the low-boiling solvent is readily removed, but the much less volatile solutes are retained in a concentrated zone or plug at the front of the column. The chromatographic separation process does not begin for the sample components until temperature programming is initiated.

Although it might be assumed that because of their low vapor pressures and high melting points (a reflection of strong intermolecular forces) the collection from a GLC column of compounds such as steroids, alkaloids, and many drugs is a relatively easy task, the condensation of such substances from a gas stream is a relatively formidable challenge. This is because compounds of this type emerge from the column in the form of an aerosol. If one attaches a length of Teflon tubing to the end of a column, leads it out of the chromatograph, and then injects an oversized sample of, e.g., a steroid, it is not difficult to observe smokelike vapors emanating from the end of the Teflon tubing. Numerous approaches and devices have been suggested and tested in a search for superior collection, and it is not the purpose of this chapter to review them (see Chapter 4). We would like to discuss one approach very briefly—one that takes advantage of the differences in boiling points of nitrogen ($-196°C$) and argon ($-186°C$; m.p. $-189°C$) (111). If one employs the latter as carrier gas, very satisfactory collection results are observed for steroids and alkaloids when the column effluent is allowed to pass into a small glass U-tube immersed in liquid nitrogen. As the argon is condensed, it carries the solute molecules down with it into a liquid phase or slush via a "rain effect." The tube containing the desired fraction is removed from the liquid nitrogen and allowed to warm to room temperature, during which time the argon evaporates, leaving the collected component deposited on the surface of the walls of the tube (111).

The temperature programming and "rain effect" techniques were employed in a preparative separation designed to provide β-sitosterol of high purity for use as an internal standard in the quantitative determination of urinary 17-ketosteroids (64,72). Satisfactory separation of β-sitosterol from another plant sterol in a soybean sterol fraction was achieved using a stainless-steel column system (65).* It appears quite likely from limited GLC data that the

---

*Although all-glass systems are generally preferable, stainless steel can be used with many compounds; see, however, Section VI, p. 317.

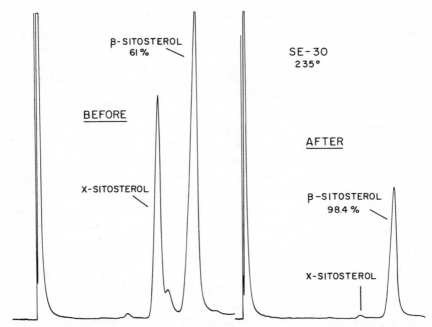

Fig. 9.5. Analysis of a sample of β-sitosterol before and after purification by preparative GLC. Analytical column conditions: 6 ft × 4 mm i.d. glass U-tube; 1% SE-30 on 100–120-mesh acid-washed and silanized Gas-Chrom P; 235°C; 20 psi. Illustration taken from Ref. 65.

companion sterol is campesterol; see Ref. 112. Figure 9.5 shows the improvement in purity of the β-sitosterol following a single preparative separation at the milligram level.

Preparative separations such as those illustrated in Figs. 9.1 and 9.5 were carried out with systems of somewhat greater column dimensions (9–12 ft × 0.5 in. i.d.) than those normally employed in analytical GLC. Increase in column diameter leads to a greater column sample capacity, a requisite for preparative work. However, too large a column diameter can result in excessive loss of resolution, detrimental to separation. Column capacity can be increased by use of heavily coated packing, but this is impractical with compounds of low volatility (many substances of interest to biochemists fall into this category), for it requires high elution temperatures. The use of thin-film packings is thus essential, but with samples of a few hundred micrograms to a milligram or more one must of necessity work under column overload conditions. A technique employed in many laboratories for the collection of GLC components is to use analytical columns and to collect (repetitively) under overload conditions. Although this may be the simplest means for obtaining microgram amounts of valuable sample, it is less satisfactory for milligram quantities. The work of VandenHeuvel and Horning (65) and Fales et al. (111) demonstrates that satisfactory column efficiencies can be achieved with hundreds of microgram- to milligram-sized components when columns somewhat larger than the usual analytical columns are employed.

There are few published data on the interrelation between sample size, column dimensions, and chromatographic behavior relating to overload, efficiency, and separation phenomena for compounds of interest to biochemists. Such information is important, of course, for the success of a preparative-scale analysis is dependent to a great extent upon column performance. Fales et al. (111) first published on the preparative-scale GLC of steroids and alkaloids and compared the separation of a mixture of steroids on two columns differing in internal diameter. The narrower column was shown, as expected, to be more efficient and to give better separation than the one of larger diameter; however, the comparison was made at only one sample level (30 μg total for a five-component mixture). A more recent study employed cholestanol (dihydrocholesterol) as a reference compound in a study of the relationship between column efficiency and diameter as a function of sample size (113). Data presented in Table 9.1 indicate that at the

TABLE 9.1

Relationship between Sample Size and Efficiency for Two OV-17 Columns[a]

Cholestanol, μg	Theoretical plates[b]	
	Analytical	Preparative
1	2900	1860
6	2860	1870
12	2460	1850
25	2280	1880
50	1670	1860
100	940	1840
250	—	1480
500	—	960

[a] Column conditions: 8 ft $\times$ 4 mm (analytical) or 11 mm (preparative) i.d. glass U-tubes; 1.8% OV-17 on 60–80-mesh acid-washed and silanized Gas-Chrom P; 240°C; 35 (analytical) or 250 (preparative) ml/min.

[b] Number of theoretical plates $= 16(t_R/w)^2$, where w is the base width of the peak with retention time (from point of injection) t_R.

Table taken from Ref. 113.

1-μg level a 4-mm-i.d. column does indeed exhibit greater efficiency than an 11-mm-i.d. counterpart (2900 vs. 1860 theoretical plates, respectively), and this level of superiority is also observed at 6 μg. The increase in sample size to 12 μg leads to a considerable decrease in the observed theoretical plate value for the analytical column although little change is noted for the preparative column. Deterioration of efficiency with sample size in the case of the 4-mm-i.d. column is further evidenced by the fact that at a sample size of 50 μg this column actually exhibits fewer theoretical plates than the 11-mm column. Indeed, one must apply approximately 500 μg of cholestanol to the larger diameter column to observe the low theoretical plate value exhibited by the analytical column at the 100-μg level. The relationship between sample size and efficiency for the OV-17 columns is presented in Fig. 9.6. The pre-

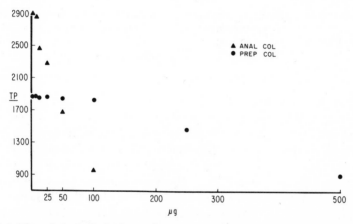

Fig. 9.6. Plot of theoretical plate values versus micrograms of cholestanol for two GLC columns differing only in internal diameter. Column conditions given in Table 9.1. Illustration taken from Ref. 113.

parative column efficiency is essentially independent of sample mass up to 100 μg and then decreases, whereas in the case of the analytical column efficiency is reduced at much lower levels. At 40–50 μg the preparative column is the more efficient of the two.

Although widely used and shown to be a very versatile stationary phase SE-30 suffers from a lack of selectivity so that closely related compounds— e.g., coprostanol (5β-cholestan-3β-ol) and cholestanol (5α-cholestan-3β-ol)— may not exhibit large differences in separation factor or retention time. This need not be objectionable, especially if the column is one of relatively high efficiency. Figure 9.7 shows the separation of a mixture of steroids with an

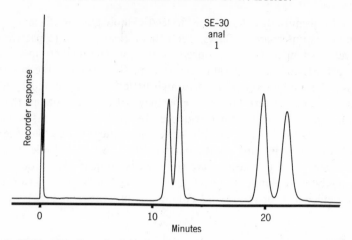

Fig. 9.7. GLC separation of a mixture of steroids [in order of elution, coprostane (0.5 µg), 5-cholestene (0.5 µg), coprostanol (1 µg), and cholestanol (1 µg)] on an SE-30 column. Column conditions: 8 ft × 4 mm i.d. glass U-tube; 2% SE-30 on 60–80-mesh acid-washed and silanized Gas-Chrom P; 250°C; 35 ml/min. Illustration taken from Ref. 113.

analytical SE-30 column. This separation is better than that observed on the SE-30 preparative column at the same sample level but, as expected, with increased sample size the situation is reversed. Thus with a sample level 200 times greater, the efficiency of the analytical column has deteriorated (Fig. 9.8) and is less satisfactory than that observed for even larger quantities on the larger-diameter column (Fig. 9.9). A comparison of Fig. 9.10 with Fig. 9.9 illustrates the improvement in separation found on the SE-30 preparative-size column for the TMSi ethers of coprostanol and cholestanol (250 µg each).

The quantity of sample that can be chromatographed successfully with an analytical column can be increased significantly by a two- or threefold increase in column diameter. A small change that should be easily accommodated by the currently available instruments makes possible the chromatography of quantities of material often sufficient for subsequent investigation by other means. Most of the applications reported in this chapter employ commercially available analytical gas chromatographs somewhat modified for collection purposes.

V. Applications

A. Steroids

The collection of GLC fractions for the isolation and identification of unknowns, the application of confirmatory techniques, and the establishment

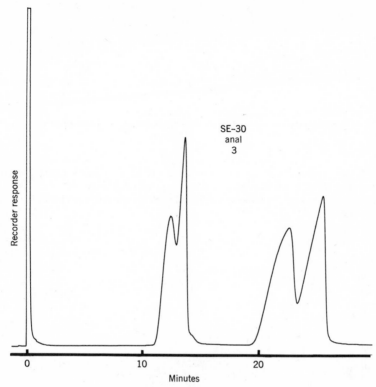

Fig. 9.8. Same as Fig. 9.7, but with higher sample load (100, 100, 200, 200 μg). Illustration taken from Ref. 113.

of purity, has been applied to many biological problems. The commonly utilized ancillary techniques have been previously cited. The following examples are intended to demonstrate specific applications to commonly encountered problems in biochemical research.

van Lier and Smith (114) reported the recovery of crystalline sterols from preparative GLC effluents. Sterols so obtained were subjected to melting-point determinations, IR and UV absorption spectrophotometry, TLC and analytical GLC for routine checks of identity and homogeneity. The authors report the collection of a variety of sterols in a state of high purity. These techniques were utilized to identify 26-hydroxycholesterol as a minor constituent sterol in isolates obtained from normal and diseased human aortas (115).

Curtius and Müller (17) utilized a slight modification of an apparatus described by Kaiser (116) for the automatic and continuous combination of GLC with TLC. They were able to obtain essentially complete resolution of a

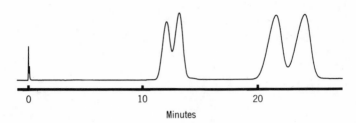

SE-30
prep
3

0   10   20

Minutes

Fig. 9.9. GLC separation of the same mixture of steroids as in Fig. 9.7 (200, 200, 400, 400 μg) with an SE-30 preparative-scale column. Column conditions: 8 ft × 11 mm i.d. glass U-tube; column packing same as in Fig. 9.7; 250°C; 250 ml/min. Illustration taken from Ref. 113.

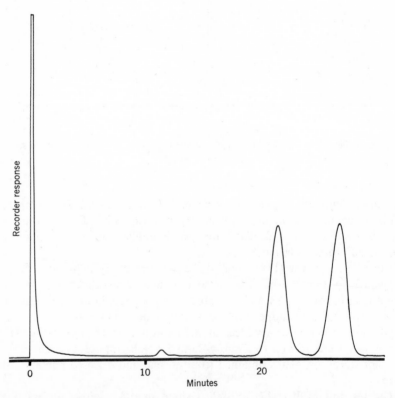

Fig. 9.10. GLC separation of a mixture of coprostanol and cholestanol as their TMSi ethers (250 μg each) with the larger-diameter SE-30 column (see Fig. 9.9).

302

mixture of 12 steroids by this technique. The TMSi ethers of the steroids were chromatographed on XE-60. The eluted components were adsorbed on a thin-layer silica gel G plate where they were converted to free steroids by spraying with 1% hydrogen chloride in methanol. The chromatoplate was developed with methylene chloride–acetone. The first three eluting peaks (allopregnanediol, pregnanediol, and androsterone), inadequately resolved as TMSi ethers by GLC, were widely separated as free steroids by the TLC technique. Similarly, vastly improved separation of pregnanetriol from 11-ketoandrosterone was obtained. This procedure utilized partition and adsorption chromatography of steroids as the nonpolar TMSi and polar free steroids, respectively. Additional characterization was obtained by spraying the developed plate with 50% sulfuric acid in methanol and utilizing the sulfuric acid chromogenicity of the steroids.

Jungman, et al. (117) described a quantitative method for the analysis of urinary 11-deoxy-17-ketosteroids and pregnanediol by GLC. After hydrolysis, the 17-ketosteroids were purified by alumina chromatography and converted to TMSi derivatives. Preliminary purification of pregnanediol was achieved by either alumina column or Girard-T separation. TLC was used for final purification. The steroids were quantitated by GLC. The recovery of ^{3}H-labeled dehydroepiandrosterone sulfate added to urine prior to hydrolysis through all manipulations of the procedure was 89%. Radioactivity content was measured by collecting the steroids in the effluent stream on p-terphenyl crystals and subsequent scintillation counting. The loss attributable to the GLC portion of the procedure was actually less than 1%. The recovery of pregnanediol diacetate-^{3}H after TLC and GLC averaged 97% in 25 determinations, with a ±2% variation. Clearly, efficient collecting systems and associated techniques served as an important tool in the validation of this procedure.

Funasaki and Gilbertson (118) isolated cholesteryl alkyl ethers from bovine cardiac muscle. Characterization was based on TLC and GLC. Cholesteryl hexadecyl ether was identified by MS and accounted for over 90% of the total components observed by GLC. These investigators collected the effluent corresponding to this major peak and removed stationary phase contamination by TLC. Subsequently, the mass spectrum of the ether was obtained by direct probe analysis with a single-focusing mass spectrometer.

The identification of 3α-hydroxyandrost-5-en-17-one in the monosulfate fraction of human urine and bile from normal subjects, as well as in blood plasma from uremic and some normal subjects, was reported by Jänne et al. (119). The monosulfate fraction was obtained by chromatography on Sephadex LH-20; after solvolysis and purification on silicic acid, the steroids were gas chromatographed. About 3 μg of the unknown was collected from the GLC effluent and subjected to further studies. These included GLC on QF-1 and SE-30 columns as free compounds, TMSi ethers, and acetates; authentic 3α-hydroxyandrost-5-en-17-one was used as a reference compound. Relative retention time values were also obtained both after hydrogenation and borohydride reduction of the unknown and the reference standard. Similar retention behavior in all cases and identical mass spectra were the basis for the identification.

Siperstein and Fagan (120) reported the marked suppression of mevalonic acid synthesis by cholesterol feeding, whereas the synthesis of β-hydroxy-β-methyl glutaric acid (β-HMG) was unaffected by this regimen. These investigators concluded that the major site of the cholesterol feedback system is located at the reaction responsible for the conversion of β-HMG to mevalonic acid. To demonstrate this effect, Siperstein et al. (121) developed a method for the separation and isolation of mevalonic acid and β-HMG by a technique employing GLC (of the methyl esters). The GLC portion of this procedure was verified by collecting peaks of the respective labeled compounds followed by liquid scintillation counting for carbon-14 assay to establish proportionality of the mass and radioactivity. The validity of isolating the de novo synthesized mevalonic acid from a rat liver system by this procedure was established utilizing acetate-2-^{14}C. The identification of the recovered labeled material was based on coincident retention times with standard methyl mevalonate. Confirmation was obtained by collecting the peak in question and converting it to the ethyl ester; the retention time of the new derivative was identical to standard ethyl mevalonate. The measurement of radioactivity in the collected column effluent showed the presence of only one peak which coincided exactly with the retention time of the ethyl mevalonate peak. The authors interpret the presence of carbon-14 in two chemical derivatives of mevalonic acid as proof that the synthesized compound was mevalonic acid.

A number of reports have appeared recently involving the role of 2,3-oxidosqualene as an intermediate in the biosynthesis of sterols from squalene (122–125). van Tamelen et al. (125), on the basis of their experiments, suggested that squalene 2,3-epoxide is a normally occurring intermediate in the enzymic cyclization of squalene. Labeled epoxide was prepared from squalene-^{14}C biosynthesized from mevalonic acid-2-^{14}C. The labeled compound was incubated with a liver enzyme system in an atmosphere of either oxygen or nitrogen. Identification of recovered sterols (incubation under nitrogen) as lanosterol and dihydrolanosterol was made with GLC and the distribution of radioactivity was determined by carbon-14 assay of collected effluents. The dihydrolanosterol contained 60% of the recovered activity, and 30% appeared in the lanosterol peak.

The major labeled product from incubations carried out in oxygen was identified as cholesterol. Samples of the recovered radioactive sterol were gas chromatographed as the methyl ether. The emergent peak contained all of the recovered radioactivity; fractions collected across the peak exhibited the same specific activity.

The conversion of plant sterols to cholesterol by insects has been the subject of several studies (97–101,126,127). Ikekawa et al. (128) reported the conversion of β-sitosterol to cholesterol in silkworm larvae and pupae. β-Sitosterol-^{3}H was introduced into larvae or pupae by appropriate means. After specified time periods sterol fractions were recovered by extraction and alumina column chromatography. Prior to GLC the sterol ester fraction was hydrolyzed. Three fractions of the effluent were collected as indicated in Fig. 9.11. The radioactivity in each fraction was determined by liquid scintillation

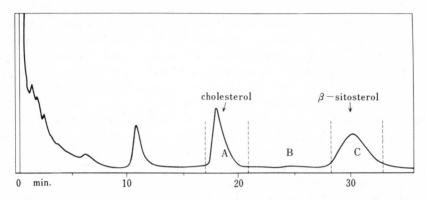

Fig. 9.11. Preparative separation of silkworm larval sterols by GLC. Three fractions (A, B, and C) were collected and radioactivity measured in a liquid scintillation counter. Column conditions: 1.8 m × 6 mm i.d. glass column; 1.5% SE-30 on 80–100-mesh Gas-Chrom P; 240°C; 100 ml/min. Illustration taken from Ref. 128.

counting. Based on their results the authors concluded that the larva is able to convert dietary β-sitosterol to cholesterol, thus showing that larval cholesterol originates from mulberry leaf β-sitosterol.

B. Lipids

The development of efficient collecting techniques has greatly increased the utility of preparative GLC as applied to the study of fats, waxes, and phospholipids. An example of obtainable recoveries is given in the report by Hammarstrand and associates (129). Average recovery was 92% for methyl palmitate-1-^{14}C and 90% for a mixture of carbon-14 labeled fatty acids containing methyl myristate, palmitate, stearate, and arachidate. The distribution of trapped carbon-14 between the methyl esters was acceptably constant, indicating suitability of the procedure for studies of the incorporation of radioactive substrates into the fatty acids of biological systems.

Nicolaides and Ansari (130) have reported the structures of dienoic fatty acids occurring in small amounts in adult human skin surface lipid. The multistep chromatographic procedure utilizes preparative GLC to obtain dienes for analytical GLC of the original and hydrogenated products, as well as for ozonolysis and subsequent GLC of the resulting aldehydes and aldesters. By these methods the authors found the C-18 and C-20 dienes to constitute over 93% of the total dienes. The $18:\Delta^{5,8}$ and $18:\Delta^{9,12}$ (linoleate) represented 81% of the C-18 dienes. Nicolaides and Ansari proposed the name sebaleic acid for $18:\Delta^{5,8}$ because it was the most abundant diene to occur in a lipid which, in their opinion, was undoubtedly of sebaceous gland origin.

Chang and Sweeley (131) established the structures of three polyenoic acids isolated from canine adrenal lipids with the aid of preparative GLC. After methanolysis of the crude lipid and extraction of methyl esters with hexane, the chromatogram shown in Fig. 9.12 was obtained on adipate polyester. Identification of peaks was based on comparisons of retention times of known standards. The chain lengths of the polyenoic acids (peaks 11, 12, and 13) were confirmed by GLC analysis of hydrogenated samples. For the isolation of the polyenoic acids by preparative GLC, 1-ml portions of a 25% solution of methyl esters in hexane were used for each injection. The collected samples were contaminated with eluted adipate polyester, which was removed by silicic acid chromatography.

GLC analyses of the purified fractions are shown in Fig. 9.13. The traces of 16–0, 16–1, 18–0, and 18–1 in fractions 12 and 13 were considered negligible for the purposes of this study. Repetition of preparative GLC resulted in the purity exhibited in fraction 11. The double bonds were located by oxidative cleavage of the olefinic bonds and GLC analysis of the (methylated) acidic fragments. The structures assigned to the polyenoic acids were 8,11,14-eicosatrienoic, 5,8,11,14-eicosatetratenoic, and 7,10,13,16-docosatetraenoic acid.

Polito et al. (132) obtained a sample of sphinga-4,14-dienine from human plasma sphingomyelin by chromatographic procedures subsequent to acid-catalyzed methanolysis of the lipid. Deduction of the structure of the sphingolipid base was based on MS data before and after osmium tetroxide oxidation, and by MS identification of sebacic acid after permanganate–periodate oxidation of the base. Samples of sphingosine and sphingadienine as the N-acetyl-O-trimethylsilyl derivatives were purified by GLC. Appropriate effluent fractions were collected for oxidation to establish the position of the olefinic groups.

Refsum's disease (heredopthia atatica polymiritiformis, HAP) (133) is an inherited neurological disorder shown by Klenk and Kahlke (134) to be associated with the accumulation of phytanic acid (3,7,11,15-tetramethylhexa-

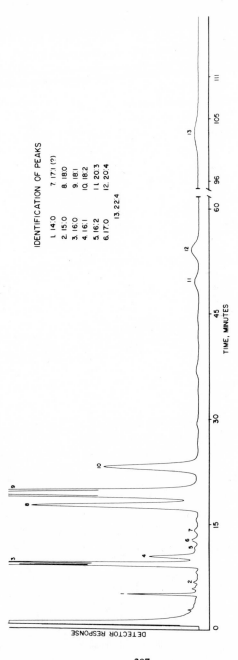

Fig. 9.12. GLC analysis of methyl esters from canine adrenal tissue. Column (6 ft × ⅛ in.) packed with 15% ethylene glycol adipate on 80–100-mesh Chromosorb W; 200°C; 20 psi. Illustration taken from Ref. 131.

IDENTIFICATION OF PEAKS

1. 14:0   7. 17:1 (?)
2. 15:0   8. 18:0
3. 16:0   9. 18:1
4. 16:1   10. 18:2
5. 16:2   11. 20:3
6. 17:0   12. 20:4
        13. 22:4

307

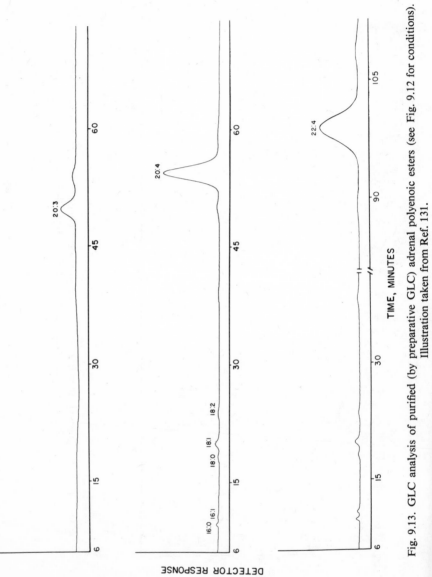

Fig. 9.13. GLC analysis of purified (by preparative GLC) adrenal polyenoic esters (see Fig. 9.12 for conditions). Illustration taken from Ref. 131.

decanoic acid) in blood and tissue. Steinberg and co-workers (135,136) demonstrated the exogenous origin of phytanic acid and that patients with HAP exhibit defective catabolism of the acid. Herndon and associates (137) recently presented results showing that the alpha oxidative pathway for phytanate oxidation, established by animal studies, is present also in human tissues and that the principal enzymic defect in Refsum's disease lies in the production of α-hydroxyphytanic acid, presumed to be the first step in phytanic acid oxidation (138–140). The conclusions of the authors were supported by the distribution of radioactivity among fatty acids in incubation mixtures with phytanic acid-U-^{14}C as a substrate utilizing control and HAP cultures. Figure 9.14 shows a gas chromatogram of recovered methylated fatty acids and the radioactivity found in collected fractions from a control culture.

Significant amounts of radioactivity were found in pristanate and 4,8,12-trimethyldecanoate, both previously identified as degradation products from a major pathway for phytanic acid metabolism in man. When the same incubation was conducted with an HAP culture, the results shown in Fig. 9.15 were obtained. Essentially only the parent phytanate peak was present. There was no evidence of breakdown into branched-chain intermediates, or of incorporation into straight-chained fatty acids.

C. Other Compounds

Preparative GLC is of course not limited to steroids and lipids in its application but has indeed been helpful in many other areas of research. Studies of carbohydrates, amino acids, alkaloids, metabolic and natural products have been conducted utilizing this technique.

Bolan and Steele (141) described a gas chromatographic method for the analysis of 10 phenolic glycosides commonly found in *Salix* species. A crystalline fraction obtained from a glycoside extract of this species was employed to test the efficiency of their procedure. The two major components of the natural mixture (as TMSi ethers) were collected from the GLC column. The TMSi ethers of salicin and picein were similarly recovered. IR and UV data obtained for pure, recovered TMSi salicin and TMSi picein corresponded to previously reported values (142,143). The two principal constituents of the natural mixture were identified as salicin and picein, based on their gas chromatographic behavior and IR and UV spectra which corresponded with those of authentic trimethylsilylated salicin and picein.

Roth and Giarman (144) conducted experiments to determine whether or not γ-aminobutyric acid (GABA) might serve as a precursor of γ-hydroxybutyric acid (GHB) in the rat brain in vivo. Interest in this problem arises from reports that γ-butyrolactone (GBL) and GHB, its hydrolytic cleavage product, can produce unconsciousness and anesthesia in man and in a variety

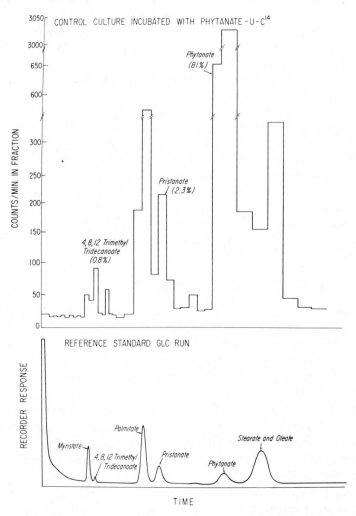

Fig. 9.14. Gas–liquid radiochromatogram of methylated fatty acids derived from a control culture incubated 48 hr with phytanic acid -U-14C. Radioactivity assayed in collected fractions is shown in upper part of figure. A standard fatty acid methyl ester mixture run under the same conditions is shown in lower part. A two-component stationary phase (7% EGS and 1.5% SE-30) on Gas-Chrom Q was employed; 175°C; 80 ml/min. The radioactive peaks corresponding to methyl phytanate, methyl pristanate, and methyl 4,8,12-trimethyltridecanoate are labeled and the percent each contributed to the total collected radioactivity is given. Illustration taken from Ref. 137.

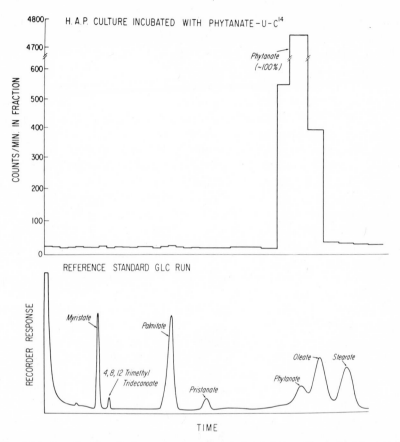

Fig. 9.15. Gas–liquid radiochromatogram of methylated fatty acids derived from HAP culture incubated 48 hr with phytanic acid U-[14]C. Column conditions same as Fig. 9.14. Illustration taken from Ref. 137.

of animals (145–147). Increased brain levels of GABA do not seem to evoke central nervous system depression (148,149). We refer to the original paper for the biological aspects of these studies. GHB-^{3}H originating from GABA-^{3}H was converted to GBL-^{3}H and gas chromatographed as such. Collections were made by bubbling effluent gas through a scintillation counting solution mixture which was ultimately counted. Radioactivity appearing in the GHB (GBL) peak of the experimental subjects in excess of radioactivity evident in the peaks originating from control animals was taken as evidence for the in vivo conversion of GABA-^{3}H to GHB-^{3}H in the rat brain.

Curtius et al. (90) separated persilylated fructose into five components by means of GLC. Preparative GLC was used to isolate individual components for characterization by IR spectrometry and MS. Of the five components, two were pyranosides, two furanosides, and one the open-chain form of fructose. An equilibrium mixture of fructose was found to contain about 33% furanoside and about 67% pyranoside.

Barnett et al. (150) described a method for the separation and estimation of pyruvate, lactate, β-hydroxybutyrate, and some Krebs cycle intermediates in rat tissues by GLC. A carbon-14 internal standard for each acid was used to allow individual corrections for losses. This was accomplished by collecting effluent peaks and measuring carbon-14 content so that adjustments could be made for losses occurring during extraction and chromatography. Column background radioactivity was monitored by trapping a peak-free segment of the effluent. The reproducibility of measurements was tested on liver extract and the authors reported a mean coefficient of variation for four experiments ranging from 3 to 7%.

Tham and Holmstedt (151,152) have reported the presence of 1-methyl-imidazole-5-acetic acid (1,5-MeIMAA) in human urine, in addition to the generally recognized histamine metabolite, 1-methylimidazole-4-acetic acid (1,4-MeIMAA). Tham (153) further pursued these studies in dogs, utilizing histamine-2-^{14}C and l-histidine-2-[14]C. After preliminary purification on ion-exchange columns, the methyl esters of the imidazolic acids were prepared and subsequently analyzed by GLC. The compounds eluting from the column were collected and, after appropriate manipulation, assayed for radioactivity. Figure 9.16 illustrates the application of this technique to urinary extracts from a dog given labeled histidine orally. Imidazoleacetic acid (ImAA) represents the bulk of both mass and radioactivity. Peaks of 1,4-MeImAA and 1,5-MeImAA were virtually free of radioactivity. The identity of the second radioactive peak (35–40 min) has not yet been established (154). Similar experiments conducted with histamine-2-^{14}C showed incorporation of the isotope into ImAA when administered orally or subcutaneously. Labeled 1,4-MeImAA was formed after subcutaneous administration of histamine; 1,5-MeImAA, although present in the chromatogram, did not contain radioactivity.

Fales et al. (111) demonstrated the feasibility of preparative-scale GLC of alkaloids with the use of thin-film analytical type packings. Figure 9.17 illustrates the separation obtained when a 7-mg alkaloid mixture derived from a plant of the Amaryllidaceae family was chromatographed. The vertical lines delineate the fraction-cutting operation. All fractions were examined by analytical GLC, IR spectroscopy, and other physical methods. The recovered components were found to be better than 98% pure as determined by analytical GLC. Compounds in fractions 2, 5, and 6 were identified as buphanisine, buphanidiene, and undulatine, respectively, known alkaloids of the Amaryllidaceae family. Compounds in fractions 1, 4, and 8 were previously unknown and first isolated by this technique.

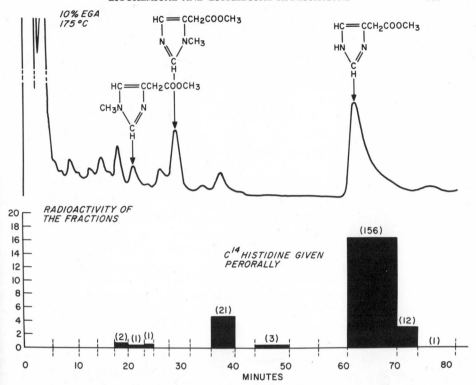

Fig. 9.16. Gas–liquid radiochromatographic analysis of a urinary extract from a dog given labeled histidine perorally. The upper panel shows the gas chromatogram; the stationary phase was 10% ethylene glycol adipate, 175°C. The lower panel shows the radioactivity of the various collected peaks; 1240 cpm of radioactivity was injected. The compound at 35–40-min retention time is unidentified. Illustration taken from Ref. 153.

The preparation damiana consists of dried leaves of *Turnera diffusa* or of *Turnera aphrodisiaca*, plants found from Texas to lower California. It was introduced into the United States in 1879 as a drug and held to possess a broad spectrum of applications: aphrodisiac, urinary tract disinfectant, and central nervous system stimulant. Auterhoff and Häufel (155) recently examined this early "wonder drug" and were able to characterize a number of its constituents with the aid of preparative GLC. The main constituents in the "low-boiling" fraction were identified as α-pinene, β-pinene, p-cymene, and 1,8-cineol by GLC retention times (obtained on three different columns), TLC, IR spectra, and optical rotation. The "higher-boiling" fraction yielded three pure components by preparative GLC, classified as sesquiterpene hydrocarbons. The chief constituent, compound F, of which 300 mg was recovered, was subjected to determination of boiling point, molecular weight, and elemental analysis, allowing the assignment of the empirical formula $C_{15}H_{24}$.

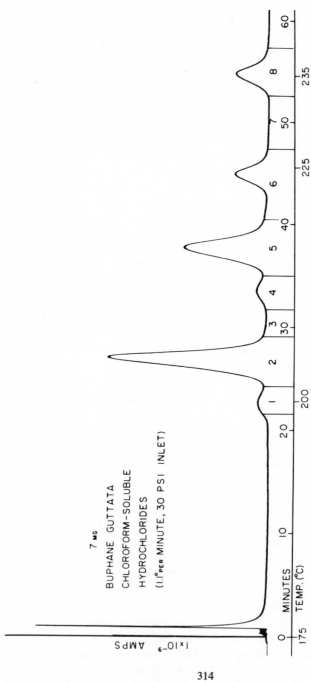

Fig. 9.17. Separation of a mixture of alkaloids. The sample was the mixed free bases from the chloroform-soluble hydrochlorides of a *Buphane guttata* alkaloid extract. Times and temperatures for the procedure are indicated. Vertical lines indicate the fraction-cutting operation. All components were found to be above 98 % purity by analytical gas chromatography. Compounds of known structure were identified as buphanisine (fraction 2), buphanidrine (fraction 5), and undulatine (fraction 6). Compounds 1, 4, and 8 are new alkaloids, isolated for the first time by this procedure. Column conditions same as in Fig. 9.1. Illustration taken from Ref. 111.

UV analysis and subsequent hydrogenation demonstrated the presence of two double bonds not in conjugation. NMR and IR spectroscopy led the authors to conclude that compound F was an isomer of commercially available guajen.

One of the early applications of preparative GLC as an integral part of an isolation and identification procedure is represented by the studies of Wharton et al. (156) dealing with the sex attractant of the American cockroach. In another report in which preparative GLC was utilized for entomological studies, Bedard et al. (157) studied the field response of the western pine beetle to its sex pheromone *exo*-brevicomin, and the possible synergism of a host terpene, myrcene. Brownlee and Silverstein (158) had previously isolated a sex attractant from the frass of the western pine beetle and using preparative GLC, a carbon skeleton determinator, and MS identified the substance as brevicomin. Myrcene is a prominent constituent of the xylem oleo resin of *Pinus ponderosa* and *Pinus coulteri*. It was obtained commercially and purified by sequential, preparative GLC. The authors reported a purity of 99.8% as determined by analytical GLC.

Rodin et al. (159) identified two components of the sex attractant of the female demerstid beetle *Trogoderma inculsum* Le Conte. The starting material for this endeavor consisted of 250,000 unmated female beetles. After a number of manipulations, 5 g of the active fraction was subjected to preparative GLC. Compound 1 (about 1 mg from 100,000 beetles) was obtained by fractionation first on an SE-30 column and then on a Carbowax-20M column. A third fractionation on a diethylene glycol succinate column was necessary to obtain about 0.2 mg of compound 2. IR, mass, and NMR spectra were obtained on the isolated compound 1; optical rotation was measured; ozonolysis, hydrogenolysis, and analytical GLC were also used to elicit structure. Compound 1 was identified as (−)-14-methyl-*cis*-8-hexadecen-1-ol and compound 2, based on IR and mass spectra, ozonolysis, and optical rotation, as (−)-methyl-14-methyl-*cis*-8-hexadecenoate. Both compounds were synthesized and found to be biologically active. Spectral data for the synthesized and isolated compounds were congruent.

Tumlinson et al. (159a) reported the isolation and identification of two terpene aldehydes and two terpene alcohols from male boll weevils *Anthonomus grandis* Boheman and their feces. Preparative GLC was utilized to obtain samples for bioassay, mass, NMR, and IR spectral studies. The two aldehydes were eluted as one peak from an SE-30 column into a drop of 2,4-dinitrophenylhydrazine reagent on a thin-layer plate. The resulting derivatives chromatographed on thin layer very similarly to standard terpene carbonyl compounds. The authors synthesized the four identified terpenes and found the mass, IR, and NMR spectra and the biological activity of the synthetic compounds to be identical to those of the corresponding natural products.

The isolation and identification of the substance causing the peculiar odor in sweat of schizophrenic patients has been reported by Smith et al. (160). Analytical GLC revealed the presence of an acid found only in schizophrenic individuals. The retention time of the acid was between that of *n*-heptanoic and *n*-octanoic acids. Preparative GLC was employed to obtain about 230 μg of the unknown acid. Mass, IR, and NMR spectra, as well as analytical GLC of the isolated substance and comparisons to a synthesized reference compound allowed the identification of the unknown as *trans*-3-methyl-2-hexenoic acid. Confirmation of the location of the double bond in both the natural and synthetic acids was obtained by cleavage across the double bond to give 2-pentanone and oxalic acid.

Taylor et al. (161) described the synthesis of *dl*-Δ^6-3,4-*trans*-tetrahydrocannabinol, the racemic modification of one of the physiologically active constituents of hashish (marijuana). Preparative GLC was utilized to separate the Δ^6-3,4-*cis* isomer from the Δ^6-3,4-*trans* isomer. The latter tetrahydrocannabinol was identical to the natural product isolated from hemp with respect to NMR, UV, and IR spectra and differed only in optical activity. The authors observed the thermal isomerization of the Δ^1-3,4-*trans* isomer to the Δ^6-3,4-*trans* isomer during GLC and suggested that the physiological effects consequent to the smoking of marijuana, now ascribed to the Δ^1-isomer, may actually be attributable to the Δ^6-isomer.

VI. Concluding Comments

If the substance the alchemists thought capable of transmuting base metals into noble metals was called the "philosophers' stone," perhaps the functional preparative GLC column could be dubbed the "philosophers' pipe," in which individual components of a mixture are purified by a gas-phase separation. Lest we moderns become too enamored of our technology, however, we should reflect upon a recent publication by Needham and Gwei-Djen (162). These authors report that as early as the 11th century A.D. Chinese iatrochemists employed rather remarkable and sophisticated fractionation procedures for producing preparations containing human urinary steroid sex hormones used in the quasiempirical therapy of that day. The final step in the process was usually sublimation in small earthenware pots. "Great attention was paid to the degree of heating, so that they should be neither too cold nor get too hot" (162). We would all agree that remarkable technological progress has been made since the days of the alchemist—although we in GLC are still greatly concerned with temperature control. There is room (indeed, need) for new and improved methodology, however, and it is on this note that we should like to conclude. Recent advances in GLC, including the development of special derivatives and also new detectors, have allowed the extension of GLC techniques to many families of compounds not normally volatile and the reduction in sample size to the nanogram to picrogram level. It is now time for the chromatographer to "look up" and extend the separation of closely related compounds of interest to the pharmaceutical chemist to the submilligram-milligram level. This may be achieved through automated procedures, improved column technology (e.g., use of the promising textured glass beads) to allow high-efficiency columns, or by as yet unexplored concepts.

It may prove possible to extend the value of preparative-scale GLC by capitalizing upon already observed phenomena. For example, although molecular alterations that occur during GLC are usually considered deleterious and to be avoided, the possibility does exist that quantitative on-column transformations may be turned to good use if the reaction product possesses a potentially useful structure. Acetyl hydroxamates of long-chain fatty acids have been shown to undergo a Lossen rearrangement to the corresponding isocyanates (163), and methanesulfonates and p-toluenesulfonates of secondary alcohols (e.g., cholestanol) undergo an elimination reaction to yield the corresponding olefin (67,164). Preparative GLC would be an excellent method for the production of pyro- and isopyrocalciferol (56). Several other examples of on-column transformations are the reaction of α-hydroxyferrocene to yield vinyl ferrocene "in a very pure state" (165) and the decarboxylation of perfluoroaromatic acids to form the corresponding hydro compounds (166). It thus appears likely that a number of reactions known to organic chemistry occur during GLC (167). One can conceive of a GLC-like system serving as a reactor tube in which the labile compound is introduced at the front and the product collected at the rear of the system, free from reagents, by-products, and starting material.

It is sometimes held, and rightly so, that the necessity of exposing the sample to elevated temperatures reduces the applicability of analytical and preparative-scale GLC. The number of authentic thermal reactions that occur during GLC is limited, however, and one must be careful to distinguish between authentic thermally induced alterations (e.g., the Lossen rearrangement (163), sulfonate eliminations (67,164), vitamin D cyclization (56), and those that involve undesired surface effects. An example of the latter is the dehydration of cholesterol during GLC that occurs with undeactivated support but which can be avoided by use of properly deactivated support (168). Surface-related decomposition during preparative-scale GLC has been observed by Bowers (169). Terpene epoxides were found to decompose when applied to a stainless-steel GLC system. Gold plating of the injection zone significantly reduced epoxide decomposition, and the additional use of gold-plated columns virtually eliminated this undesirable phenomenon. This regal solution indicates the kind of ingenuity and technology that will extend the applicability of preparative-scale GLC.

A promising approach for the separation of substances of very low volatility (sugars, amino acids, nucleosides, nucleotides, polymers) without derivatization involves the use of a gas above its critical point as a mobile phase (at pressures up to 2000 atm most gases become fluids with liquidlike properties, hence can serve as solvents) (170). To move a step further away from conventional GLC, recent developments in the field of liquid chromatography such as high-speed (171,172) and high-efficiency (171,173) systems, and the use of new stationary phases and support materials [e.g., the "brushes of Halasz" (174), and controlled surface porosity supports (172)] make liquid chromatography increasingly attractive to persons working with complex mixtures of nonvolatile materials.* The gas chromatographer need not fear these new techniques—they should prove to be helpful, challenging, and complementary.

One must have access to a variety of methods to solve a variety of problems successfully, and must be able to choose for a given problem the approach that provides the solution with the least expenditure of money and effort. There is no need to activate an expensive piece of laboratory equipment when two recrystallizations from a commonly available solvent can effect the desired purification. When simpler approaches prove unsuccessful, however, more sophisticated techniques such as preparative GLC may be the method of choice. It is the judgment of the scientist, his judicious choice of approach, that is the key step in any procedure. We hope this chapter has brought the capabilities, potential, and limitations of preparative-scale GLC in biochemical and biomedical research to the attention of scientists working in these vital areas.

Acknowledgments

It is a pleasure to acknowledge the generous cooperation of H. M. Fales, L. J. Goad, E. C. Horning, N. Ikekawa, D. Steinberg, C. C. Sweeley, L. Swell, and R. Tham in supplying us with glossy prints or photographic negatives of illustrations for use in this chapter.

References

1. A. J. P. Martin and R. L. M. Synge, *Biochem. J.*, **35**, 1358 (1941).
2. S. R. Lipsky and R. A. Landowne, *Ann. Rev. Biochem.*, **29**, 649 (1960).
3. W. J. A. VandenHeuvel, C. C. Sweeley, and E. C. Horning, *J. Am. Chem. Soc.*, **82**, 3481 (1960).
4. H. A. Lloyd, H. M. Fales, P. F. Highet, W. J. A. VandenHeuvel, and W. C. Wildman, *J. Am. Chem. Soc.*, **82**, 3791 (1960).

*However, it is equally limited to small samples and the extension to preparative use is fraught with most of the same problems as preparative gas chromatography.

5. E. C. Horning and W. J. A. VandenHeuvel, *Ann. Rev. Biochem.*, **32**, 709 (1963).
6. M. G. Horning, E. A. Boucher, and A. M. Moss, *J. Gas Chromatog.*, **5**, 297 (1967); L. T. Sennello, F. A. Kummerow, and C. J. Argoudelis, *J. Hetero. Chem.*, **4**, 295 (1967); A. R. Prosser, A. J. Sheppard, and D. A. Libby, *J. Assoc. Offic. Agri. Chem.*, **50**, 1348 (1967).
7. M. G. Horning, E. A. Boucher, A. M. Moss, and E. C. Horning, *Anal. Letters*, **1**, 713 (1968).
8. R. L. Hancock, *J. Chromatog. Sci.*, **7**, 366 (1969).
9. F. Eisenberg, Jr., and A. H. Bolden, *Anal. Biochem.*, **29**, 284 (1969).
10. R. Binks, J. MacMillan, and R. J. Price, *Phytochemistry*, **8**, 271 (1969).
11. F. Vane and M. G. Horning, *Anal. Letters*, **2**, 357 (1969); E. Granstrom and B. Samuelsson, *J. Am. Chem. Soc.*, **91**, 3398 (1969); M. Hamberg, *European J. Biochem.*, **6**, 135 (1968).
12. K. Tsuji and J. H. Robertson, *Anal. Chem.*, **41**, 1332 (1969).
13. W. E. Wilson and J. E. Ripley, *Anal. Chem.*, **41**, 810 (1969).
14. G. Casparrini, M. G. Horning, and E. C. Horning, *Anal. Letters.*, **1**, 481 (1968).
15. B. J. Gudzinowicz, *Gas Chromatographic Analysis of Drugs and Pesticides*, Marcel Dekker, New York, 1967; H. H. Wotiz and S. J. Clark, *Gas Chromatography in the Analysis of Steroid Hormones*, Plenum Press, New York, 1966; H. V. Street, *J. Chromatog.*, **41**, 358 (1969).
16. C. J. W. Brooks, L. Hanaineh, A. McCormick, G. Steel, and J. S. Young, in *Gas Chromatography of Hormonal Steroids*, R. Scholler and M. F. Jayle, Eds., Gordon and Breach, New York, 1968; F. A. Vandenheuvel and A. S. Court, *J. Chromatog.*, **39**, 1 (1969).
17. H. -C. Curtius and M. Müller, *J. Chromatog.*, **32**, 222 (1968).
18. R. Kaiser, *Chem. Brit.*, **5**(2), 54 (1969).
19. J. A. McCloskey, in *Methods in Enzymology*, Vol. 14, *Lipids*, J. M. Lowenstein, Ed., Academic Press, New York, 1969; E. C. Horning, C. J. W. Brooks, and W. J. A. VandenHeuvel, in *Advances in Lipid Research*, R. Paoletti and D. Kritchevsky, Eds., Vol. 6, Academic Press, New York, 1968.
20. L. Swell, *Anal. Biochem.*, **16**, 70 (1966).
21. P. T. Russell, R. T. van Aller, and W. R. Nes, *J. Biol. Chem.*, **242**, 5802 (1967).
22. R. O. Martin, *Anal. Chem.*, **40**, 1197 (1968).
23. S. Smith, R. Watts, and R. Dils, *J. Lipid Res.*, **9**, 52 (1968).
24. W. R. Supina, in *Biomedical Applications of Gas Chromatography*, H. A. Szymanski, Ed., Plenum Press, New York, 1964.
25. C. T. Bishop, in *Methods of Biochemical Analysis*, D. Glick, Ed., Vol. X, Interscience, New York, 1962.
26. W. J. A. VandenHeuvel and E. C. Horning, *Biochem. Biophys. Res. Commun.*, **4**, 399 (1961).
27. C. C. Sweeley, W. W. Wells, and R. Bentley, in *Methods of Enzymology*, Vol. 8, S. P. Colowick and N. O. Kaplan, Eds., Academic Press, New York, 1966.
28. B. Weinstein, in *Methods of Biochemical Analysis*, D. Glick, Ed., Vol 14, Interscience, New York, 1966.
29 R. Ruhlmann and W. Giesecke, *Angew. Chem.*, **73**, 113 (1961).
30. D. Stalling, C. W. Gehrke, and R. W. Zumwalt, *Biochem. Biophys. Res. Commun.*, **31**, 616 (1968).
31. W. J. A. VandenHeuvel, J. L. Smith, and J. S. Cohen, *J. Chromatog. Sci.*, **8**, 567 (1970).

32. C. W. Gehrke, D. Roach, R. W. Zumwalt, D. L. Stalling, and L. L. Wall, *Quantitative Gas-Liquid Chromatography of Amino Acids in Proteins and Biological Substances*, Analytical Biochemistry Laboratories, Columbia, Missouri, 1968; D. Roach and C. W. Gehrke, *J. Chromatog.*, **43**, 303 (1969).

33. A. Del Favero, A. Darbre, and M. Waterfield, *J. Chromatog.*, **40**, 213 (1969); A. Islam and A. Darbre, *J. Chromatog.*, **43**, 11 (1969); E. Gelpi, W. A. Koenig, J. Gilbert, and J. Oro, *J. Chromatog. Sci.*, **7**, 604 (1969).

34. J. J. Pisano, W. J. A. VandenHeuvel, and E. C. Horning, *Biochem. Biophys. Res. Commun.*, **7**, 82 (1962).

35. J. J. Pisano, in *Theory and Application of Gas Chromatography in Industry and Medicine*, H. S. Kroman and S. R. Bender, Eds., Grune and Stratton, New York, 1968.

36. R. E. Harman, J. L. Patterson, and W. J. A. VandenHeuvel, *Anal. Biochem.*, **25**, 452 (1968).

36a. B W. Melvas, *Acta Chem. Scand.*, **23**, 1679 (1969).

37. M. Makita and W. W. Wells, *Anal. Biochem.*, **5**, 523 (1963); W. J. A. VandenHeuvel and K. L. K. Braly, *J. Chromatog.*, **31**, 9 (1967).

38. J. Sjövall, in *Biomedical Applications of Gas Chromatography*, H. A. Szymanski, Ed., Plenum Press, New York, 1964; J. Sjövall, in *The Gas-Liquid Chromatography of Steroids*, J. K. Grant, Ed., Cambridge University Press, Cambridge, 1967).

39. W. J. A. VandenHeuvel, *J. Chromatog.*, **25**, 29 (1966).

40. C. E. Dalgliesh, E. C. Horning, M. G. Horning, K. L. Knox, and K. Yarger, *Biochem. J.*, **101**, 792 (1966).

41. W. W. Wells, T. Katagi, R. Bentley, and C. C. Sweeley, *Biochim. Biophys. Acta*, **82**, 408 (1964).

42. P. I. Jaakonmaki, K. A. Yarger, and E. C. Horning, *Biochim. Biophys. Acta*, **137**, 216 (1967); W. J. A. VandenHeuvel, *J. Chromatog.*, **28**, 406 (1967).

43. M. G. Horning, A. M. Moss, E. A. Boucher, and E. C. Horning, *Anal. Letters*, **1**, 311 (1968); J. Vessman, A. M. Moss, M. G. Horning, and E. C. Horning, *Anal. Letters*, **2**, 81 (1969); E. Anggard and G. Sedvall, *Anal. Chem.*, **41**, 1250 (1969).

44. M. G. Horning, A. M. Moss, and E. C. Horning, *Biochim. Biophys. Acta*, **148**, 497 (1967); W. J. A. VandenHeuvel, *J. Chromatog.*, **36**, 354 (1968).

45. H M. Fales and T. Luukkainen, *Anal. Chem.*, **37**, 955 (1965)

46. M. G. Horning, A. M. Moss, and E. C. Horning, *Anal. Biochem.*, **22**, 284 (1968).

47. J. F. Klebe, H. Finkbeiner, and D. M. White, *J. Am. Chem. Soc.*, **88**, 3390 (1966).

48. A. E. Pierce, *Silylation of Organic Compounds*, Pierce Chemical Co., Rockford, Illinois, 1968.

49. E. M. Chambaz and E. C. Horning, *Anal. Biochem.*, **30**, 7 (1969).

50. E. C. Horning and W. J. A. VandenHeuvel, in *Advances in Chromatography*, Vol. 1, J. C. Giddings and R. A. Keller, Eds., Marcel Dekker, New York, 1965.

51. W. J. A. VandenHeuvel, C. C. Sweeley, and E. C. Horning, in *Drugs Affecting Lipid Metabolism*, S. Garattini and R. Paoletti, Eds., Elsevier, Amsterdam, 1961; unpublished results.

52. W. J. A. VandenHeuvel, J. Sjövall, and E. C. Horning, *Biochim. Biophys. Acta.*, **48**, 596 (1961).

53. W. J. A. VandenHeuvel, *J. Chromatog.*, **26**, 396 (1967).

54. R. A. Okerholm, P. I. Brecher, and H. H. Woltiz, *Steroids*, **12**, 435 (1968).

55. C. J. W. Brooks, E. C. Horning, and J. S. Young, *Lipids*, **3**, 391 (1968).

55a. J. Meinwald, Y. C. Meinwald, and P. H. Mazzocchi, *Science*, **164**, 1174 (1969).

56. H. Ziffer, W. J. A. VandenHeuvel, E. O. A. Haahti, and E. C. Horning, *J. Am. Chem. Soc.*, **82**, 6411 (1960).

56a. W. J. A. VandenHeuvel and E. C. Horning, *Biochem. Biophys. Res. Commun.*, **3**, 356 (1960).

57. E. Minini, J. C. Orr, R. C. Gibb, and L. L. Engel, in *Gas Chromatography of Hormonal Steroids*, R. Scholler and M. F. Jayle, Eds., Gordon and Breach, New York, 1968.

58. C. J. W. Brooks, *Anal. Chem.*, **37**, 636 (1965).

59. M. A. Kirscher and H. M. Fales, *Anal. Chem.*, **34**, 1548 (1962).

60. W. L. Gardiner and E. C. Horning, *Biochem. Biophys. Acta.*, **115**, 524 (1966)

61. W. J. A. VandenHeuvel, J. L. Patterson, and K. L. K. Braly, *Biochim. Biophys. Acta*, **144**, 691 (1967).

62. G. M. Anthony, C. J. W. Brooks, I. Maclean, and I. Sangster, *J. Chromatog. Sci.*, **7**, 632 (1969).

63. R. W. Kelly, *Steroids*, **13**, 507 (1969); *J. Chromatog.*, **43**, 229 (1969).

64. E. C. Horning, W. J. A. VandenHeuvel, and B. G. Creech, in *Methods of Biochemical Analysis*, Vol. 11, D. Glick, Ed., New York, 1963.

65. W. J. A. VandenHeuvel and E. C. Horning, in *Biomedical Applications of Gas Chromatography*, H. A. Szymanski, Ed., Plenum Press, New York, 1964.

66. E. C. Horning, K. C. Maddock, K. V. Anthony, and W. J. A. VandenHeuvel, *Anal. Chem.*, **35**, 526 (1963).

67. W. J. A. VandenHeuvel, R. N. Stillwell, W. L. Gardiner, S. Wikstrom, and E. C. Horning, *J. Chromatog.*, **19**, 22 (1965).

68. R. C. Dutky, W. E. Robbins, T. J. Shortino, J. N. Kaplanis, and H. E. Vroman, *J. Insect Physiol.*, **13**, 1501 (1967).

69. W. J. A. VandenHeuvel, W. L. Gardiner, and E. C. Horning, *J. Chromatog.*, **27**, 85 (1967).

70. E. O. A. Haahti and H. M. Fales, *J. Lipid Res.*, **8**, 131 (1967).

71. K. Tanaka, M. G. Horning, and E. C. Horning, unpublished results, 1964.

72. W. J. A. VandenHeuvel, B. G. Creech, and E. C. Horning, *Anal. Biochem.*, **4**, 191 (1962).

73. W. J. A. VandenHeuvel, *Separation Sci.*, **3**, 151 (1968).

74. J. Sjövall and K. Sjövall, *Steroids*, **12**, 359 (1968).

75. G. T. Brooks and A. Harrison, *Chem. Ind.* (*London*), 1414 (1966).

76. C. J. W. Brooks, E. M. Chambaz, and E. C. Horning, *Anal. Biochem.*, **19**, 234 (1967).

77. C. J. W. Brooks and J. Watson, *J. Chromatog.*, **31**, 396 (1967).

78. J. E. Karkkainen, E. O. A. Haahti, and A. P. Lehtonen, *Anal. Chem.*, **38**, 1316 (1966).

79. R. Ryhage, *Anal. Chem.*, **36**, 759 (1964).

80. J. T. Watson and K. Biemann, *Anal. Chem.*, **36**, 1135 (1964).

81. S. R. Lipsky, C. G. Horvath, and W. J. McMurray, *Anal. Chem.*, **38**, 1585 (1966).

82. C. Merritt, Jr., M. L. Bazinet, and W. G. Yeomans, *J. Chromatog. Sci.*, **7**, 122, 1969.

83. D. R. Black, R. A. Flath, and R. Teranishi, *J. Chromatog. Sci.*, **7**, 284, 1969.

84. C. C. Sweeley, W. H. Elliott, I. Fries, and R. Ryhage, *Anal. Chem.*, **38**, 1549 (1966).

85. P. M. Krueger and J. A. McCloskey, *Anal. Chem.*, **41**, 1930 (1969).

86. R. Tham, L. Nystrom, and B. Holmstedt, *Biochem. Pharmacol.*, **17**, 1735 (1968).

87. T. A. Bryce, G. Eglinton, R. J. Hamilton, M. Martin-Smith, and G. Subramanian, *Phytochemistry*, **6**, 727 (1967).

87a. P. I. Jaakonmaki, K. A. Yarger, and E. C. Horning, *Biochim. Biophys. Acta*, **137**, 216 (1967).

88. M. Greer, C. E. Hutcheson, and C. M. Williams, *Clin. Chim. Acta*, **22**, 461 (1968).

89. A. J. Glazko, T. Chang, J. Baukema, W. A. Dill, J. R. Goulet, and R. A. Buchanen, *Clin. Pharm. Therap.*, **10**, 498 (1969); G. R. Wilkinson and E. L. Way, *Biochem. Pharm.*, **18**, 1435 (1969); K. H. Palmer, B. Martin, B. Baggett, and M. E. Wall, *Biochem. Pharm.*, **18**, 1845 (1969); C. -G. Hammar, W. Hammer, B. Holmstedt, B. Karlen, F. Sjoquist, and J. Vessman, *Biochem. Pharm.*, **18**, 1549 (1969).

90. H. -C. Curtius, M. Müller, and J. A. Völlmin, *J. Chromatog.*, **37**, 216 (1968).

91. P. I. Jaakonmaki, K. L. Knox, E. C. Horning, and M. G. Horning, *European J. Pharmacol.*, **1**, 63 (1967).

92. C. -G. Hammar, B. Holmstedt, and R. Ryhage, *Anal. Biochem.*, **25**, 532 (1968); C. -G. Hammar, I. Hanin, B. Holmstedt, R. J. Kitz, D. J. Jenden, and B. Karlen, *Nature*, **220**, 915 (1968).

93. A. Karmen, L. Giuffrida, and R. L. Bowman, *J. Lipid Res.*, **3**, 44 (1962).

94. P. J. Thomas and H. J. Dutton, *Anal. Chem.*, **41**, 657 (1969).

95. B. Kliman and C. Briefer, Jr., in *The Gas-Liquid Chromatography of Steroids*, J. K. Grant, Ed., Cambridge University Press, Cambridge, 1967.

96. L. J. Goad and T. W. Goodwin, *Eruopean J. Biochem.*, **7**, 502 (1969).

97. W. E. Robbins, R. C. Dutky, R. E. Monroe, and J. N. Kaplanis, *Ann. Entomol. Soc. Am.*, **55**, 102 (1962).

98. C. H. Schaeffer, J. N. Kaplanis, and W. E. Robbins, *J. Insect Physiol.*, **11**, 1013 (1965).

99. J. N. Kaplanis, W. E. Robbins, R. E. Monroe, T. J. Shortino, and M. J. Thompson, *J. Insect Physiol.*, **11**, 251 (1965).

100. J. N. Kaplanis, R. E. Monroe, W. E. Robbins, and S. J. Louloudes, *Ann. Entomol. Soc. Am.*, **56**, 198 (1963).

101. J. A. Svoboda, M. J. Thompson, and W. E. Robbins, *Life Sci.*, **6**, 395 (1967).

102. G. Popjak, A. E. Lowe, D. Moore, L. Brown, and F. A. Smith, *J. Lipid Res.*, **1**, 29 (1959); G. Popjak, A. E. Lowe, and D. Moore, *J. Lipid Res.*, **3**, 364 (1962).

103. A. Karmen, *J. Assoc. Offic. Agr. Chem.*, **47**, 15 (1964).

104. A. Karmen, I. McCaffrey, and R. L. Bowman, *J. Lipid Res.*, **3**, 372 (1962).

105. A. Karmen, *J. Gas Chromatog.*, **5**, 502 (1967).

106. A. T. James, in *New Biochemical Separations*, A. T. James and L. J. Morris, Eds., Van Nostrand, London, 1964.

107. A. Karmen, in *Methods in Enzymology*, Vol. 14, J. M. Lowenstein, Ed., Academic Press, New York, 1969.

108. M. A. Kirschner and M. B. Lipsett, *J. Lipid Res.*, **6**, 7 (1965).

109. J. A. McCloskey, A. M. Lawson, and F. A. J. M. Leemans, *Chem. Commun.*, 285, (1967).

110. R. Bentley, N. C. Saha, and C. C. Sweeley, *Anal. Chem.*, **37**, 1118 (1965).

111. H. M. Fales, E. O. A. Haahti, T. Luukkainen, W. J. A. VandenHeuvel, and E. C. Horning, *Anal. Biochem.*, **4**, 296 (1962).

112. M. J. Thompson, W. E. Robbins, and G. L. Baker, *Steroids*, **2**, 505 (1963).

113. W. J. A. VandenHeuvel and G. W. Kuron, *J. Chromatog. Sci.*, **7**, 651 (1969).

114. J. E. van Lier and L. L. Smith, *J. Chromatog.*, **36**, 7 (1968).

115. J. E. van Lier and L. L. Smith, *Biochem.*, **6**(10), 3269 (1967).

116. R. Kaiser, *Chromatographie in der Gas Phase*, Band IV, Quantitative Auswertung, Bibliographisches Institut, Mannheim, 1965.

117. R. A. Jungman, E. Calvary, and J. S. Schweppe, *J. Clin. Endocrinol. Metab.*, **27**, 355 (1967).

118. H. Funasaki and J. R. Gilbertson, *J. Lipid Res.*, **9**, 766 (1968).

119. O. Jänne, T. Laatikainen, J. Vainio, and R. Vihko, *Steroids*, **13**, 121 (1969).

120. M. D. Siperstein and V. M. Fagan, *J. Biol. Chem.*, **241**, 602 (1966).

121. M. D. Siperstein, V. M. Fagan, and J. M. Dietschy, *J. Biol. Chem.*, **241**, 597 (1966).

122. E. J. Corey, W. E. Russey, and P. R. Ortiz de Montellano, *J. Am. Chem. Soc.*, **88,** 4750 (1966).
123. E. J. Corey and W. E. Russey, *J. Am. Chem. Soc.*, **88,** 4751 (1966).
124. J. D. Willett, K. B. Sharpless, K. E. Lord, E. E. van Tamelen, and R. B. Clayton, *J. Biol. Chem.*, **242,** 4182 (1967).
125. E. E. van Tamelen, J. D. Willett, R. B. Clayton, and K. E. Lord, *J. Am. Chem. Soc.*, **88,** 4752 (1966).
126. W. Bergmann, *J. Biol. Chem.*, **107,** 527 (1934).
127. A. J. Clark and K. Bloch, *J. Biol. Chem*, **234,** 2589 (1959).
128. N. Ikekawa, M. Suzuki (nee Saito), M. Kobayashi, and K. Tsuda, *Chem. Pharm. Bull.*, **14,** 834 (1966).
129. K. Hammarstrand, J. M. Juntunen, and A. R. Hennes, *Anal. Biochem.*, **27,** 172 (1969).
130. N. Nicolaides and M. N. Ansari, *Lipids*, **4,** 79 (1969).
131. T-C. Lo Chang and C. C. Sweeley, *J. Lipid Res.*, **3,** 170 (1962).
132. A. J. Polito, T. Akita, and C. C. Sweeley, *Biochemistry*, **7**(7), 2609 (1968).
133. S. Refsum, *Acta Psychiat. Scand. Supply.*, **38,** 9 (1946).
134. E. Klenk and W. Kahlke, *Z. Physiol. Chem.*, **333,** 133 (1963).
135. D. Steinberg, J. Avigan, C. E. Mize, L. Eldjarn, H. Try, and S. Refsum, *Biochem. Biophys. Res. Commun.*, **19,** 783 (1965).
136 D. Steinberg, F. Q. Vroom, W. K. Engel, J. Cammermeyer, C. E. Mize, and J. Avigan, *Ann. Intern. Med.*, **66,** 365 (1967).
137. J. H. Herndon, Jr., D. Steinberg, B. W. Uhlendorf, and H. M. Fales, *J. Clin. Invest.*, **48,** 1017 (1969).
138. J. Avigan, D. Steinberg, A. Gutman, C. E. Mize, and G. W. A. Milne, *Biochem. Biophys. Res. Commun.*, **24,** 838 (1966).
139. C. E. Mize, D. Steinberg, J. Avigan, and H. M. Fales, *Biochem. Biophys. Res. Commun.*, **25,** 359 (1966).
140. S. -C. Tsai, J. H. Herndon, Jr., B. W. Uhlendorf, H. M. Fales, and C. E. Mize, *Biochem. Biophys. Res. Commun.*, **28,** 571 (1967).
141. M. Bolan and J. W. Steele, *J. Chromatog.*, **36,** 22 (1968).
142. H. Thieme, *Pharmazie*, **19,** 471 (1964).
143. C. S. C. Wong, Thesis, University of Manitoba, 1966; see Ref. 141.
144. R. H. Roth and N. J. Giarman, *Biochem. Pharmacol.*, **18,** 247 (1969).
145. H. Laborit, J. M. Jouany, J. Gerard, and P. Fabiani, *Neuropsychopharmacology*, **2,** 490 (1961).
146. B. A. Rubin and N. J. Giarman, *Yale J. Biol. Med.*, **19,** 1017 (1947).
147. H. Laborit, J. M. Jouany, J. Gerard, and F. Fabiani, *Presse Med.*, **68,** 1867 (1960).
148. N. J. Giarman and K. F. Schmidt, *Brit. J. Pharmacol. Chemotherap.*, **20,** 563 (1963).
149. Ch. Mitoma and S. E. Neubauer, *Experientia*, **24,** 12 (1968).
150. D. Barnett, R. D. Cohen, C. N. Tassopoulos, J. R. Turtle, A. Dimitriadau, and T. R. Fraser, *Anal. Biochem.*, **26,** 68 (1968).
151. R. Tham, *Life Sci.*, **4,** 293 (1965).
152. R. Tham and B. Holmstedt, *J. Chromatog.*, **19,** 286 (1965).
153. R. Tham, *J. Chromatog.*, **22,** 245 (1966).
154. R. Tham, *personal communication*, 1969.
155. H. Auterhoff and H. P. Häufel, *Arch. Pharm.*, **301,** 537 (1968).
156. D. R. A. Wharton, E. D. Black, G. Merritt, M. L. Wharton, M. Bazinet, and J. T. Walsh, *Science*, **137,** 1062 (1962).
157. W. D. Bedard, P. E. Tilden, D. L. Wood, R. M. Silverstein, R. G. Brownlee, and J. O. Rodin, *Science*, **164,** 1284 (1969).
158. R. G. Brownlee and R. M. Silverstein, *Anal. Chem.*, **40,** 2077 (1968).

159. J. O. Rodin, R. M. Silverstein, W E. Buckholder, and J. E. Gorman, *Science*, **165**, 904 (1969).

159a. J. H. Tumlinson, D. D. Hardee, R. C. Gueldner, A. C. Thompson, P. A. Hedin, and J. P. Minyard, *Science*, **166**, 1010 (1969).

160. K. Smith, G. F. Thompson, and H. D. Koster, *Science*, **166**, 398 (1969).

161. E. C. Taylor, K. Lenard, and Y. Shvo, *J. Am. Chem. Soc.*, **88**, 367 (1966).

162. J. Needham and L. Gwei-Djen, *Endeavour*, **27**, 130 (1968).

163. P. R. Vagelos, W. J. A. VandenHeuvel, and M. G. Horning, *Anal. Biochem.*, **2**, 50 (1961).

164. W. J. A. VandenHeuvel, *J. Chromatog.*, **43**, 215 (1969).

165. O. E. Ayers, T. G. Smith, J. D. Burnett, and B. W. Ponder, *Anal. Chem.*, **38**, 1606 (1966).

166. R. J. DePasquale and C. Tamborski, *Chem. Ind. (London)*, 771 (1968).

167. V. G. Berezkin, *Analytical Reaction Gas Chromatography*, Moscow, 1966. Translated by L. S. Ettre, Plenum Press, New York, 1968.

168. W. R. Supina, R. S. Henly, and R. F. Kruppa, *J. Am. Oil Chem. Soc.*, **43**, 202A (1966).

169. W. S. Bowers, private communication, 1969

170. J. C. Giddings, M. N. Meyers, and J. W. King, *J. Chromatog. Sci.*, **7**, 276 (1969).

171. J. F. K. Huber, *J. Chromatog. Sci.*, **7**, 85 (1969).

172. J. J. Kirklahd, *J. Chromatog. Sci*, **7**, 361 (1969).

173. S. T. Sie and N. van den Hoed, *J. Chromatog. Sci.*, **7**, 257 (1969).

174. I. Halasz and S. Sebastian, *Angew. Chem., Intern. Ed. Engl.*, **8**, 453 (1965); *J. Chromatog. Sci.*, **7**, 129 (1969).

CHAPTER 10

Continuous Chromatographic Techniques

P. E. Barker, *Professor of Chemical Engineering, University of Aston, Birmingham, England*

I. Introduction

Continuous chromatographic refining (CCR) is a mass transfer separation technique which originated soon after the inception of analytical gas chromatography (GC) by James and Martin in 1952. Whereas analytical GC emphasizes the separation of minute quantities of material, CCR attempts to optimize the use of the main components of GC, namely, an inert gas, a solid carrier, and a relatively nonvolatile liquid, for the separation of much larger quantities of feed mixtures ranging from light hydrocarbons to essential oils, metal chelates, etc. More recently, the CCR technique has been extended to include the continuous separation of nonvolatile mixtures such as enzymes, proteins, and carbohydrate polymers such as dextran, etc., in which the system now consists of, e.g., an inert mobile liquid phase and a chromatographic medium such as controlled pore size silica beads.

In optimizing the chromatographic type system for larger-scale separations, as the scale increases from grams per hour to kilograms per hour operation, not only is it of importance to design the most efficient separation system, but the most efficient system at minimum overall operating and capital equipment cost. Different viewpoints concerning the way in which a chromatographic system can best be scaled up have been taken over the last 15 years, and these fall into two main categories: (1) batch (2) continuous.

In (1) a direct scale-up of the analytical process is attempted by using larger-diameter packed beds, incorporating perhaps a baffling system which gives strong radial unit mixing but minimum longitudinal mixing. In (2) the chromatographic bed is moved countercurrent to the inert mobile phase so that by judicious choice of flow rates of bed and inert phase a component or group of components moves preferentially with the inert mobile phase while the other component or remainder of the components moves in the direction of the chromatographic bed.

The batch process was the first to be tried in 1953–1954, and the technique has been actively developed over the intervening years so that commercial units of 12-in. diameter or greater are now available.

The continuous mode of processing was introduced soon after (1955–1956) and has taken several forms. In particular may be mentioned the vertical moving-bed technique in which attrition of particles occurs and, later, the rotating-circular-column technique which eliminated the mechanical handling and attrition of chromatographic particles. While scientific publications have only given details of small diameter continuous units, we have reason to believe that continuous units comparable to the batch units are in commercial operation.

Those investigators who choose the continuous type processing do so primarily because:

1. The sheer volume occupied by large-scale injections of feed into a batch column reduces the efficiency of the column although to some extent this can be offset by periodic injections of smaller quantities of feed.

2. With the continuous process loss of mass transfer efficiency with increasing column size is not as critical, since column length can be increased and purity of products readily achieved. Increasing column length in the batch process past a certain point causes more back mixing and even poorer product purities.

3. To obtain band shift between two components in the batch process requires a greater number of plates than in the continuous process, although some or all of the advantage may be lost if a mixture containing several components has to be separated, since $N - 1$ columns are required to separate an n component mixture completely as in other continuous mass transfer processes.

4. Continuous processing is in general favored in the chemical industry in the majority of mass transfer separation processes, since experience has shown that greater throughputs, higher purities, and lower costs are normally possible than in an equivalent batch process.

II. The Principle of Continuous Chromatographic Refining

This method, similar to other processes for separating different compounds, is dependent upon the components distributing themselves differently between two phases, one mobile the other stationary. The mobile phase may be gas or liquid. The stationary phase may be a nonvolatile solvent absorbed on the surface of a suitable solid such as kieselguhr, the solid acting as a carrier and taking no part in the separation or, alternatively, a solid phase such as an adsorbent, molecular sieve, ion-exchange resin, etc.

Each phase has a preferential selectivity for one or other of the solute components. Movement of two phases countercurrently produces in each phase a relatively low concentration of the other, so that given a sufficient height of column complete separation is possible.

The presence of the stationary nonvolatile solvent phase in continuous chromatography acts in a similar manner to the solvent in extractive distillation, whereby gas–liquid equilibria are modified by the presence of the solvent, making a physical separation of the components of the mixture easier. Moreover, by distributing these solvents as very thin films on specially prepared solids with a high surface-to-volume ratio, high mass transfer rates are obtained. When a solid phase alone is used, a variety of surface phenomenon which may be adsorbtive, diffusive, or electronic come into play, thereby producing a selective retention of one component over another.

Most of the work to date in continuous chromatography has involved gas–liquid systems, and for this reason the following theoretical treatments refer to such systems. However, in the subsequent treatment it is hoped the reader will not forget the generality of the technique and its application to other two-phase systems.

The suitability of a nonvolatile solvent for a given separation together with conditions of operation may be judged from a chromatogram (Fig. 10.1) of the binary mixture (AB) obtained on a GC apparatus.

The linear velocity of component A through the column is L/t_{RA}, where L is the length of the column and t_{RA} the retention time of component A. Similarly, the velocity of B is L/t_{RB}. Thus if the column packing is moved countercurrently to the carrier gas stream at a velocity greater than L/t_{RB} but less than L/t_{RA}, component B will be carried in the direction of the packing absorbed in the liquid phase. Component A will be eluted in the carrier gas.

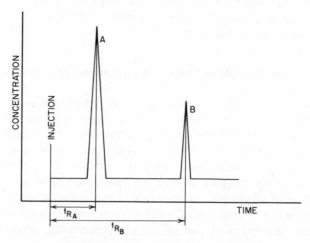

Fig. 10.1. Chromatogram of binary mixture.

It is more convenient to express the conditions of separation in terms of the volume flow of solvent and carrier gas rather than linear velocities.

The partition coefficient K for a particular component may be defined as: *Weight of component per unit volume of solvent phase/weight of component per unit volume of carrier gas phase.* Let the partition coefficient of components A and B at the temperature of the rectification section be K_A^R and K_B^R, respectively (Fig. 10.2). Let the volume flow rates of solvent and carrier

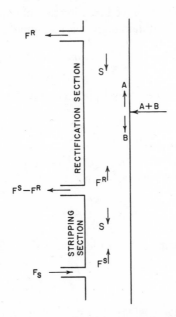

Fig. 10.2. Gas and solvent flows.

gas in the rectification section be S and F^R, respectively. These may be considered constant when the temperature of the column is kept constant, and the pressure drop across the column is small.

A material balance on component A around the feed point gives

$$M_A = F^R Y_A + S X_A \qquad (10.1)$$

where M_A is the feed rate of component A to the column, $F^R Y_A$ denotes the amount of component traveling upwards in the gas stream, and $S X_A$ the amount traveling downward in the solvent stream. Y_A and X_A are the weights of component A per unit volume of carrier gas phase and solvent phase, respectively.

The condition for resultant movement of component A up the column is that

$$F^R Y_A > S X_A \tag{10.2}$$

i.e.,

$$\frac{F^R}{S} > \frac{X_A}{Y_A} \tag{10.3}$$

Now assuming the amount of feed entering the column is very small, so that infinite dilution conditions obtain in the gas and liquid phases.

$$\frac{F^R}{S} > K_A^R \tag{10.4}$$

Similarly, for component B to travel preferentially down the column

$$\frac{F^R}{S} < K_B^R \tag{10.5}$$

Hence separation occurs if the ratio of gas to liquid flows is between the partition coefficients under investigation

$$K_A^R < \frac{F^R}{S} < K_B^R \tag{10.6}$$

Component B will be stripped from the solvent if

$$K_B^S < \frac{F^S}{S} \tag{10.7}$$

where K_B^S is the partition coefficient of component B at the temperature of operation of the stripping section and F^S is the carrier gas flow rate in that section.

III. Effect of Finite Concentration of Solutes in the Mobile Phase

The preceding relations are true only at infinite dilution and for an infinite number of plates in the packed section. As neither of these conditions is realized in practice, the real operating ranges are slightly different.

It was shown by Barker and Lloyd (1) for hydrocarbon systems that the effect of finite concentration can either increase or decrease the partition coefficient depending on the system and the concentration, but in general for chromatographic systems it is found either to have negligible effect or to increase the value of the partition coefficient.

Assuming K to increase with component concentrations, the actual operating range from Eq. 10.6 is, according to Fitch et al. (2)

$$(K_A^R + Y_A) < \frac{F^R}{S} < (K_B^R + Y_B) \tag{10.8}$$

where Y_A, Y_B are factors accounting for the effect of finite concentrations on the partition coefficient. The effect of a finite column length, i.e., a finite number of theoretical plates is to narrow the operating range.

$$(K_A^R + Y_A + \delta_A) < \frac{F^R}{S} < (K_B^R + Y_B - \delta_B) \tag{10.9}$$

where δ_A, δ_B are factors accounting for the effect of finite column length on the partition coefficient.

IV. Probabilistic Approach

Sciance and Crosser (3) have developed a relationship between the required column length, the feed location, and the degree of separation of a binary mixture based on the probability theory. Assuming no longtudinal diffusion, and that the rates of adsorption and desorption are in equilibrium (i.e., their ratio is approximated by the partition coefficient), the relations arrived at for a column having the feed point at its center are

$$\ln (\mu_z)_A = \frac{LR_A}{2F^R} (K_A^R - \phi) \tag{10.10}$$

$$\ln [(1 - \mu_z)_B] = \frac{-LR_B}{2F^R} (K_B^R - \phi) \tag{10.11}$$

where $(\mu_z)_A$ = mass ratio component A in bottoms/feed
$1 - (\mu_z)_B$ = mass ratio component B in tops/feed
ϕ = operating F^R/S
L = column length
R_A, R_B = rate constants of desorption

If the system is designed to give equal product purity, Eqs. 10.10 and 10.11 can be combined and simplified.

$$\frac{R_B}{R_A} = \frac{\phi/K_A^R - 1}{SF - \phi/K_A^R} \tag{10.12}$$

where SF = separation factor (K_B^R/K_A^R).

To find the required column length for a given separation, knowledge of the values R_A and R_B is necessary. The only way these values can be determined is by curve fitting on the elution curve obtained from an analytical column (39). The same theoretical operating conditions are obtained by this probabilistic approach for the separation of a binary mixture as by the method used to derive Eq. 10.6.

V. General Description of a Moving-Bed Apparatus for Binary Separations

A typical apparatus for moving-bed chromatography is shown in Fig. (10.3). The vertical column, which may be of metal or glass and usually about 1-in. diameter is fed by chromatographic type packings from a hopper A, the flow of solids being controlled by a variable orifice B at the column base and removed by a feed table C. The solids pass through a rectification

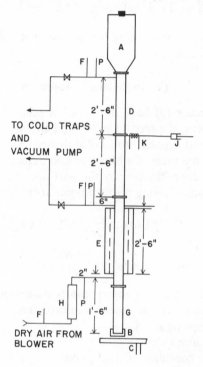

Fig. 10.3. Continuous column.

section D and into a stripping section E, which is surrounded by an electrically heated air jacket. The section between the base of the stripper and the orifice enables the solid to cool down prior to passing to the atmosphere or, alternatively, to be gas-lifted back to hopper A. The column is vibrated to ensure a continuous steady stream of solids. The lengths of D and E are shown in Fig. (10.3) to be 5 ft and 2 ft 6 in., respectively, although D could be any length depending on the difficulty of separation. A dry gas enters a heater H, passing to the column at a pressure slightly above atmospheric to prevent leakage of moist air into the column.

The liquid or gas mixture to be separated is injected into the column by a micropump *J*. The more strongly absorbed component passes down in the liquid phase and is removed in the heated stripper by the gas. The less strongly absorbed component passes up in the gas stream. Both products pass through a recovery system, which usually consists of cold traps containing solid carbon dioxide and acetone, although with less volatile systems water or even air-cooled condensers are adequate. Rarely, is it possible to achieve greater than 90% recovery of the components by heat-exchange methods for volatile solutes, although this is usually regarded as adequate for laboratory purposes. To recover the remaining amount of component from the gas stream, adsorbents or molecular sieves are necessary. Such methods are only used when recirculating inert carrier gases on production-scale chromatographic equipment.

The gas used is normally nitrogen, although Barker et al. (8,11,20) have successfully used air for 12 years for nondecomposing separations. Air is used for cheapness but should only be used on small-scale equipment, taking particular care that all the equipment is properly bonded and earthed, and that no source of ignition can occur on the equipment. Heated surface temperatures should be kept below the ignition temperature of the components being separated. If in doubt, use nitrogen.

The jacket temperature of the stripper is usually kept about 50°C above the temperature of *D*, which in the diagram is shown unjacketed and therefore at ambient conditions. For light hydrocarbon and volatile solvent mixtures, such an arrangement is satisfactory, but for comparatively nonvolatile mixtures *D* is jacketed and operated at a temperature approximating that at which successful separation in an analytical GC apparatus could be achieved.

Although liquid feeds may be injected into the column it is preferable to vaporize the feed by a heater *K* to achieve more efficient column operation.

VI. The Separation of Binary Mixtures by Using Moving-Bed Chromatography

One of the advantages of this process over simple distillation is that separations can be achieved by differences in polarity as well as boiling point. Polar molecules are those in which the charge density is not uniform over the surface of the molecule but is concentrated at one or two points. The method can therefore be appropriately demonstrated by separating two liquids with approximately the same boiling point but of different polarity. Such a system is benzene (b.p. 80.1°C) and cyclohexane (b.p. 80.7°C) using a polyglycol derivative (polyoxyethylene 400 diricinoleate) as stationary phase absorbed on 10–20-mesh kieselguhr. Typical operating conditions on a 1-in.-diameter column when separating a 50/50 *v/v* 30 ml/hr liquid feed mixtures are shown in Table 10.1, while concentration profiles within the column corresponding to the three runs are given in Fig. 10.4.

The ideal separation range for cyclohexane and benzene is for values of F^R/S from 750 to 300, which are the approximate partition coefficients of benzene and cyclohexane, respectively, at 20°C. The actual range of separation is slightly narrower than the predicted value; see Fig. 10.5. The theoretical range of flow ratios corresponds to

$$K^R_{\text{cyclohexane}} < \frac{F^R}{S} < K^R_{\text{benzene}} \tag{10.13}$$

which is true at infinite dilution and for an infinite number of theoretical plates. It was found in the work on equilibrium determinations by Barker and Lloyd (1) that the partition coefficients of benzene were unaffected by

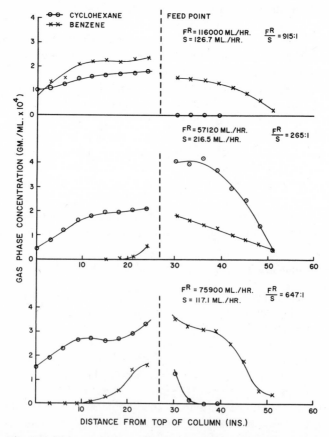

Fig. 10.4. Concentration profiles for the system cyclohexane/benzene.

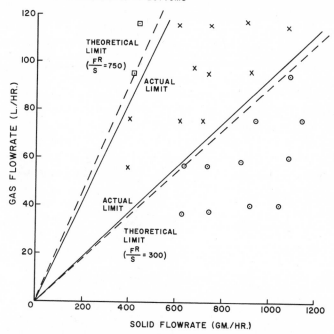

Legend:
□ BENZENE IN TOPS
× COMPLETE SEPARATION
○ CYCLOHEXANE IN BOTTOMS

Fig. 10.5. Separation requirements for cyclohexane and benzene.

TABLE 10.1
Operating Data for Separation of Cyclohexane–Benzene

	Run number		
	1	2	3
Rectification outlet pressure, mm Hg	730.5	749.3	725.5
Stripper outlet pressure, mm Hg	749.7	758.1	737.9
Mean column pressure, mm Hg	740.1	753.7	731.7
Flow rate of pure air, rectification section, ml/hr			
At NTP	105,200	52,800	67,800
At column conditions	116,000	57,120	75,900
Flow rate of pure air, stripping section at NTP, ml/hr	92,800	93,500	91,900
Flow rate of solids, g/hr	432	739	400
Flow rate of solvent, ml/hr	126.7	216.5	117.1

335

P. E. BARKER

the presence of cyclohexane or benzene. Cyclohexane behaved differently in that the partition coefficients were increased by the presence of benzene but not cyclohexane, while the effect of a finite number of mass transfer stages is to narrow the operating range to give an actual range according to Eq. 10.9.

A further illustration of the technique when separating the cyclohexane–methylcyclohexane system is shown in Figs. 10.6 and 10.7, with the corresponding operating data in Table 10.2. It is observed that the operating range for this system is narrower than for the benzene–cyclohexane system. The reason for this is that none of the three solutes has permanent dipoles, but benzene is more polarizable per unit volume than either cyclohexane or methylcylohexane because of the presence of the highly mobile electrons in

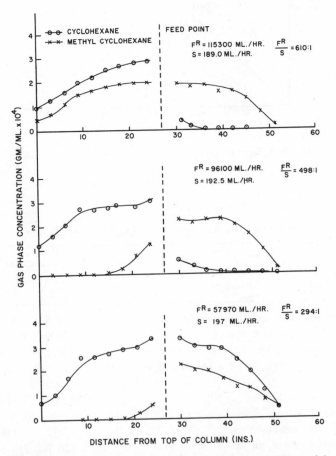

Fig. 10.6. Concentration profiles for the system cyclohexane/methyl cyclohexane.

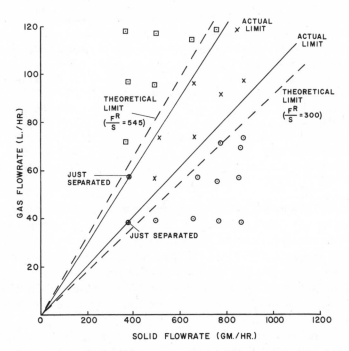

Fig. 10.7. Separation requirements for methylcyclohexane and cyclohexane.

the molecular structure. Thus in the case of a polar solvent of the type used, a dipole is induced by the energy field of the solvent molecules. This effect is greater for benzene, making it less volatile, and explains the large difference in partition coefficient values between cyclohexane and benzene, although they boil at virtually the same temperature.

Schultz (4), using a single column from which both products are collected, evaluated in terms of velocities the operating conditions of the column. He investigated the separation of *cis*- and *trans*-butene-2 on a column 100 cm long and 1 cm in diameter, using 30% dibutyl phthalate on 0.3–0.4 mm Sterchamol particles. At a feed rate of 78 ml/hr consisting of 37.6% by volume of *trans* and 62.4% *cis*, the purity of *trans* 32 cm above the feed was 99.73% and the purity of *cis* at 32 cm below the feed was 99.4% by volume. The column was operated at 22°C which was well above the boiling point of both materials. Using a larger column, 138 cm long and 2.6-cm diameter,

TABLE 10.2
Operating Data for Separation of Cyclohexane–Methylcyclohexane

	Run Number		
	1	2	3
Rectification outlet pressure, mm Hg	720.8	721.8	726.6
Stripper outlet pressure, mm Hg	739.6	739.2	735.4
Mean column pressure, mm Hg	730.2	730.5	731.0
Flow rate of pure air, rectification section, ml/hr			
At NTP, ml/hr	103,800	86,200	52,000
At column conditions, ml/hr	115,300	96,100	57,970
Flow rate of pure air, stripping section at NTP, ml/hr	90,600	92,300	92,100
Flow rate of solids g/hr	645	657	674
Flow rate of solvent m/hr	189.0	192.5	197.0

21 g/hr of a 38.8 mole % 2.2 dimethylbutane and 61.2% mole % cyclopentane mixture was separated using 7.8 benzoquinoline as a nonvolatile liquid, with the column operating at the boiling point of the materials. Product purities in excess of 99.999 mole % were claimed to have been obtained.

Tiley and co-workers (2), using a 2-cm-diameter column, 4 ft long and a packing of 20% w/w dinonyl phthlate on 44–60 BSS mesh Celite, investigated the separation of 1.1 binary feed mixtures of diethyl ether, dimethoxymethane, and dichloromethane.

The feed material was vaporized into a stream of nitrogen before entering the column, and although this resulted in low feed rates of 3–5 ml/hr, it gave rapid attainment of equilibrium around the feed point.

Scott (5) also used the technique to separate aromatic hydrocarbons from coal gas.

VII. The Evaluation of Column Efficiency in Terms of HTUs

Since moving-bed chromatography is a steady-state countercurrent differential process, the efficiency may be expressed by the transfer unit concept proposed by Chilton and Colburn (6).

A material balance for a component over a column height dL gives.

$$S\, dx = F^R\, dy \tag{10.14}$$

$$= K_G\, a\, (y - y_e)\, dL \tag{10.15}$$

where $K_G a$ = overall gas phase mass transfer coefficient

      y = solute concentration in gas phase (grams per milliliter)

      y_e = equilibrium value of solute concentration in gas phase (grams per milliliter)

The height of the column may be calculated

$$L = \frac{F^R}{K_G a} \int_{y_2}^{y_1} \frac{dy}{y - y_\epsilon} \tag{10.16}$$

This may be written

$$L = H_{OG} \times N_{OG} \tag{10.17}$$

where H_{OG} is the height of a transfer unit $(= F^R/K_G a)$ and N_{OG} is the number of transfer units

$$\int_{y_2}^{y_1} \frac{dy}{(y - y_\epsilon)}$$

Where it is possible to assume that the operating and equilibrium lines are straight it can be shown that

$$\int_{y_2}^{y_1} \frac{dy}{y - y_\epsilon} = \frac{y_1 - y_2}{(y - y_\epsilon)_1 - (y - y_\epsilon)_2} \ln \left\{ \frac{(y - y_\epsilon)_1}{(y - y_\epsilon)_2} \right\} = \frac{L K_G a}{F^R} \tag{10.18}$$

This may be written:

(1) For investigations above the feed point

$$N_{OG} = \frac{1}{F^R/(K^F S - 1)} \ln \left[\frac{M_T/K^F S - y_2(F^R/K^F S - 1)}{M_T/K^F S - y_1(F^R/K^F S - 1)} \right] \tag{10.19a}$$

where K^F = partition coefficient at finite concentrations on solute-free basis

(2) For investigations below the feed point

$$N_{OG} = \frac{1}{(1 - F^R/K^F S)} \ln \left[\frac{M_B/K^F S - y_2(1 - F_R/K^F S)}{M_B/K^F S - y_1(1 - F_R/K^F S)} \right] \tag{10.19b}$$

where M_T = flow rate of solute leaving in the top product stream
M_B = flow rate of solute leaving in the bottom product stream

Applying equations 10.19a and 10.19b to the runs shown in Figs. 10.4 and 10.6 gave results as shown in Tables 10.3a and 10.3b. The values of H_{OG} obtained for cyclohexane above the feed point and for methylcyclohexane and benzene below the feed point are shown in Fig. 10.8. A logarithmic plot of H_{OG} against liquid flow rate of solvent flow yielded a straight-line relationship for each of the three components.

The corresponding empirically determined equations are:

For cyclohexane $\qquad\qquad H_{OG} = \dfrac{60.82}{S^{0.514}} \tag{10.20a}$

TABLE 10.3a

Investigation of Top 9 in. of Column

Gas flow rate, solute-free (F^R), ml/hr	Liquid flow rate, solute-free (S), ml/hr	Column temp., °C	Mean solute concentration, g/ml × 10⁴	Partition coefficient		Weight of solute in tops (M_T) g	Gas phase concentrations		N_{og}	H_{og}, in.	H_{og}, cm	Solute investigated
				Solute basis	Solute-free K		At top of column (y_2), g/ml × 10⁴	9 in. from top of column (y_1), g/ml × 10⁴				
					Cyclohexane–benzene							
116,000	126.7	20.3	1.21	297	327	11.73	1.011	1.560	2.025	4.44	11.29	Cyclohexane
116,000	126.7	20.3	1.30	769	845	8.41	0.80	2.14	2.043	4.40	11.19	Benzene
57,120	216.5	20.8	1.12	291	295	2.36	0.41	1.611	2.558	3.52	8.94	Cyclohexane
75,900	117.1	21.0	2.10	289	293	11.57	1.52	2.67	2.05	4.39	11.15	Cyclohexane
					Cyclohexane–methylcyclohexane							
115,300	189.0	19.0	1.52	328.8	329	11.71	1.016	2.04	2.288	3.93	9.98	Cyclohexane
115,300	189.0	19.0	1.00	572	574	5.50	0.413	1.47	2.366	3.80	9.66	Methyl-cyclohexane
96,100	192.5	19.8	1.90	294	300	11.52	1.20	2.66	2.513	3.58	9.09	Cyclohexane
57,970	197.0	19.5	1.10	296.9	302	4.37	0.755	1.59	2.491	3.61	9.16	Cyclohexane

TABLE 10.3*b*

Investigation of Section $42\frac{1}{4}$–$51\frac{1}{4}$ in. from Top of Column

Gas flow rate, solute-free (F^R), ml/hr	Liquid flow rate, solute-free (S), ml/hr	Column temp., °C	Mean solute concentration, g/ml $\times 10^4$	Partition coefficient Solute basis	Partition coefficient Solute-free K	Weight of solute in bottoms (M_B), g	Gas phase concentrations $42\frac{1}{4}$ in. from top of column (y_1), g/ml $\times 10^4$	$51\frac{1}{4}$ in. from top of column (y_2), g/ml $\times 10^4$	N_{og}	H_{og}, in.	H_{og}, cm	Solute investigated
						Cyclohexane–benzene						
116,000	126.7	20.3	0.74	769	790	5.00	1.406	0.15	2.04	4.40	11.19	Benzene
57,120	216.5	20.8	1.51	295	312	9.15	2.84	0.32	2.323	3.87	9.84	Cyclohexane
57,120	216.5	20.8	0.62	758	802	13.27	1.03	0.381	2.97	3.03	7.70	Benzene
75,900	117.1	21.0	1.31	764	835	12.81	2.360	0.284	2.09	4.31	10.93	Benzene
						Cyclohexane–methylcyclohexane						
115,300	189.0	19.0	0.85	603	624	7.04	1.595	0.264	2.369	3.80	9.66	Methyl-cyclohexane
96,100	192.5	19.8	1.06	585	620	11.61	2.03	0.296	2.375	3.79	9.60	Methyl-cyclohexane
57,970	197.0	19.5	0.75	591	630	11.60	1.356	0.329	2.405	3.74	9.48	Methyl-cyclohexane
57,970	197.0	19.5	1.40	319.8	340	7.31	2.413	0.350	2.32	3.88	9.86	Cyclohexane

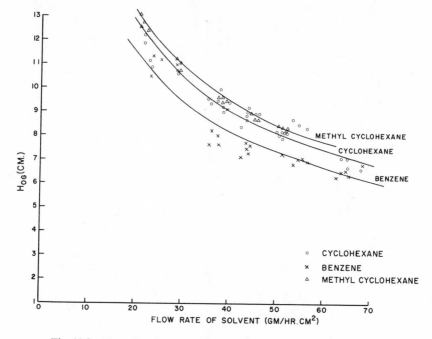

Fig. 10.8. H_{OG} values for cyclohexane, benzene and methylcyclohexane.

For methylcyclohexane
$$H_{OG} = \frac{60.24}{S^{0.501}}$$
(10.20b)

For benzene
$$H_{OG} = \frac{50.47}{S^{0.495}}$$
(10.20c)

VIII. Location of Resistance to Mass Transfer

The general equations for mass transfer is

$$\frac{1}{K_G a} = \frac{1}{k_g a} + \frac{1}{K^F k_L a}$$
(10.21)

where K^F = partition coefficient at finite concentration on solute-free basis

$1/K_G a$ = total resistance to mass transfer

$1/k_g a$ = gas phase resistance to mass transfer

$1/K^F k_L a$ = liquid phase resistance to mass transfer

Now $1/k_g a = f(F)$, where F is the flow rate of the gas, since the gas film thickness, hence the gas film resistance varies with gas flow rate. We assume that $1/k_L a$ is not a function of gas flow rate, i.e., that the gas flow does not tend to mix the liquid. This is reasonable if the solvent is situated mainly in the pores of the packing.

Thus the above equation can be arranged:

$$\frac{1}{K_G a} = \frac{1}{f(F)} + \frac{1}{K^F k_L a} \tag{10.22}$$

At infinite gas velocity there is no resistance in the gas phase, i.e., when $1/F = 0$. Barker and Lloyd (40) in a plot of $1/K_G a$ against $1/F$ at constant liquid flow S^1, found it gave a straight-line relation which extrapolated back to the origin as shown in Fig. 10.9. A statistical treatment of their results indicated a negligible interception on the ordinate axis. This is to be expected since the values of H_{OG} showed no measurable dependence on gas flow rate. Thus as $1/F = 0$, $1/K_G a = 1/Kk_L a = 0$. This is assumed to be constant since $k_L \neq f(F)$ and S^1 is constant.

It appears therefore that for moving-bed columns under the conditions used by Barker and Lloyd in their studies that the resistance to mass transfer lies in the gas film.

IX. The Evaluation of Column Efficiency in Terms of Overall HETPs

Separation of a binary mixture by continuous GC is a process similar to two-solvent countercurrent extraction, in which one solvent is the inert carrier gas and the other is the nonvolatile stationary phase. The only difference is that in chromatographic separations the heavier component is stripped from the stationary phase before the latter leaves the column.

Following the approach of Alders (7) in the treatment of liquid–liquid extraction in which the process is considered as a large number of discrete stages, a relationship for the number of equivalent theoretical plates necessary for the separation of a binary mixture into products of a given purity can be derived. Tiley and co-workers (2), assuming a column consisting of a large number of plates with a central feed point and equal product purities, have obtained an approximate relation for the number of plates in the column. A more general relationship which does not assume equal product purities nor restrict the application to a 1:1 feed mixture has been derived by Barker and Huntington (8,9).

Considering a two-solvent countercurrent extraction column in which a binary feed mixture enters at a point between the two ends of the column; one component to one offtake in the first solvent (gas stream), while the other component travels to the other offtake in the second solvent (non-

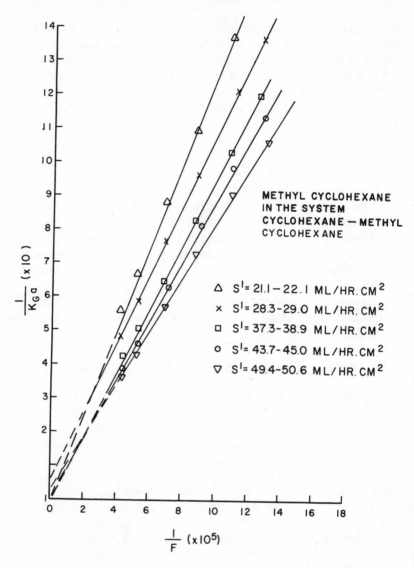

Fig. 10.9. Plot of $1/K_Ga$ against $1/F$ at constant liquid flow.

344

volatile liquid). From a binary feed mixture containing 1 part of component i, assume that x parts of i are removed in the gas stream, $1 - x$ parts are removed in the solvent stream, and the fraction of i not extracted $\psi = x$.

If in unit time y parts of i pass from the feed to the $(m + 1)$-th stage and z parts pass to the $(m - 1)$-th stage, where m is the feed stage, then $E_1 = z/y$, hence

$$\frac{x}{y} = \frac{E_1 - 1}{E_1^n - 1} \tag{10.23}$$

$$\frac{1 - x}{z} = \frac{1/E_2 - 1}{1/E_2{}^m - 1} \tag{10.24}$$

where E_1 and E_2 are the extraction factors for each component above and below the feed, n and m are the number of stages above and below the feed, respectively.

$$\frac{1 - x}{z} = \frac{[(1/E_2) - 1]/(1/E_2{}^m) - 1] \, (x/y)}{(E_1 - 1)/(E_1{}^n - 1)} \tag{10.25}$$

from which

$$x = \psi = \frac{(E_1 - 1) \, (E_2^m - 1)}{E_1 \, E_2^{m-1} \, (E_2{}^m - 1) \, (E_1{}^n - 1) + (E_1 - 1) \, (E_2{}^m - 1)} \tag{10.26}$$

If $E_1 = E_2 = E$, i.e., at infinite dilution

$$\psi = \frac{E^m - 1}{E^{m+n} - 1} \tag{10.27}$$

If the column contains a high number of plates, such that $m + n$ approximates to $m + n - 1$, which is the total number of plates in the column, and the number of plates above and below the feed are the same, i.e., $m = n$

$$\psi = \frac{1}{E^n + 1} \tag{10.28}$$

Considering components A and B under conditions such that $E_A < 1 < E_B$, then if $n \gg 1$, using Eq. 10.27,

$$\psi_A = \frac{E_A{}^n - 1}{E_A^{2n} - 1} \rightarrow 1 - E_A^n \tag{10.29}$$

$$\psi_B = \frac{1 - E_B{}^n}{1 - E_B^{2n}} \rightarrow E_B^{-n} \tag{10.30}$$

For a $1:1$ by weight feed mixture $\psi_A = 1 - \psi_B$, hence $E_A \, E_B = 1$.

As $E = K/\phi$, the required flow ratio ϕ is $(K_A K_B)^{\frac{1}{2}}$. If in unit time the feed mixture contains W_A parts of component A and W_B parts of component B then

$$\psi_A = \frac{W_A - \delta W_A}{W_A} = 1 - \frac{\delta W_A}{W_A} \tag{10.31}$$

$$\psi_B = \frac{\delta W_B}{W_B} \tag{10.32}$$

where ψ_A and ψ_B are the fractions not extracted at the top of the column. δW_A is the amount of component A that is extracted (i.e., appears in the bottom product) and δW_B is the amount of component B that is not extracted (i.e., appears in the top product).

Combining Eqs. 10.29 and 10.31

$$1 - E_A{}^n = 1 - \frac{\delta W_A}{W_A} \tag{10.33}$$

$$1 - (K_A/\phi)^n = M_{AT}/W_A \tag{10.34}$$

Where M_{AT} is the collection rate of component A at the top of the column

$$n \log K_A - n \log \phi = \log (1 - M_{AT}/W_A) \tag{10.35}$$

Similarly, combining Eqs. 10.30 and 10.32

$$E_B^{-n} = \delta W_A/W_B \tag{10.36}$$

$$(K_B/\phi)^{-n} = M_{BT}/W_B \tag{10.37}$$

Where M_{BT} is the collection rate of component B at the top of the column

$$\simeq n \log K_B + n \log \phi = \log M_{BT}/W_B \tag{10.38}$$

For a specified purity of products there is a range of flow ratio, depending on the number of theoretical plates. The maximum is represented by Eq. 10.38, and the minimum by Eq. 10.35.

As the ratio of partition coefficients is likely to vary to a lesser extent than the actual values, a combined equation is more accurate

$$\log \frac{\phi_{\max}}{\phi_{\min}} = \log \frac{K_B}{K_A} + \frac{1}{n}\left[\log\left(1 - \frac{M_{AT}}{W_A}\right) + \log \frac{M_{BT}}{W_N}\right] \tag{10.39}$$

In the operation of the vertical moving-bed column, a single operating value ϕ is used, i.e., $\phi_{\max} = \phi_{\min}$ and the equation reduces to

$$\log SF = \frac{-2}{N_c}\left[\log\left(1 - \frac{M_{AT}}{W_A}\right) + \log \frac{M_{BT}}{W_B}\right] \tag{10.40}$$

where N_c is the total number of theoretical plates ($=2n$) in the continuous column. For a 1:1 feed mixture and equal product purities

$$N_c = \frac{-4}{\log SF} \log I \tag{10.41}$$

where I = product impurity fraction.

In the case of the circular chromatographic column described later in the chapter for which the operating values of ϕ vary slightly above and below the feed owing to the bleed of carrier gas injected at the feed point, a more accurate estimate of N_c is obtained by taking the value of ϕ above the feed as ϕ_{max} and the value below as ϕ_{min}.

X. Separation of Three-Component Mixtures

A more common problem of separation is one in which one component or group of components must be removed from components appearing on a chromatogram on either side of the single component or group of components.

A patented apparatus (Fig. 10.10) has been devised by Barker and Lloyd (10) for carrying out such separations, while Barker and Huntington (11) report experimental findings using such equipment.

The column used in their experimental work was constructed from copper tubing with brass flanges and Neoprene gaskets.

The packing was placed in the hoppers A and G from which it passed by gravity through the columns and the controlling orifice B onto the rotating table C, finally being removed by a scraper arm onto the pan of a weighing machine. In some initial experiments 13% w/w polyoxyethylene 400 diricimoleate on 14–18 B.S. mesh Johns-Manville C22 firebrick was used.

The control of the solids in the side arm was effected by the intrusion of a 1.27-cm-diameter side arm into the 2.54-cm-diameter main column. The position of the side arm R altered the effective cross-sectional area allotted to each part of the column, hence the ratio of the solid flows in the main column to the side arm. It was found necessary to have the intrusion, otherwise no solids flow was obtained in the side arm.

The solids leaving the main hopper A passed through the rectification section MNP and then into the stripping section E. The distances M, N, P, and E were 68.6, 63.5, 68.6, and 52.1 cm, respectively. The solids leaving the hopper G served to reflux back the heaviest component entering the side arm R, thus permitting a pure side product to be obtained. The effective length of L for mass transfer was 82.5 cm. The stripper E was surrounded by an electrically heated air jacket.

Air was supplied to the column below the stripping section by a blower and was drawn off under a slight vacuum from three product offtake points.

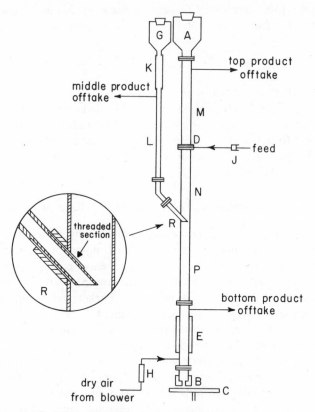

Fig. 10.10. Moving-bed column for ternary separations.

The air was dried and preheated to 50°C in a heater *H*. It entered the column at a pressure slightly above atmospheric to prevent leakage of moist air into the system through the orifice *B*.

A 1:1:1 *v/v* liquid mixture of benzene, cyclohexane, and methylcyclohexane was injected unheated into the column at the feed point *D* by a micropump *J*. All the product streams were passed through cold traps containing carbon dioxide and acetone to condense out most of the separated products.

Gas sampling points were situated at 7.62-cm intervals in the rectification sections of the columns.

These side tubes were fitted with serum caps so that gas samples could be taken and analyzed in a conventional GLC column.

XI. General Operating Conditions for Moving-Bed Ternary Separations

When using the equipment as a single column and without the side arm in use, product purities of 99.5% were obtained at the top and bottom of the column and maximum purities of 78.6% by volume of methylcyclohexane contaminated by benzene from the middle section.

Table 10.4 shows complete operating data for six of the runs when the side arm was brought into use. A typical concentration profile is shown in Fig. 10.11. These data show how benzene entering the side arm can be refluxed back, thus enabling a pure side stream to be produced. Because of insufficient column length below the junction R, a pure bottoms product of benzene could not be obtained simultaneously with high purity overhead and side stream products.

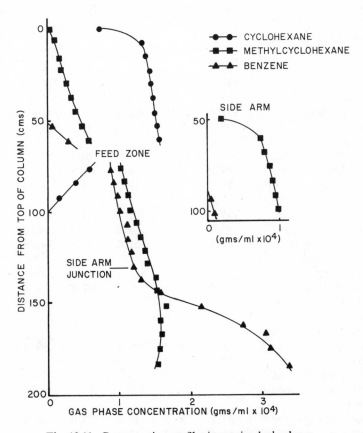

Fig. 10.11. Concentration profiles in moving-bed column.

TABLE 10.4

Operating Data for Ternary Separation with Side Arm in Operation

	Run number					
	1	2	3	4	5	6
Upper column						
Pure air flow at column conditions (F^R/S), liters/hr	40.6	43.8	52.3	49.8	52.2	60.5
	455	475	557	532	559	641
Side arm						
Pure air flow at column conditions (F^R/S), liters/hr	49.3	45.8	44.9	39.5	38.6	36.7
	787	764	703	658	644	605
Middle column						
Pure air flow at column conditions (F^R/S), liters/hr	89.9	89.6	97.2	89.3	90.8	97.2
	592	587	630	581	592	628
Mean column						
Pressure, mm Hg	731.4	721.9	735.4	713.0	726.1	722.9
Temperature, °K	295.2	295.0	295.2	295.3	295.2	295.2
Upper column						
Solids flow rate, g/hr	705	736	712	740	738	745
Solvent flow, ml/hr	89.2	92.8	90.3	93.7	93.3	94.5
Solvent flow, g/hr cm²	18.75	19.52	18.98	19.69	19.65	19.86
Side arm						
Solids flow rate, g/hr	495	474	503	474	474	478
Solvent flow, ml/hr	62.7	60.0	63.8	60.0	60.0	60.0
Solvent flow, g/hr cm²	51.1	48.8	52.0	48.8	48.8	49.4
Mole percentages						
Top product						
Cyclohexane	99.5	99.5	97.3	97.5	98.7	91.6
Methylcyclohexane	—	—	2.7	2.5	1.3	8.4
Side-arm product						
Cyclohexane	34.2	—	—	—	—	—
Methylcyclohexane	53.3	93.7	99.5	99.5	99.5	99.5
Benzene	12.5	6.3	—	—	—	—
In gas phase at point above bottoms product offtake						
Methylcyclohexane	33.1	31.3	28.2	30.0	26.4	20.7
Benzene	66.9	68.7	71.8	70.0	73.6	79.3

Extending the principles determined for binary separations (Eqs. 10.6 and 10.7) to ternary systems, in the sections M and N of the column the ideal separating range for cyclohexane to be carried up the column and the methylcyclohexane and benzene downward corresponds to:

$$K^R_{\text{cyclohexane}} < \frac{F_1^R}{S_1^R} < K^R_{\text{methylcyclohexane}} < K^R_{\text{benzene}} \qquad (10.42)$$

where subscript 1 refers to flows in Section M and N of the column.

In the side arm for benzene to be refluxed back down the column the ideal separating range corresponds to

$$K^R_{\text{methylcyclohexane}} < \frac{F_2^R}{S_2^R} < K^R_{\text{cyclohexane}} < K^R_{\text{benzene}} \qquad (10.43)$$

where subscript 2, refers to flows in Section L of the column.

Similarly, in section P of the column

$$K^R_{\text{methylcyclohexane}} < \frac{F_1^R + F_2^R}{S_1^R + S_2^R} < K^R_{\text{benzene}} \qquad (10.44)$$

One of the limiting factors in the operation of the column with the side arm is the ratio F_2^R/S_2^R must be of the same order as the ratio

$$\frac{F_1^R + F_2^R}{S_1^R + S_2^R}$$

i.e., both must be between the partition coefficients of methylcyclohexane and benzene. These ideal operating ranges are modified by the presence of the other components and by using columns of finite length as previously described for binary separations.

The principle could be extended to multicomponent separations or, alternatively, a series of single columns in series operating at different F^R/S ratios could be used to separate multicomponent systems.

The efficiency data in terms of HTUs are shown in Table 10.5 for the series of runs given in Table 10.4. This table shows that the mass transfer efficiency of the column is reduced when the percent of liquid phase is reduced, presumably because of the poorer coverage of brick giving a reduced surface area for mass transfer.

The H_{OG} values obtained for methylcyclohexane in the side arm given in Table 10.5 illustrate the considerable beneficial effect of using smaller-diameter tubes. The effect of tube size reduction from 2.54 to 1.27 cm more than offsets the increase in H_{OG} attributable to decreasing the liquid phase percentage from 29.6 to 13%. It is recalled that the empirical Eq. 10.20b

TABLE 10.5
HTU Data for Operating Conditions in Table 10.4

Rectification section

Gas phase concentration of cyclohexane, g/ml $\times 10^4$					
22.9 cm from top	1.478	1.151	1.174	1.123	0.916
Partition coefficient of cyclohexane	277	279	275	276	285
HTU cyclohexane 0–22.9 cm	13.21	14.40	15.60	14.81	14.02
From Eq. 10.20a $H_{OG} = 60.82/S^{0.514}$ for 1-in.-diameter columns	13.31	13.40	13.20	13.17	13.08

Side arm $\frac{1}{2}$-in.-diameter columns

Gas phase concentration of methycyclohexane, g/ml $\times 10^4$					
At top of column	0.282	0.220	0.271	0.235	0.280
22.9 cm from top	0.803	0.725	0.945	0.855	0.963
Partition coefficient of methylcyclohexane	534	535	537	542	538
HTU methylcyclohexane 0–22.9 cm	6.48	5.69	6.40	6.37	7.30
From Eq. 10.20b $H_{OG} = 60.24/S^{0.501}$ for 1-in.-diameter columns	8.58	8.32	8.58	8.58	8.55

(namely, $H_{OG} = 60.24/S^{0.501}$) was determined for 1-in.-diameter columns separating methylcyclohexane from benzene and using 29.6% polyglycol liquid phase on C22 Johns-Manville firebrick.

XII. Circular Columns for Continuous Chromatographic Refining

It is inevitable that with the moving-bed column previously described, some breakdown of the solid through attrition will occur leading to loss of brick and solvent on elutriation. To eliminate this loss and to give greater ease of operation without the problem of physically moving solids, circular chromatography columns that enable separations to be carried out without the disadvantages mentioned have been proposed by a variety of inventors.

The operation of the circular column is based on the moving-bed principle. The column, which is in the form of a circle, rotates passed fixed inlet and outlet ports. This means that the main disadvantages of the moving-bed process, viz attrition and solids flow control, are obviated by having no relative movement between the packing and the column wall. Only sufficient packing to fill the column is required as no recirculation is necessary. The use of this type of column was first suggested by the schematic diagrams of Pichler (12) and in the patent granted the Gulf Research and Development Company, (13). The first actual equipment constructed on these lines was by Luft (14).

In all these arrangements the flow of carrier gas within the column must be controlled by pressure drop, hence much of the column, which could be used for separation, must be employed just in creating a pressure drop. In a recent article Glasser (15) has described a similar piece of equipment in which again flow is controlled by pressure drop. In the scheme proposed by Barker (16), a column was designed that gave a undirectional flow of carrier fluid by using cam-operated valves and thereby giving efficient use of column length and carrier fluid.

In the first three arrangements shown in Fig. 10.12, the effective separation length cannot exceed half of the total circumference, and there is an

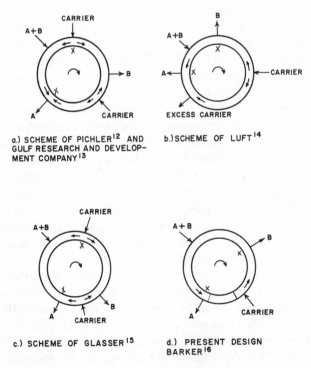

Fig. 10.12. Circular columns for continuous separations.

excess of carrier gas traveling in the same direction as the column rotation, thereby reducing the offtake concentration. Glasser (15), in arrangement *c*, uses only about 20% of the total carrier gas input for stripping the absorbed component or passing through the column used. Littlewood (17) points out that excess gas in the column reduces the concentration and may permit higher throughputs, as the throughput is limited by gas phase saturation. In fact, in this scheme the excess air reduces the concentration only in the cocurrent section of the column and not in the separating section itself. In scheme *d*, the design proposed by Barker (16), as the carrier gas travels only in one direction, the length of the column that can be used for separation purposes is limited only by the amount required for stripping and to provide a gas seal.

XIII. General Operating Principle of the Barker-5-Ft-Diameter Circular Chromatographic Machine

The prototype illustrated in Fig. 10.13 and shown schematically in Fig. 10.14 consists essentially of a 1½-in. square cross-section chamber in the form of a circle of 5-ft diameter.

The chamber is divided into eight equal sections connected by means of external valves, each section containing a copper helix through which can be passed a heating or cooling medium. By suitable cam arrangements, as the chamber (or column) rotates, different temperatures can be achieved in the packed sections so that parts may act as a stripping section. A hot, dry inert carrier gas enters the column at *D* and passes countercurrent to the direction of rotation of the column, leaving through the offtake ports *B* and *C*. To prevent the carrier gas from traveling from *D* to *B*, the valve chamber segments, which is in the arc *DB*, is always closed. The feed mixture to be separated enters continuously at *A*, the lighter (less strongly absorbed) component traveling in the gas stream toward product offtake port *B*, while the heavier (more strongly absorbed) components absorbed in the nonvolatile liquid with which the support is coated travel in the direction of rotation of the column and are stripped by heat in the section *CD* and leave through the product offtake port *C*.

To permit the gas flow into and out of the column through the inlet and outlet ports there are 180 gas passages equally spaced over the chamber face, each gas passage being closed by a self-sealing valve. These valves are automatically opened when they pass under one of the inlet or offtake ports. Sealing is achieved on the face of the chamber by means of O-rings that cover at least three of the 3/64-in.-diameter gas passages at any time under each port. The O-rings are sealed against the face by means of spring-loaded plates.

Fig. 10.13. 5 ft-diameter chromatographic machine.

The feed mixture entering the column at A is vaporized before entry, and to ensure that any that condenses on entry does not remain in the valve chambers a "bleed" of carrier gas is fed in just after the feed port at E. Sampling points are incorporated in the lines between the chamber sections so that the concentration profile within the packed bed can be obtained. The circular packed bed is rotated by a toothed wheel driving a chain attached to the periphery of the column, at speeds between 1 and 10 rph.

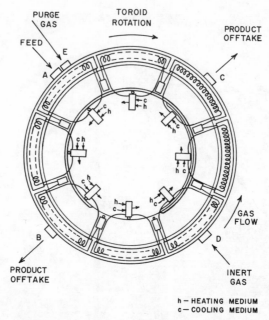

Fig. 10.14. Circular chromatography column.

The commercial machine manufactured by the Universal Fisher Group Ltd., Crawley, England (Fig. 10.15) has many improvements over the prototype, including separate entry and exit ports for feed and product off-take ports, respectively. Also, electrically heated elements in lieu of the copper helices give higher temperature capability and easier temperature control and a refined method of sealing between the stationary ports and the moving wheel.

XIV. Separations Using the 5-Ft-Diameter Circular Chromatograph. Hydrocarbon Separations—97% Cyclopentane Purification

The purification of a 97% pure cyclopentane containing five detectable impurities as shown in Fig. 10.16 illustrates the high throughput capability of this type of machine. The machine was packed with a Johns-Manville C22 firebrick coated with 30% polyoxyethylene 400 diricinoleate. Although more favorable stationary phases might have been chosen, Barker and Huntington (18) chose the phase because it had been used in earlier work on 1-in. diameter moving-bed columns and they were interested in observing the relationship between the two types of equipment. This phase, however, did have the advantage of bringing all the detectable impurities to one side of the main cyclo-

Fig. 10.15. 5 ft-diameter commercial chromatographic machine.

pentane peak, so that the impurities could be removed in one pass through the machine rather than two.

The operating behavior of this type of equipment is similar to the moving-bed column, so the gas and liquid rates to achieve separation can again be approximately determined from the partition coefficients of the two adjacent components of interest in the separation. For the stationary phase used in these experiments, the values of partition coefficient at 20°C were $K_{\text{cyclopentane}}$

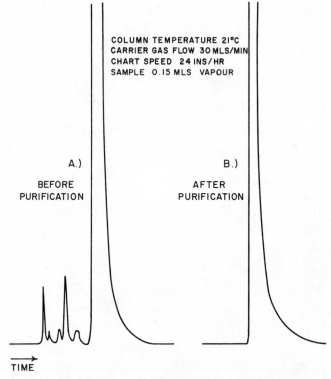

COLUMN TEMPERATURE 21°C
CARRIER GAS FLOW 30 MLS/MIN
CHART SPEED 24 INS/HR
SAMPLE 0.15 MLS VAPOUR

A.)

BEFORE
PURIFICATION

B.)

AFTER
PURIFICATION

TIME

Fig. 10.16. Chromatogram of cyclopentane before and after purification.

$= 119.1$ and $K_{\text{impurity}} = 78.5$, hence the separation factor given by the ratio of these K values is 1.518.

Table 10.6 shows the evaluation of the operating conditions in terms of gas flow that just allow a pure bottom product (cyclopentane) to be realized. In all these runs it is observed that the top product, which ideally should just be the impurities, always contains a large percentage of cyclopentane. This is because of the small amount of impurity initially present in the mixture and the limited effective length of 107 in. for mass transfer between the two offtake ports.

As the feed rate is increased, the gas flow also must be slightly increased to maintain pure cyclopentane as bottom product. Table 10.7 shows the proportion of the feed leaving at each offtake port for varying feed rates. The saturation limit of the carrier gas, which was air, had been exceeded at all the feed rates used in Table 10.7 and accounts for the high proportion of cyclopentane in the bottom product. This indicates that the column can be oper-

TABLE 10.6

Evaluation of Operating Conditions for Cyclopentane Purification[a]

Flow at ambient conditions, liters/hr			Mean column pressure, psig		F^R/S,		Top product, percent of cyclopentane (on inert free gas basis)	Bottom product, cyclopentane
Upper offtake flow rate	Lower offtake flow rate	Bleed carrier flow rate	Upper column	Lower column	Upper column	Lower column		
291	504	45	9.2	11.3	160	124	96.4	Pure; limit 34 in. from offtake
235	504	47	9.8	12.0	126	92.7	93.1	Pure; limit 8 in. from offtake
199	504	47	10.3	12.1	104	76.2	90.8	Pure
164	504	51	11.1	12.7	84	56.7	88.1	Trace impurity
131	504	56	11.6	13.1	67	37.6	68.2	Trace impurity 99% pure

[a] Column operating temperature was 20°C and feed rate was 72 ml/hr.

TABLE 10.7

Variation of Cyclopentane-Feed Distribution with Feed Rate[a]

Feed rate, ml/hr	Flows at ambient conditions, liters/hr			Mean column pressure, psig		F^R/S		Amount of feed mixture leaving, %	
	Upper offtake flow rate	Lower offtake flow rate	Bleed carrier flow rate	Upper column	Lower column	Upper column	Lower column	Top product offtake	Bottom product offtake
120	208	504	47	10.3	12.1	109	80	17.4	82.6
202	208	504	47	10.3	12.1	109	80	16.2	83.8
300	214	504	47	10.2	12.1	112	83	17.5	82.5
410	232	504	47	9.9	12.0	122	92	15.2	84.8

[a] Column operating temperature was 20°C and at all feed rates pure bottom product (cyclopentane) was obtained.

ated successfully above the saturation limit, but to prevent the occurrence of liquid pockets within the column it is preferable to increase the gas flow rate so that saturation does not occur at such low feed rates.

XV. 99% Cyclohexane Purification

The 99% cyclohexane fraction chosen had one detectable impurity (Fig. 10.17) when using a 6 ft long × 4 mm i.d. column packed with 52–60-mesh Johns-Manville C22 firebrick coated with 6% w/w of polyoxyethylene 400 dircinoleate and a flame-ionization–type detector. The partition coefficients for the cyclohexane and impurity at 20°C are $K_{cyclohexane} = 306$ and $K_{impurity} = 193.6$, hence the separation factor is 1.582. Although the separation factor is of an order similar to that of the cyclopentane system, it can be seen that the actual gas flow requirements are much greater, by a factor of about 3.

Table 10.8 shows the proportion of the feed leaving the column at both off-take ports, maintaining a pure bottom product, but in this instance the column was not saturated at any of the feed throughputs. Table 10.8 also shows a high loss of the cyclohexane product into the impurity fraction, necessitating the rerunning of this fraction through the machine to reduce excessive wastage.

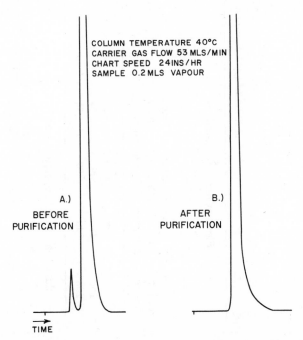

COLUMN TEMPERATURE 40°C
CARRIER GAS FLOW 53 MLS/MIN
CHART SPEED 24 INS/HR
SAMPLE 0.2 MLS VAPOUR

A.)

BEFORE
PURIFICATION

B.)

AFTER
PURIFICATION

TIME

Fig. 10.17. Chromatogram of cyclohexane before and after purification.

TABLE 10.8

Variation of Cyclohexane-Feed Distribution with Feed Rate[a]

Feed rate, ml/hr	Flows at ambient conditions, liters/hr			Mean column pressure, psig		F^R/S		Amount of feed mixture leaving, %		Limit of impurity from bottom port, in.
	Upper offtake flow rate	Lower offtake flow rate	Bleed carrier flow rate	Upper column	Lower column	Upper column	Lower column	Top offtake	Bottom offtake	
108	648	600	216	3.5	8.0	467	249	40.5	59.5	6
160	780	492	254	3.0	7.6	583	310	55.3	44.7	22
208	840	492	254	2.8	7.5	623	345	55.5	44.5	14

[a] Column operating temperature was 21°C.

XVI. The Separation of the Azeotropic System Cyclohexane–Benzene

The cyclohexane–benzene system has been extensively studied by Barker and Huntington (8) and their results for this system show the effect of column temperature, feed rate, and gas/liquid ratio on the behavior of a circular continuous chromatography machine.

The four column temperatures employed in their experiments were 20–21, 31–32, 35–36 and 43–44°C. Feed rates were varied from 66 to 100 ml/hr, while F_R/S ratios varied from 168 to 892. In all cases the feed was heated to about 87°C before entering the column so that it entered as a vapor.

The operating conditions of each of the runs are presented in Table 10.9. The upper column refers to the section between the top offtake and the feed, and the lower column to the section between the bottom offtake and the feed zone.

All the runs were operating in the laminar region for gas flow except for the column operation at 20–21°C when the maximum value of the Reynolds number based on the empty column was 13.6, which is in the transition region (*Re* 10–30).

Plots of concentration against distance show the same characteristic features as those obtained for vertical moving-bed columns, and are illustrated for the runs at 20–21°C in Fig. 10.18. There is a sharp drop in concentration at the top of the column, attributable to the incoming fresh packing, and the concentration of benzene below the feed increases as the column is descended. The partition coefficients of cyclohexane and benzene at this temperature are 302 and775, respectively, giving a separation factor of 2.57.

At 20–21°C the concentration level within the column is much less than in the vertical column. When plotted in terms of mole fractions, again similar profiles are observed (Fig. 10.19). As the ratio increases, the length of column necessary for the separation increases, particularly below the feed zone. In Fig. 10.19 it would be expected that curve *A7* should lie between *A8* and *A9* from F_R/S considerations, but it can be seen that because of the higher feed rate (110 ml/hr compared with 82 ml/hr) an impure bottom product was in fact obtained. The maximum feed rate that can be used to give pure products is about 90 ml/hr. The maximum and minimum operating values of F_R/S in the experiments that permitted realization of pure products were about 712 and 485, respectively, but it must be remembered that these are the mean values over the sections.

As the column temperature increases, the partition coefficients of the components decrease, so that the gas flow must also decrease as the column is being operated at a constant liquid rate. However, the separation factor also decreases and makes the separation a little more difficult.

TABLE 10.9

Operating Conditions for the Separation of Cyclohexane–Benzene at Increasing Temperatures

Temperature and run number	Feed rate, ml/hr	Solvent flow, ml/hr	Offtake carrier flow at ambient conditions, liters/hr			Mean column pressure, psig		Mean column operating F^R/S	
			Upper	Bleed	Lower	Upper	Lower	Upper	Lower
20–21°C									
A1	80	1111	690	188	560	2.7	5.6	524	326
A2	80	1111	778	186	546	2.5	5.3	596	390
A3	80	1778	828	192	546	2.7	5.8	892	582
A4	80	778	634	211	560	4.8	8.2	616	352
A5	82	767	740	200	557	2.4	5.5	836	485
A6	82	767	664	200	557	4.1	6.4	674	420
A7	82	767	717	198	613	3.6	6.1	748	475
A8	110	767	731	196	602	3.2	5.8	780	497
A9	110	767	697	202	597	4.0	6.3	712	449
31–32°C									
B1	66	758	512	67	762	3.9	7.5	553	403
B2	66	758	446	71	762	4.5	7.9	464	334

B3	66	758	358	73	762	5.3	8.5	362	248
B4	66	758	388	73	688	5.1	8.4	397	274
B5	81	758	398	73	732	4.9	8.5	407	282
B6	89	758	388	72	732	5.0	8.2	398	277
B7	89	758	378	70	732	4.7	7.8	391	273
B8	100	758	388	72	732	5.0	8.4	398	274
35–36°C									
C1	66	767	316	102	567	5.1	8.4	321	189
C2	65	767	395	99	602	4.9	8.3	405	260
C3	65	767	363	104	602	5.4	8.7	362	226
C4	86	767	274	94	573	5.5	8.7	274	158
C5	83	758	318	96	877	4.7	8.0	378	244
C6	74	758	294	98	874	4.8	8.1	340	212
C7	74	758	330	107	877	4.3	7.6	398	245
43–44°C									
D1	94	758	502	98	563	4.7	7.0	539	388
D2	94	758	343	107	626	5.5	8.6	350	211
D3	94	758	291	105	612	6.0	9.0	289	163
D4	64	758	322	99	585	5.4	7.9	332	204
D5	64	758	297	103	573	5.7	8.4	308	174

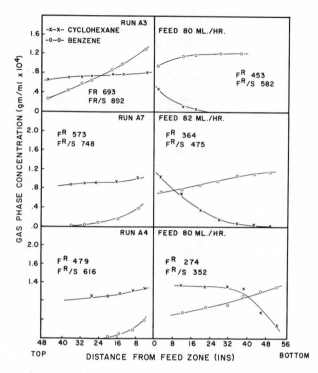

Fig. 10.18. Concentration profile for the separation of cyclohexane/benzene at 20–21°C.

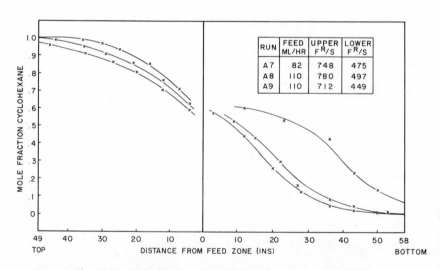

Fig. 10.19. Mole fraction of cyclohexane in column at 20–21°C.

366

Decreasing the gas flow rates increases the concentration level within the column for a constant feed rate. As the temperature increases, the gas phase saturation also increases quite rapidly, so that it is not expected that reducing the gas flow rate because of increased temperature will cause column saturation.

At column temperatures of 31–32°C and 35–36°C, the concentration profiles were of the same form as at the lower temperature.

The highest column temperature used was 43–44°C, at which temperature the partition coefficients of cyclohexane and benzene are 145 and 345, respectively, giving a separation factor of 2.38. The gas flow rate used to perform the separation at this temperature was very low, giving much higher concentrations within the column (Fig. 10.20). Run *D5* was the only run to give two

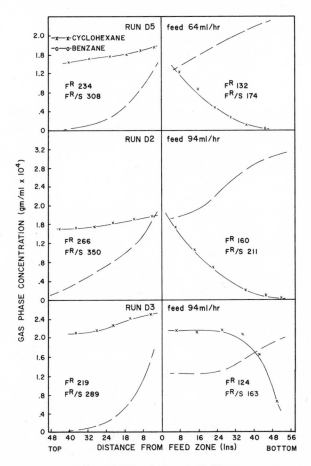

Fig. 10.20. Separation at 43–44°C.

pure products and indicates that the maximum feed rate is about 70 ml/hr. The maximum value of F^R/S giving pure products is about 308 and the minimum 174, which shows that the theoretical range is slightly narrowed in practice. It also appears that the range is lessened as the feed rate is increased.

XVII. Column Efficiency

Column efficiency for the benzene–cyclohexane system has been expressed in terms of both HTU and the overall HETP for the runs in which operating conditions satisfy the relationships. For the evaluation of the HTU, a mean value of F^R/S was employed in each of the sections and the value of HTU determined over 5-in. sections along the column (Table 10.10).

TABLE 10.10
Efficiency Based on Benzene above the Feed Zone

| | HTU, in. | | | | | |
| | Section above the feed zone, in. | | | | | |
Run number	5–10	10–15	15–20	20–25	25–30	Mean
A4	1.11	1.02	—	—	—	1.06
A6	0.93	0.85	—	—	—	0.89
A9	0.82	0.90	0.85	—	—	0.86
B3	2.92	2.36	2.24	—	—	2.51
B4	3.62	2.82	2.99	2.86	2.29	2.91
B5	3.08	3.20	3.34	3.28	2.78	3.14
B6	2.70	3.14	3.00	2.51	2.16	2.70
B7	4.72	4.35	3.50	3.15	2.06	3.55
C1	3.97	3.27	2.83	—	—	3.36
C4	2.86	2.49	—	—	—	2.67
C6	3.55	3.07	2.92	2.86	2.16	2.91
D3	1.79	1.63	1.70	1.87	1.83	1.76
D5	1.88	1.53	1.38	1.48	1.14	1.48

At 20–21°C the mean value HTU is about 1 in., the same as that found in the vertical moving-bed columns. At a temperature of 31–32°C, the HTU value is in general between 2 and 3.5 in. The same range is observed at a column temperature of 35–36°C, but at 43–44°C the value drops to a mean value of about 1.5 in. The reason for the low values at 20–21°C may be that turbulence was beginning to occur in the bed, which promotes mass transfer.

The calculated HTU values below the feed are shown in Table 10.11. The HTU values are in general much higher than those observed for benzene above the feed, although it can be seen that as the distance from the feed zone decreases the value of HTU gradually decreases and approaches the value obtained for benzene. The value of benzene below the feed zone is considered to result from the equilibrium being established more slowly below the feed than above, because as the column operating temperature is below the boiling point of the mixture some condenses on entry and is carried down the column as liquid.

TABLE 10.11

Efficiency Based on Cyclohexane below the Feed Zone

	HTU, in.					
	Section below the feed zone, in.					
Run number	5–10	10–15	15–20	20–25	25–30	30–35
A3	6.71	4.92	—	—	—	—
A5	8.15	7.02	7.12	5.72	4.37	—
A7	11.2	8.48	6.83	7.04	5.44	4.59
A8	15.6	14.0	10.8	8.06	7.48	5.25
B1	13.3	12.6	9.45	5.98	—	—
B2	7.63	6.25	5.75	4.13	—	—
B3	4.77	4.0	3.33	2.21	1.98	1.44
B4	4.07	3.57	3.44	2.93	2.45	—
B5	4.90	4.29	3.55	2.72	2.96	—
B6	5.25	4.66	3.45	3.18	3.24	2.76
C2	8.24	6.73	6.08	6.33	5.23	5.12
C3	6.42	5.74	4.72	4.21	3.44	3.73
C5	6.17	5.40	5.01	4.27	4.06	4.13
C7	5.31	4.50	4.61	5.62	4.45	4.07
D1	9.75	9.31	7.82	—	—	—
D2	18.3	8.55	8.55	6.80	5.62	4.95
D4	6.97	5.87	5.15	4.75	3.82	3.09
D5	4.21	4.03	3.18	2.56	2.38	2.02

The overall mass transfer efficiency of the circular column in terms of HETP for those operating conditions satisfying Eq. 10.40 are presented in Table 10.12, where it is seen that the mean value for a feed rate of 65 ml/hr is about 2 in. From investigations on analytical columns, it would be expected that the efficiency should rise with temperature to reach a maximum about 10°C below the boiling points of the components being separated. In the operation of the circular column, this has not been observed probably because of the use of a bleed stream of gas which must be kept above a certain minimum value. This gas stream becomes increasingly important as the temperature rises because the gas flow in the upper section is reduced. The flow in the lower column is different from that in the upper section by an amount equal to this bleed stream.

TABLE 10.12

Overall Efficiency of the Separation of Cyclohexane–Benzene

Run number	Feed rate, ml/hr	Distance from feed zone, in.		Total, in.	ϕUpper/ ϕlower	N_C	HETP, in.
		Above	Below				
A6	82	30	58	88	1.604	44.2	1.99
A7	80	40	47	87	1.575	42.6	2.40
B3	66	28	52	80	1.46	37.2	2.15
B4	66	40	36	76	1.45	36.8	2.06
B5	81	49	39	88	1.44	36.4	2.42
B6	89	46	46	92	1.44	36.4	2.53
B7	89	44	51	95	1.434	36.1	2.63
C1	66	42	58	100	1.698	51.7	1.93
C3	65	48	56	104	1.602	43.2	2.41
D5	64	47	55	102	1.770	59.9	1.70

XVIII. The Separation of the Azeotropic System Diethyl Ether–Dichloromethane

This system is a further example of the ease with which azeotropic systems separate into their constituents using the CCR technique. The partition coefficients at 20°C using the polyglycol phase previously mentioned are $K_{\text{diethyl ether}} = 101$ and $K_{\text{dichloromethane}} = 482$, giving a separation factor of 4.77. The boiling points of the two materials are 34.8 and 41°C, respectively.

The actual operating conditions are shown in Table 10.13, which indicates the effective use of the column at different rates, with the upper and lower sections of the column being used to an equal extent. It can be seen that the maximum throughput consistent with pure products is about 220 ml of feed stock per hour under these conditions. Products were defined as pure at

TABLE 10.13

Evaluation of the Operating Conditions for Separating Diethyl Ether–Dichloromethane[a]

Feed rate, ml/hr	Flows at ambient conditions, liters/hr			Mean column pressure, psig		F^R/S		Product purity and limit of the impurity	
	Upper offtake flow rate	Lower offtake flow rate	Bleed carrier flow rate	Upper column	Lower column	Upper column	Lower column	Top offtake DEE	Bottom offtake DCM
64	437	348	85	8.2	11.0	252	179	Pure limit 31 in. from offtake	Pure limit 38 in. from offtake
117	464	348	85	7.6	11.1	268	193	Pure limit 25 in. from offtake	Pure limit 22 in. from offtake
176	488	336	80	7.2	11.2	282	202	Pure limit 9 in. from offtake	Pure limit 10 in. from offtake
202	437	348	86	8.0	11.0	252	178	Pure	Pure
220	451	348	86	8.0	11.0	260	186	Pure	Pure
267	437	348	86	7.7	11.0	274	200	99% pure	98.7% pure
267	476	348	86	7.3	11.2	274	200	92.6% pure	Pure

[a] Column operating temperature was 20°C.

$>99.5\%$. Table 10.14 shows the values of HETP for this separation, indicating an increase in HETP with feed rate, presumably the result of higher concentration levels within the column.

TABLE 10.14

Overall Efficiency for the Separation of Diethyl Ether–Dichloromethane

Feed rate, ml/hr	Distance from feed zone, in.		Total, in.	ϕUpper/ ϕlower	N_C	HETP, in.
	Above	Below				
64	18	20	38	1.41	17.45	2.18
117	24	36	60	1.39	17.20	3.59
176	40	48	88	1.395	17.27	5.10
202	49	58	107	1.42	17.51	6.11
220	49	58	107	1.398	17.29	6.19

XIX. The Separation of the Close-Boiling Dimethoxymethane–Dichloromethane Mixture

The boiling points of the two constituents are 42 and 41°C, respectively and, using the polyglycol phase previously mentioned, the partition coefficients at 20°C are $K_{DMM} = 172$ and $K_{DCM} = 482$, giving a separation factor of 2.80. This system illustrates another advantage of the CCR technique, how by suitable choice of stationary phase the vapor liquid equilibria of a system can be modified, making the system much easier to separate than if no third phase were used (Table 10.15). Used in this way, the technique is analogous to extractive distillation, however, on a small scale extractive distillation columns need to be carefully and expensively instrumented and are normally more difficult to operate successfully than the equipment described here.

The efficiency of the circular column in terms of HETP is shown in Table 10.16 which again shows the reduction in the efficiency of the column with increasing feed rate.

XX. General Operating Principle of the Barker Compact Circular Chromatographic Machine

One of the main deficiencies of the 5-ft-diameter circular chromatographic machine was the limited separating power resulting from the insufficient length of mass transfer zone available for separation purposes (9 ft). To maintain the basic concept of a circular machine and still obtain long mass trans-

TABLE 10.15

Evaluation of the Operating Conditions for Separating Dimethoxymethane–Dichloromethane[a]

Feed rate, ml/hr	Flows at ambient conditions, liters/hr			Mean column pressure, psig		F^R/S		Product purity and limit of the impurity	
	Upper offtake flow rate	Lower offtake flow rate	Bleed carrier flow rate	Upper column	Lower column	Upper column	Lower column	Top offtake DMM	Bottom offtake DCM
93	600	504	89	5.0	8.9	396	282	92% pure	Pure limit 28 in. from offtake
48	443	432	76	5.6	9.2	290	202	Pure limit 33 in. from offtake	Pure limit 10 in. from offtake
48	525	432	72	5.2	9.0	347	250	Pure limit 18 in. from offtake	Pure limit 16 in. from offtake
70	525	432	72	5.2	9.0	347	250	Pure	Pure limit 8 in. from offtake
88	525	432	72	5.2	9.0	347	250	Pure	Pure
130	525	432	72	5.2	9.0	347	250	98% pure	Pure
172	488	432	72	5.2	9.0	323	230	Pure	97.5% pure

[a] Column operating temperature was 20°C.

TABLE 10.16

Overall Efficiency for the Separation of Dimethoxymethane–Dichloromethane

Feed rate, ml/hr	Distance from feed zone, in. Above	Below	Total, in.	ϕUpper/ ϕlower	N_C	HETP, in.
48	31	42	73	1.39	30.2	2.42
70	49	50	99	1.39	30.2	3.28
88	49	58	107	1.39	30.2	3.54

fer zones makes the machine very cumbersome. The difficulty of compactness has been overcome by Barker (19) by forming a closed-loop cylindrical nest of straight tubes held between a top and bottom ring, as shown in the prototype model in Fig. 10.21 and schematically in Fig. 10.22a.

The circular chromatography machine consists essentially of 44 1-in.-bore stainless-steel tubes, 9 in. in length forming a closed-loop cylindrical nest of tubes disposed between top and bottom rings. This tube bundle can be rotated at speeds between 0.2 and 2.0 rph by a variable-speed drive. Four stationary ports are located around the circumference of the upper ring available for service with clockwise rotation of the tube bundle as carrier gas inlet (I) and faster fraction outlet (IV).

To permit a undirectional flow of inert gas countercurrent to the rotation of the bundle, poppet valves are closed in turn by a cam as they pass between the carrier gas inlet (I) and faster product outlet (IV).

The tube bundle and ports are housed in an electrically heated thermostatically controlled oven operating to 200°C. The oven is vented and continuously purged with preheated air as a precaution against the accumulation of explosive vapors in the event of accidental leakage. An explosion disc is also fitted in the top of the oven as an extra precaution. A photograph of the commercial model manufactured by the Universal Fisher Group Ltd., Crawley, England is shown in Fig. 10.22b.

The principle of operation is similar to that described for the 5-ft-diameter circular machine. A hot, dry, inert carrier gas enters at port I, passing up and down tubes in a general direction countercurrent to the rotation of the tube bundle and leaving offtake ports II and IV. The feed mixture to be separated enters continuously at III, the lighter (less strongly absorbed) components traveling preferentially in the gas stream toward product port IV. The heavier (more strongly absorbed) components travel in the direction of rotation of

Fig. 10.21. Compact circular chromatographic machine.

the tube bundle, being stripped off by the hot, inert gas in the zone between ports *I* and *II* and leaving through the offtake port *II*.

XXI. The Production of High-Purity Hydrocarbons. Cyclopentane Purification

The improved separation capability of the compact circular chromatography machine compared with the 5-ft circular chromatograph is demonstrated by the results published by Barker and Al-Madfai (20) on cyclopentanes

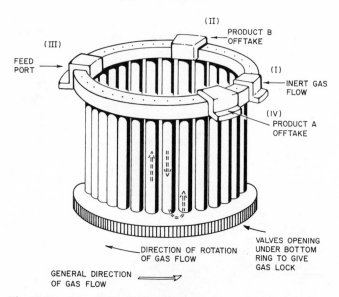

Fig. 10.22a. Schematic of compact circular chromatographic machine.

and cyclohexane purifications. A 97% cyclopentane with four detectable impurities as shown in Fig. 10.23 was used. The nearest impurity to the cyclopentane was identified as 2,2-dimethylbutane, having a partition coefficient of 78.5 compared with 119.1 for cyclopentane. Hence the separation factor was 1.518. The tubes of the machine were packed with 14–18-mesh Johns-Manville C22 firebrick coated with 30% w/w polyolyethylene 400 diricinoleate. The total weight of support was 2956.7 g, of which 887.0 g were liquid phase.

Table 10.17 shows the operating data required to achieve the column performance given in Table 10.18, the experiments being designed to obtain pure cyclopentane, with minimum loss of this material in the impurity fraction. Very high purity (99.999%) with virtually no loss of cyclopentane in the impurity fraction was achieved (Fig. 10.23). It should be noted that the purities recorded throughout this chapter refer to detectable purity based on GC with flame ionization detectors, rather than absolute purity.

The reduced throughput of the compact circular machine is to be expected because the cross-sectional area of the packed section is only one-third that of the packed section used in the 5-ft-diameter circular machine. It is seen that throughput is approximately proportional to cross-sectional area, while the length of mass transfer section (14 ft) compared to 9 ft for the 5-ft-diameter circular machine considerably improves the purity and yield of the desired products.

Fig. 10.22b. Commercial compact circular chromatographic machine.

XXII. Cyclohexane Purification

Further evidence of the superior separating power of the compact circular machine for the purification of 99% cyclohexane is shown in Fig. 10.24 and Table 10.19. The one detectable impurity was identified mainly as methylcyclopentane having a partition coefficient of 193.6 at 20°C. Hence compared to 306 for cyclohexane, the separation factor is 1.582.

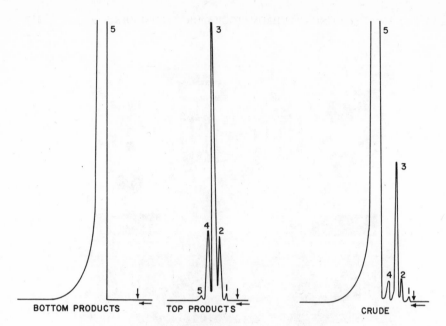

Fig. 10.23. Purification of cyclopentane.

I. IMPURITY
2. CYCLOHEXANE

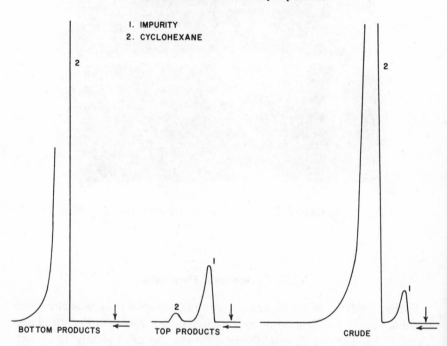

Fig. 10.24. Purification of cyclohexane.

TABLE 10.17

Operating Conditions for Cyclopentane Purification[a]

Run	Feed rate, ml/hr	Carrier gas flow rates at ambient conditions, liters/hour			Mean column pressure, psig		F^R/S	
		Upper offtake	Lower offtake	Feed purge	Upper column	Lower column	Upper column	Lower column
1	26.9	94.9	167	19.9	5.5	6.0	80.3	61.9
2	41.6	94.9	167	19.9	6.1	6.3	77.8	60.9
3	60.3	94.9	167	19.9	6.1	6.3	77.8	60.9
4	154.1	94.9	360	19.9	3.5	4.5	89.9	67.1
5	154.4	112.9	360	19.9	3.5	4.5	105.85	83.2

[a] All runs made at 1.0 RPH = 0.8625 liters/hr stationary phase, temperature = 22°C.

TABLE 10.18

Column Performance for Cyclopentane Purification

Run	Feed rate, ml/hr	Recovery of cyclopentane in lower offtake, %	Purity of lower offtake product, %	Limit of impurities from lower offtake, ft	Effective mass transfer length used in resolving mixture (14 ft available), %
1	26.9	100	99.999	12	30
2	41.6	99.84	99.999	9	40
3	60.3	99.84	99.999	3	50
4	154.1	99.9	99.18	—	100
5	154.4	99.9	99.999	1	90

XXIII. The Separation of n-Hexane from Crude Hexane

The separation of n-hexane from a crude hexane containing four main impurities illustrates the ability of the compact circular machine to separate isomers and also the separation of a component from impurities distributed on either side of the desired product as shown in Fig. 10.25.

A GC analysis of the crude n-hexane showed four main peaks on the chromatogram; these peaks were identified and taken to be in order of elution from the column as 2-methylpentane (2mp), 3-methylpentane (3-mp), n-hexane (n-hex) and methylcyclopentane (mcp). The partition coefficients of

TABLE 10.19

Operating Conditions for the Cyclohexane Purification[a]

	Feed rate, ml/hr	Carrier gas flow rates at ambient conditions, liters/hr			Mean column pressure, psig		F^R/S, liters/hr	
Run		Upper offtake	Lower offtake	Feed purge	Upper column	Lower column	Upper column	Lower column
1	36.8	242	648	29	2.4	3.5	241.7	200
2	64	242	648	29	2.4	3.5	241.7	200
3	76.4	242	648	29	3.5	4.5	226.7	197.7
4	152	242	648	29	2.4	3.5	241.7	200
5	126.7	343.2	648	31.2	3.3	5.1	324.6	269

[a] All runs made at 1.0 RPH = 0.8625 liters/hr; temperature = 21°C.

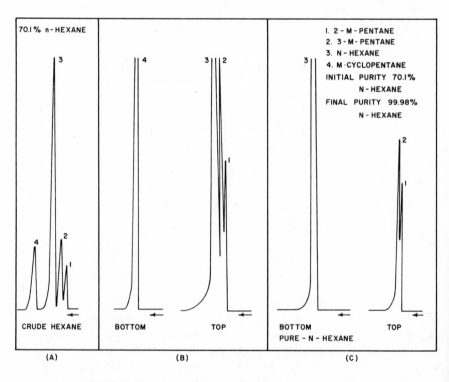

Fig. 10.25. Separation of pure *n*-hexane from crude hexane.

these compounds at 20°C were found to be $K_{2mp} = 101.8$, $K_{3mp} = 119.1$, $K_{n\text{-hex}} = 142.0$, and $K_{mcp} = 200.8$, giving separation factors of 1.170 for 3mp/2mp, 1.192 for n-hex/3mp and 1.414 for mcp/n-hex.

In the first series of runs (1 to 4 in Table 10.20), methylcyclopentane was removed from the crude n-hexane with comparative ease (Fig. 10.25b) because of the separation factor of 1.414 for mcp/n-hex and the long length of column available for mass transfer. The material collected from the top product port, i.e., n-hex, 2mp, and 3mp was then rerun to separate the n-hex from the 2mp and 3mp (Fig. 10.25c). Run 5 shows typical conditions for a feed rate of 12.8 ml/hr. The separation now is between two isomers n-hex/3mp where the separation factor is 1.192, hence the lower feed rate compared with runs 1 to 4.

Runs 6 and 9 show the versatility of the machine in being able to cut at any chosen point on a chromatogram.

Typical component distributions within the column at steady-state conditions are shown in Figs 10.26a and 10.26b.

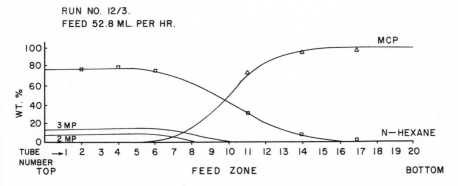

Fig. 10.26a. Separation between n-hexane and methylcyclopentane.

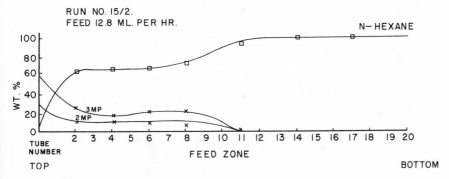

Fig. 10.26b. Separation of pure n-hexane from 2- and 3-methyl pentanes.

TABLE 10.20
The Separation of Crude Hexane into its Main Constituents

Run no.	Feed rate, ml/hr	Flow at ambient conditions, liters/hr			Mean column pressure, psig		F^R/S		Column temp, °C	RPH	Upper offtake products	Lower offtake products
		Upper offtake	Lower offtake	Feed purge	Upper column	Lower column	Upper column	Lower column				
Separation between n-hexane and methylcyclopentane												
1	12.7	160.8	251	0	2.5	3.5	159.5	150.8	20	1.00	2mp, 3mp, n-hex	mcp Complete separation
2	24.7	213	252	57	2.5	3.5	211.1	147.3	18	1.00	2mp, 3mp, n-hex mcp < 1%	mcp
3	52.8	179.9	228	20.9	3.5	4.5	168.6	132.4	19.5	1.00	2mp, 3mp, n-hex	mcp Complete separation
4	78.3	307	285	37.1	3.6	5.2	167.1	135.3	21	1.71	2mp, 3mp, n-hex, mcp	n-hex, mcp Loss 7% n-hex from lower offtake
Separation of pure n-hexane												
5	12.8	241.4	186	28.4	5.7	7.3	117.9	92.27	17	1.71	2mp, 3mp, n-hex Pure n-hexane	Loss 3% n-hex from upper offtake
Separation between 2-methylpentane and 3-methylpentane												
6	11.6	99	156	15	3.5	4.5	92.9	74.8	17.5	1.00	2mp	3mp, n-hex, mcp Complete separation
7	21.2	78	186	0	4.3	4.8	70.2	68.5	22.5	1.00	2mp	3mp, n-hex, mcp Complete separation
Separation between 3-methylpentane and n-hexane												
8	14.8	115	246	15	5.25	5.75	98.5	83.46	17	1.00	2mp, 3mp	n-hex, mcp Complete separation
9	26.1	109.8	186	0	4	4.6	109.2	105.8	21.5	1.00	2mp, 3mp	n-hex, mcp Complete separation

XXIV. The Separation of Linalol from Rosewood Oil

A chromatogram of the rosewood oil (Fig. 10.27) showed the following compounds or groups of compounds in order of elution from an analytical chromatogram: terpene hydrocarbons, cineole, benzaldehyde, *cis-* and *trans-*linalol oxides, the main peak of linalol followed by terpene alcohols. The aim of these experiments was to try to separate compounds such as linalol without degradation of the mixture. Table 10.21 shows operating conditions and performance data for the experiments made with this oil.

In run 3 complete removal of linalol oxides and light hydrocarbons was achieved (Fig. 10.27). For in run 4, Table 10.21 shows conditions when cutting on the other side of the linalol peak.

These experiments showed the ability of a compact chromatograph to handle high-boiling, heat-sensitive liquid mixtures, giving products that had suffered no detectable decomposition effects. It is considered that this type of separation technique will find particular application in handling heat-sensitive liquid mixtures because the presence of the inert gas acts in a similar manner to low-vacuum operation, while by operating primarily in the gaseous state, residence times within the equipment are small, thereby reducing the chance for significant decomposition to occur. Materials of construction of the equipment obviously play an important role in determining decomposition effects. Currently, 18/8 stainless steel and Graflon, i.e., PTFE, and carbon are used, although glass equipment is contemplated for those separations requiring it.

XXV. Separation of α-Pinene from U. S. Grade Spirit of Turpentine

Turpentine, which contains primarily α- and β-pinenes together with a few percent of camphene has been quoted frequently in the literature as a yardstick for separations by batch-operated gas chromatographs. It is therefore of interest to compare the quoted batch-operated throughput achievements with those for compact circular chromatographs.

Abcor (43) has reported the separation of α-pinenes from β-pinenes using crude turpentine. They reported a maximum feed rate of 22 liters (41 lb) per 24 hr when using a 4-in.-diameter packed column 9 ft long fitted with special baffles to obtain maximum column efficiency. The reported recovery of 98.6% pure α-pinenes was 90%, and the equivalent feed rate through a 1-in.-diameter column corresponds to 55.4 ml/hr.

In the investigation by Barker and Al-Madfai reported in Table 10.22 and Fig. 10.28, a 99% pure α-pinene at 91.3% recovery was obtained by CCR techniques at a feed rate of 66.5 ml/hr. This feed rate is 20% more than reported by Abcor for an equivalent 1-in.-diameter column. With higher temperatures nearer to the boiling points of the pinenes (ca. 152°C), Barker and

TABLE 10.21
Separation of Linalol from Rosewood Oil

Run no.	Feed, ml/hr	RPH of column	Stationary phase ml/hr	Carrier gas at ambient conditions, liters/hr			Mean column pressures, psig		F_R/S, at mean column cond.	Column temp., °C	Upper offtake products	Lower offtake products
				Upper column offtake	Lower column offtake	Feed purge	Upper lower	Column lower lower				
1	8	0.1755	60.3	72	282	0	1	3	1373	110	Light hydrocarbons, oxides, no linalol	Some oxides, linalol, and higher alcohols
2	8	0.1755	60.3	102	300	0	1.6	4.6	1890	130	Light hydrocarbons, oxides, a little linalol	Linalol and higher alcohols only
3	3.9	0.476	164.4	165	402	0	2.2	6.8	1046	130	Light hydrocarbons, oxides only	Linalol, higher alcohols
4	9.7	0.2375	81.6	147	374	0	1.8	5.2	2006	135	Light hydrocarbons, oxides, linalol, trace heavy alcohols	Trace linalol, higher alcohols

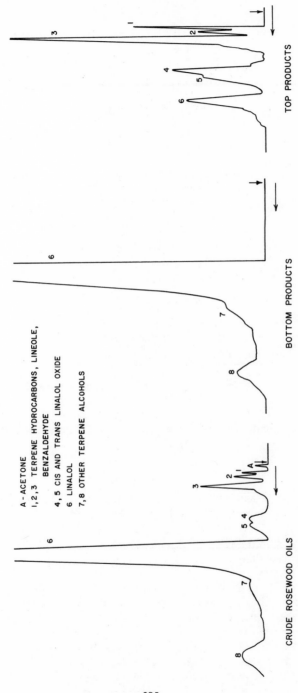

A - ACETONE
1,2,3 TERPENE HYDROCARBONS, LINEOLE,
 BENZALDEHYDE
4,5 CIS AND TRANS LINALOL OXIDE
6 LINALOL
7,8 OTHER TERPENE ALCOHOLS

TOP PRODUCTS

BOTTOM PRODUCTS

CRUDE ROSEWOOD OILS

Fig. 10.27. Separation of linalol from rosewood oils.

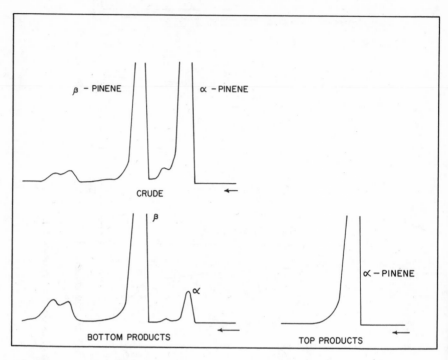

Fig. 10.28. Separation of alpha pinenes from U.S. grade spirit of turpentine.

Al-Madfai anticipate even higher percentage feed differences over the static bed results.

TABLE 10.22

Separation of α- and β-Pinenes from U. S. Commercial Turpentine Crudes

	Carrier gas flow at ambient conditions, liters/hr						Postulated feed rate through 4-in.-
				Average			
			Oven	column		Re-	diameter
Feed	Upper	Lower	tempera-	pressure,	Purity,	covery,	column
ml/hr	offtake	offtake	ture, °C	psi	%	%	liters/24 hr
33.5	165	342	110	5	99+	98.3	12.91
66.5	165	264	113	4.8	99.5	91.3	21.65
109.8	189	240	120	5	99	62.0	42.17

The overall percentage recovery indicated in Table 10.22 refers to the product actually recovered by a two-stage cooling process, which is not necessarily the maximum that could be recovered with more sophisticated equipment.

XXVI. Other Equipment Achieving Continuous Chromatographic Separations

A. Radial Flow Columns

Equipment attributable to Mosier (21) in which the feed travels from the center to the circumference of an annular packing is described in the literature. By relative rotation of the packing and the feed inlet, the paths taken by different components of the feed are different depending on the retention volume of the component, and so continuous separations can be achieved. The equipment shown in Fig. 10.29 requires that either the feed injection and

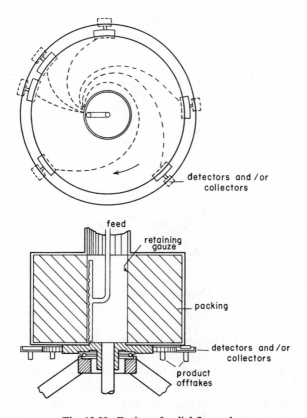

Fig. 10.29. Design of radial flow column.

collection system can be rotated with the packing stationary, or that the packing can be rotated with the feed system static, the latter method being preferred.

B. Helical Flow Columns

This type of column first proposed by Martin (22) is based on an annular packed column, or a circular tube array which is rotated. The feed enters at the top, and the paths traveled by the different components are in the form of helices.

Dinelli et al. (23,24) constructed such a column (Fig. 10.30) from 100 columns (6-mm-in. diameter, 1.2 m long), arranged in the form of a circle. The

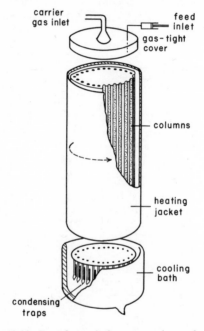

Fig. 10.30. Rotating unit for preparative-scale GC.

column bundle rotates at speeds of from 1 to 50 rph, and both the feed inlet port and product offtake receivers are stationary.

The carrier gas travels down the tubes and because of the rotation the different components of the feed, by virtue of their different retention times, each follow a different path and so can be collected in separate traps. Knowledge of the speed of rotation, the carrier flow, and retention times of the components permits evaluation of the receiver at which the component should appear.

$$L = 2\pi r Q t_R \tag{10.45}$$

where L = distance of receiver from feed position
   r = cylinder radius
   Q = speed of revolution
   t_R = retention time of component

Optimum operating conditions have also been determined (25). By using this column a mixture of heptane and toluene has been separated at 60 ml/hr, heptane appearing in receivers 8 to 22 and toluene in receivers 72 to 4 (compared with 9 to 24 and 70 to 3 as calculated).

As the feed rate is increased, the number of receivers in which the component appears increases, and eventually the components begin to overlap, so there is a maximum feed rate that can be used to give products of a required purity. For separation of cyclohexane–benzene on tricresyl phosphate at 80°C at purities of more than 99.9%, a theoretical maximum value of 220 ml/hr was calculated, although in practice it was found to be 200 ml/hr.

The unit was enlarged using 36 tubes of 2.4-m length in the form of an inverted "U" and in this form has been used to separate isomers (26) and in the purification of hydrocarbon fractions for which higher throughputs and purities are more readily obtainable than with an Oldershaw column.

The same type of column, but using a packed annulus of multiple tubes has also been proposed (27,28) although no operating data has been published. It is considered that the mechanical problems involved in making a perfectly symmetrical annulus would be greater than in the construction using multiple tubes, although the latter requires that the column be matched to give the same elution characteristics (29).

Ingenious as these helical flow schemes are, they are still batch-operated columns, suffering the inherent drawbacks outlined earlier in the chapter when attempting to separate large quantities of feed.

Turina et al. (30) used an apparatus consisting of two parallel glass plates 1 mm apart containing the inert support. Carrier gas is introduced uniformly along one vertical edge and emerges at the opposite edge through 26 uniformly spaced exhaust ports. The nonvolatile liquid flows through the packing vertically and leaves at the base. If the feed mixture is introduced continuously into the packing at the corner situated between the entrance of both carrier and nonvolatile liquid, the individual components will travel at an angle from the horizontal depending on the retention time and appear at different positions in either the existing carrier gas or in the nonvolatile liquid phase. This has been applied to continuous TLC using triangular plates (31).

C. Sinusoidal Flow Columns

Thompson (32,33) developed a semicontinuous separation using a single analytical column in which the inlet stream varied from 100% carrier gas to 100% feed mixture in a continuous sinusoidal manner so that the concentration of all the components in the feed mixture oscillated sinusoidally in phase with one another. As the concentration waves traveled through the column, they were differentially retarded and attentuated by the stationary phase. Hence by adjustment of the frequency and flow rate selected, component waves can be caused to emerge from the column out of phase with one another so that the valves, operating at the same frequency as the wave generator, can direct the effluent to different condensing traps for collection of the material. Using a 1:1 feed mixture of ethane and propane with benzyl ether as the stationary phase and carbon dioxide as the carrier gas, Thompson separated products of 70% pure ethane and 74% propane at a total feed rate of 82 ml/min and a frequency of 40 cpm.

D. Flowing Stationary Phase Columns

In all the vertical moving-bed equipment described earlier, the solid support flowed under gravity down the column, but the same basic principle can be achieved by having a solid support held stationary, as in a conventional packed column, and allowing the nonvolatile stationary phase liquid to flow down the column at a known rate over the support.

Tiley and co-workers (34) investigated this principle. Using a 1-in.-diameter column to separate 5 ml/hr of 1:1 feed mixture of diethyl ether and dichloromethane, with dinonyl phthalate as the nonvolatile flowing liquid, they found that product purities in excess of 99.9% could be achieved although the efficiency was much less (HETP about 2.5 min) than when the moving-bed column was used.

Although this scheme is simpler to construct than moving-bed or circular columns, poorer efficiencies and low throughputs are in fact obtained. This is to be expected since the nonvolatile liquids flowing over the packing pieces usually have high viscosities giving low diffusion rates, an effect that is counteracted in moving-bed chromatographs by distributing the liquid as thin films over solid supports of very large surface area. Also, packed columns suffer from liquid maldistribution effects and have a much lower surface area/unit volume ratio than when solid supports are used.

Kuhn et al. (35,36) used the temperature dependence of the partition coefficients in flowing liquid columns for the separation of multicomponent mixtures. By having sections of the column at different temperatures and constant gas and liquid flow rates, the different components of a mixture collect at different specific places in the column depending on the partition coefficient.

The column used consisted of five sections each maintained at a different temperature, decreasing as the column was descended. The nonvolatile liquid paraffin oil containing 10% stearic acid flowed over steel spirals countercurrent to the flowing carrier gas. The ternary mixture of propionic, *n*-butyric, and *n*-valeric acids was introduced into the top of the first section (hottest) at the base of the column. All three components rose through section two, and the column temperature gradient was arranged so that *n*-valeric acid concentrated between sections 2 and 3, *n*-butyric acid between 3 and 4, and propionic acid between 4 and 5. The product could be continuously withdrawn, and purities of 95% were achieved.

Other flowing-liquid film equipment papers have been reported in the literature (37,38).

XXVII. The Continuous Separation of Nonvolatile Organic Materials by Liquid–Solid Chromatography

Whereas most of the published work using CCR techniques relates to gas chromatographic type systems, the liquid–solid chromatographic counterpart is in fact practiced commercially using batch techniques for separating, e.g., glucose from fructose, reducing the molecular distribution of carbohydrate polymers such as dextran, etc.

Using rotating circular column techniques similar to those described previously for the separation of volatile organic substances, preliminary studies into the separation of carbohydrates have been completed (42) and work now continues in evaluating the technique for the separation of enzymes, antibiotics, hormones, carbohydrate polymers, etc.

The equipment being used is a small laboratory model consisting of a circular bundle made up of 44 tubes, 1 ft long and 3/8 i.d. packed with, e.g., porous silica beads in order to effect separations on the basis of molecular size (gel permeation chromatography).

With degassed, deionized water as eluant, dextran, the blood plasma substitute having a molecular weight range 40,000–200,000, has been fractionated continuously into samples with narrower molecular weight distributions.

By using a multihead micropump, the flows of liquid into and out of the apparatus have been controlled precisely, so that continuous night and day operation for periods of up to 1 month have been made.

XXVIII. Comparison between Batch and Continuous Chromatographic Refining

No experimental and economic data for processing identical systems exists in the literature to make a true comparison between production-scale batch and continuous chromatographic processing. Abcor (41) has published some capital and processing cost data for batch processing, while Barker and Huntington (8) published cost data relating to two particular hydrocarbon separations assuming throughputs of 200 gal/hr of feed. The more comprehensive data of Abcor gave operating costs varying from 3¢/lb for relatively easy and volatile systems (at the 5,000,000 lb/year rate) to $15/lb for difficult separations and low volatile systems (at the 8000 lb/year rate). Barker (8) reported processing costs for volatile systems varying from easy to average difficulty of from 1.5 to 3.5 c/lb (at the 14,000,000 lb/year rate). At one end of the scale, processing costs of a similar order of magnitude are being quoted, but the accuracy is inadequate to give a clear decision. Further work is obviously needed in this area.

With most new developing techniques it is impossible to say with certainty whether batch or continuous processing will finally predominate, although the author and his colleagues who have worked for 15 years on the latter favour the continuous process.

Symbols

F^R	Gas flow rate in rectification section of column
H_{OG}	Height of an overall gas phase transfer unit
K^R	Partition coefficient of a solute at infinite dilution in the rectification section of the column
K^F	Partition coefficient of a solute at finite concentration on a solute-free basis
K^∞	Partition coefficient of a solute at infinite dilution
$K_G a$	Overall gas phase mass transfer coefficient
L	Column length
M	Solute flow rate
N_{OG}	Number of overall gas phase transfer units
N_c	Number of theoretical plates (steady-state conditions)
S	Solvent rate
SF	Separation factor $= K_B/K_A$
t_R	Retention time of component
W_A, W_B	Feed rate of components A and B, respectively
Y_A, X_A	Weight of component A per unit volume of carrier phase and solvent phase, respectively
Subscripts	A, B, etc. refer to solutes A, B, etc.
Superscript S	refers to stripping section
Superscript R	refers to rectification section
Y_A, Y_B	Factors accounting for effect of finite concentration on the partition coefficient
δ_A, δ_B	Factors accounting for effect of finite column length on the partition coefficient

References

1. P. E. Barker and D. I. Lloyd, *J. Inst. Petrol.*, **49**, 73 (1963).
2. G. R. Fitch, M. E. Probert, and P. F. Tiley, *J. Chem. Soc.*, 4875 (1962).
3. C. T. Sciance and O. K. Crosser, *J. Am. Inst. Chem. Eng.*, **12**, 100 (1966).
4. H. Schultz, *Gas Chromatography 1962*, M. Van Swaay, Ed., Butterworths, 1963, p. 225.
5. R. P. W. Scott, *Gas Chromatography 1958*, D. H. Desty, Ed., Butterworths, London, 1958, p. 189.
6. T. H. Chilton and A. P. Colburn, *Ind. Eng. Chem.*, **27**, 255 (1935).
7. L. Alders, *Liquid-Liquid Extraction*, 2nd ed., Elsevier, Amsterdam, 1959, Chapter V.
8. P. E. Barker and D. H. Huntington, *Dechema Monographien*, Vol. 62, p. 153, 1968, published by Dechema.
9. D. H. Huntington, Ph. D. Thesis, Birmingham University, England (1967).
10. P. E. Barker and D. I. Lloyd, U. S. Patent 3,338,031.
11. P. E. Barker and D. H. Huntington, *Advances in Gas Chromatography 1965*, A. Zlatkis and L. S. Ettre, Eds., Preston Technical Abstracts Co., Evanston, Illinois, 1966, p. 162.
12. H. Pichler and H. Schultz, *Brennstoff Chem.*, **39**, 48 (1958).
13. Gulf Research and Development Co., U. S. Patent 2,893,955 (1959).
14. Mine Safety Appliances Co. (L. Luft), U. S. Patent 3,016,107 (1962).
15. D. Glasser, *Gas Chromatography 1966*, A. B. Littlewood, Ed., Institute of Petroleum, London 1967, 119. p.
16. P. E. Barker and Universal Fisher Engineering Co., Ltd., Crawley, Sussex, Pat. Appl. No. 33630/65 and 43629/65. Patent issued under No. 114596. Also foreign patents.
17. F. W. Willmott and A. B. Littlewood, *J. Gas Chromatog.*, **4**, 401 (1966).
18. P. E. Barker and D. H. Huntington, Proceedings of the Sixth International Symposium on Chromatography, Rome, 1966, Institute of Petroleum, London.
19. P. E. Barker and Universal Fisher Eng. Co., Ltd., Crawley, Sussex, Pat. Appl. No. 5764/68 and 44375/68.
20. P. E. Barker and S. Al-Madfai, Proceedings of the Fifth International Symposium on Advances in Chromatography, Las Vagas, 1969. Preston Technical Abstracts Co., Evanston, Illinois, p. 123.
21. Cities Service Research and Development Co. L. C. (Mosier), U. S. Patent 3,078,647 (1963).
22. A. J. P. Martin, *Discussions Faraday Soc.*, **7**, 332 (1949).
23. D. Dinelli, S. Polezzo, and M. Taramasso, *J. Chromatog.*, **7**, 447 (1962).
24. D. Dinelli, M. Taramasso, and S. Polezzo, U. S. Patent 3,187,486 (1965).
25. S. Polezzo and M. Taramasso, *J. Chromatog.*, **11**, 19 (1963).
26. M. Taramasso and D. Dinelli, *J. Gas Chromatog.*, **2**, 150 (1964).
27. Deutsche-Erdol-Aktiengesellschaft, German Patent 1,033,638 (1958).
28. Deutsche-Erdol-Aktiengesellschaft, British Patent 810,767 (1959).
29. Anonymous, *Chem. Eng. News*, 37, **40**, 74 (1962).
30. S. Turina, V. Krajovan, and T. Kostomaj, *Z. Anal. Chem.*, **189**, 100 (1962).
31. S. Turina, V. Krajovan, and M. Obradovic, *Anal. Chem.*, **36**, 1905 (1964).
32. Canadian Patients and Developments, Ltd. (D. W. Thompson), U. .S Patent 3,136,616 (1964).
33. D. W. Thompson, *Trans. Inst. Chem. Engrs.*, **39**, 19 (1961).
34. B. J. Bradley and P. F. Tiley, *Chem. Ind. (London)*, **18**, 743 (1963).
35. W. Kuhn, E. Narten, and M. Thurkauf, 5th World Petroleum Congress, 1958. Section V, Paper 5, p. 45.

36. W. Kuhn, E. Narten, and M. Thurkauf, *Helv. Chim. Acta*, **41,** 2135 (1958).
37. O. Grubner and E. Kucera, *Collection Czech. Chem. Commun.*, **29,** 722 (1964).
38. Phillips Petroleum Co. (B. O. Ayers), U. S. Patent 3,162,036 (1964).
39. J. C. Giddings and H. Eyring, *J. Phys. Chem.*, **59,** 416 (1955).
40. P. E. Barker and D. I. Lloyd, Proceedings of the Symposium on Less Common Means of Separation, April 1963. Institution of Chemical Engineers, London, p. 68.
41. J. M. Ryan, R. S. Timmins, and J. F. O'Donnell, *Chem. Eng. Prog.*, **64,** 53 (1968).
42. P. E. Barker, S. A. Barker, B. W. Hatt, and P. J. Somers, *Chemical & Process Eng.*, **52,** No. 1, 64 (1971).
43. Anonymous, *Chem. Eng. News* (May 23, 1966).

Index

395